AF552934

Domestic and Wild Animals Trichology
A Forensic Approach

NIPA® GENX ELECTRONIC RESOURCES & SOLUTIONS P. LTD.
New Delhi-110 034

Domestic and Wild Animals Trichology
A Forensic Approach

Rupali Y. Charjan
Assistant Professor
Department of Veterinary Anatomy and Histology
Nagpur Veterinary College
Maharashtra Animal and Fishery Sciences University
Nagpur, Maharashtra

Sanjay B. Banubakode
Professor and Head
Department of Veterinary Anatomy and Histology
Nagpur Veterinary College
Maharashtra Animal and Fishery Sciences University
Nagpur, Maharashtra

Naresh C. Nandeshwar
Professor
Department of Veterinary Anatomy and Histology
Nagpur Veterinary College
Maharashtra Animal and Fishery Sciences University
Nagpur Maharashtra

NIPA® GENX ELECTRONIC RESOURCES & SOLUTIONS P. LTD.
New Delhi-110 034

NIPA® GENX ELECTRONIC
RESOURCES & SOLUTIONS P. LTD.

101,103, Vikas Surya Plaza, CU Block
L.S.C. Market, Pitam Pura, New Delhi-110 034
Ph : +91 11 27341616, 27341717, 27341718
E-mail: newindiapublishingagency@gmail.com
Website: www.nipabooks.com

For customer assistance, please contact
Phone: + 91-11-27 34 17 17
Fax: + 91-11-27 34 16 16
E-Mail: feedbacks@nipabooks.com

Print ISBN: 978-93-58879-00-1

ebook ISBN: 978-93-58872-93-4

Composed and Designed by NIPA®.

महाराष्ट्र पशु व मत्स्य विज्ञान विद्यापीठ
फुटाळा तलाव मार्ग, नागपूर - 440001
MAHARASHTRA ANIMAL AND FISHERY SCIENCES UNIVERSITY
Futala Lake Road, Nagpur – 440001, Maharashtra (India)
Phone: +91-0712 2511088, 0712 2511282
Email: vcmafsu@gmail.com, vc@mafsu.in Web: www.mafsu.in

Dr. Niteen V. Patil
Ph.D., FADS, FNAVS, FANSI, FANA (Animal Nutrition)
Vice Chancellor

डॉ. नितीन व. पाटील
पिएच.डी, एफ.ए.डी,एस., एफ.एन.ए.व्हि.एस. एफ.ए.एन.एस.आय., एफ.ए.एन.ए. (पशुपोषण शास्त्र)
कुलगुरू

Foreword

It is an honour for me to contribute this foreword for the unique kind of book entitled "Domestic and Wild Animals Trichology : A forensic Approach". All the three authors are well recognized at National level as outstanding Anatomists. The compilation of book has skilfully done by the authors. I am very happy to note that the book is written to extend scientifically complied data, which is useful to government organizations such as wild life division of forest and police department to solve the vetero-legal cases.

It is with great pleasure and anticipation that I introduce you to the captivating world of "Domestic and Wild Animal Trichology: A Forensic Approach." In the intricate tapestry of forensic sciences, the study of animal hair is often-overlooked yet it permits vital thread, and in this exceptional work, the author unravels the mysteries and nuances of trichology with unprecedented depth.

As we venture into the realm of forensic investigation, the exploration of animal hair becomes an indispensable tool for understanding crime scenes, unravelling wildlife mysteries, and discerning the complexities of human-animal interactions. In this comprehensive volume, the authors guide us through the science of trichology, offering profound insights into the significance of hair analysis in forensic contexts.

Dr. Niteen V. Patil
Vice-Chancellor
Maharashtra Animal and Fishery Sciences University
Nagpur

Preface

The study of structural details of hair for various purposes is called as Trichology. The hair morphology is useful for species identification as well as for study of evolution and domestication of various animal species in Zoology, Veterinary Science and Forensic sciences. Unlike biological samples such as bone pieces, blood stains, skin, and viscera. The hair, are chemically most stable component, do not under go decomposition, and structural changes even after enzymatic activity, and can not be digested. Due to these peculiar characteristics, hair can be used as biological evidence in forensic investigations in solving vetero-legal cases.

After the crime has been committed, some or the other biological material such as hair remains behind as an evidence which can prove to be conspicuous enough to lock up criminals, such recovered samples from animals are helpful in identification of species by studying its gross anatomical structure, cuticular, medullary and cross sectional microscopic structure, scanning electron microscopic structure and sequencing of DNA. Therefore, an attempt has been made to publish the results of the author's work in for of photographs of hair at gross, microscopic and scanning electron microscopic level as well as the sequence of DNA of various species of domestic, and wild animals. These photographs and legends will help the scientist to understand the details of hair to identify the species of animals by comparing the photographs of the book by visual conceptualization. This book liberally uses the most routine methods of examinations, so that the scientists are able to appreciate the details of photographic interpretation. The colour images of structural details of hair will allow the scientific workers to draw conclusions from the impressions instantly. The diagrammatic representations are designed to give reader a ready resource of orientation when needed.

In general, the book is designed to intensify the experience of budding scientific workers who are beginners and have no or little experience.

Authors

Acknowledgment

First and foremost, I place on record the yeoman contribution of Dr. V. R. Bhamburkar, Retired Professor and Head, Department of Veterinary Anatomy and Histology and Associate Dean, Nagpur Veterinary College, Nagpur, who have given me the idea of undertaking the work on identification of species of domestic and wild animals. It was his dream project to help government organizations like wild life division of Forest Department and Police Department to solve the vetero-legal cases so as to put the criminals behind bar. With profound joy and happiness, I express my heartfelt thanks to him.

I am grateful to Dr. N. V. Patil, Hon'ble Vice-chancellor, Maharashtra Animal and fishery Sciences University, Nagpur and Dr. N. V. Kurkure, Director of Research, Maharashtra Animal and Fishery Sciences University, Nagpur for their constant encouragement and providing all types help in successful completion of publication of this book in a very short and stipulated time.

I pay honor with deep sense of respect to, Dr. A. P. Somkuwar, Associate Dean, Nagpur Veterinary College, Nagpur, for his keen interest, necessary suggestions and wholehearted cooperation during the research work.

I am also thankful to Dr. N. C. Nandeshwar, Professor, Department of Veterinary Anatomy and Histology, Nagpur Veterinary College, Nagpur and my Ph. D. Guide and Co-author for his constant encouragement and valuable suggestion.

I gratefully acknowledge Dr, S. B. Banubakode, Professor and Head, Department of Veterinary Anatomy and Histology, Nagpur Veterinary College, Nagpur and Co-author for his valuable guidance in strategic planning, organizing, and designing the construction of book. The completion of task of writing this book would not have been possible without his precious help.

I am delighted to express my sincere thanks to ever helpful personality to Professor S. N. Gawande, University Librarian, Maharashtra Animal and Fishery Sciences University, Nagpur for helping me to plan the basic framework of book and providing all kind of facilities to search and collect the scientific literature.

I owe my deep sense of gratitude to Dr. S. W. Kolte, Professor and Head, Department of Veterinary Parasitology for providing excellent laboratory support throughout the course of study and unparallel guidance.

Authors

Contents

1

Introduction

India is well known for its rich flora and fauna and is referred as one of the richest biodiversity nations of the world. The mammalian population of India is a combination of Palaearctic, Oriental and Afro-tropical fauna (Sahajpal *et al.* 2009). It is the home of 400 different mammalian species, 129 of these species are protected under Wildlife (Protection) Act 1972 (India) and an arbitrator to Convention on International Trade in Endangered Species (CITES) (Nameer, 2015). Illegal trade of the wildlife for preparation of various products is the common issue in the conservation and is also responsible for regional extinction of the species.

According to the findings of previous researchers, the wildlife biological materials are the third most common illegally traded items after narcotics and firearms (Hanfee, 1998). The most common wildlife products, which are illegally traded includes meat, antler or horns, nails, bones, musk, ivory, biles, pods, skin fur, claws, teeth, wool etc. These biological materials are traded in different forms like processed leather, brushes, frames, ornaments, showpieces and also as Traditional Chinese Medicines (Sahajpal *et al.*, 2009].

The identification of species from processed and unprocessed wildlife trade comes under forensic science. The forensic science play an important role in solving vetero-legal cases. It became very difficult to investigate and identify the species on the basis of small sized samples like meat, bone pieces, organ parts, skin pieces, hair and various processed and unprocessed wildlife trades. The hair are the most common biological evidence encountered during the course of vetero-legal investigation, due to its properties such as

1. Hair are the chemically most stable component of the body.
2. Do not undergo decomposition.
3. Do not change the structural characters even after enzymatic activity.
4. Cannot be digested etc.

Hence, hair are most commonly recovered biological evidence at the site of crime and sent for forensic investigations particularly for the species identification.

The hair morphology is useful for species identification as well as for study of evolution and domestication of various animal species, in zoology, phylogenetic, taxonomic, textile testing, archaeological studies and forensic sciences (Farag *et al.*, 2015). The study of structural details of hair for various purposes is known as trichology.

Hair, whether from human or from wild or domestic animals, can be an important physical evidence for solving vetero-legal cases. Hair normally falls from the body over the course of a period. It will stick to a various materials, especially fabric, clothing and injuries. Hair are not easily destroyed, even with exposure to moisture, environment and decomposition of accompanying tissue.

Hair are typical skin products of mammals. They are fibrous outgrowths of the mammalian skin, produced by hair follicles. The mammalian pelage, except human and sheep consists of three different hair types, viz; strong, mostly straight primary hair / guard hair, thin, soft and curly, secondary hair / wool hair / fur hair and very thick, stiff hair, tactile hair or vibrissae etc. Each hair can be divided into three parts as root, shaft and tip. The part of the hair present under the skin is the root, the longest part above the skin is the shaft and pointed part of hair is the tip. The hair shaft, which is the outer and longest part of the hair, present over the skin and it perform various functions. The most important functions of hair are: insulation, protection against UV radiation and mechanical hazards, but hair also has functions in sense of touch as mechanical receptors (Brunner and Coman, 1974).

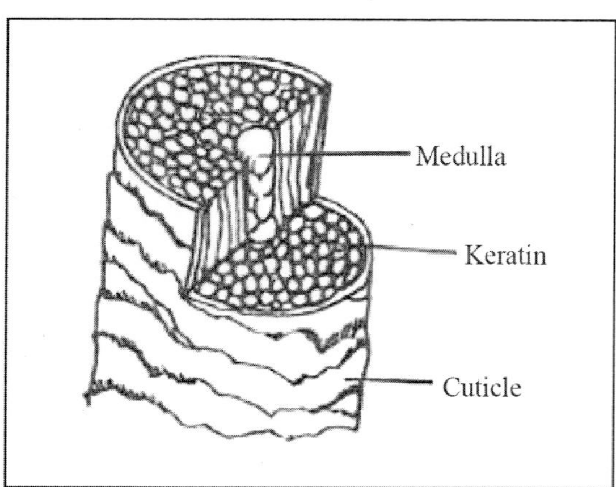

Fig. 1. Anatomy of hair: three principal layers of hair

The shaft of guard hair is composed of three concentric layers of cells from inner to outer as medulla, cortex and cuticle (Fig. 1.). The central medulla comprised of shrunken, pigmented or non pigmented air filled cells. The cortex, composed of dead cells, which are packed in homogenous, hyaline like mass. The outer hair cuticle, which consists of a layer of flat, strongly cornified, transparent cells, arranged in overlapping manner, the scales (Brunner and Coman, 1974).

The hair shaft can be classified in three regions: i) pars apicalis, ii) pars intermedia and iii) pars basalis. The part of the shaft towards tip of the hair is called as pars apicalis or apical part or distal part, which in several mammalian groups has a spindle shaped to strongly flattened thickening and referred as shield. The next part is termed as pars intermedia or medial part or middle part, which forms about one quarter of the hair shaft length and shows a more or less constant hair cuticle pattern. The part of the shaft towards the root is known as pars basalis or basal part or proximal part, forms the one third to half of the hair shaft and generally shows varying hair cuticle pattern (Kuhn, 2009).

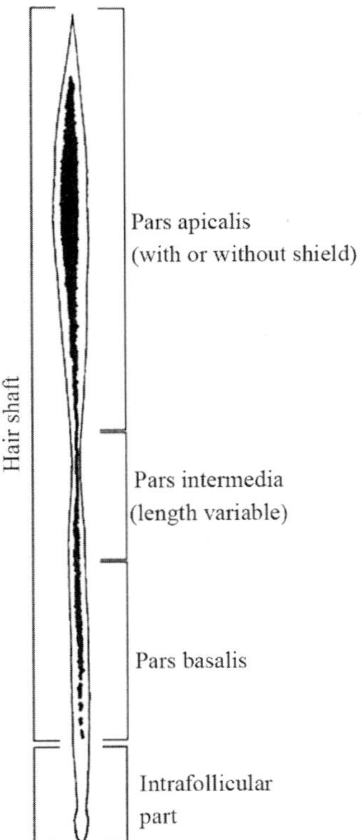

Fig. 2. Three subdivions of hair shaft along the length of hair

The structure of guard hair varies from species to species or groups of related species. Hence, it is possible to identify a group or even a species of animal on the basis of structural details of guard hair. The microscopic structure of hair shows variation from species to species. Upon scanning of available literature, it has been observed that the scanty information is available on the comparative study of domestic and wild animal hair. The present study could represent a useful tool in Veterinary Forensic investigation and forensic biology allowing for curbing the practices of illegal transport of slaughtered animals, poaching or wildlife crime and fraud in textile and fur industry.

The identification and comparison of domestic and wild animal hair can be helpful for solving legal cases. There are many structural differences in hair of various species of animals. These differences are often descriptive type. The studies on various parameters of hair have been carried out by various scientists. The hair are non invasive samples and can be useful as an alternative to blood or tissue as genetic material for species identification. In addition to morphological and chemical analysis, nuclear and mitochondrial DNA investigation can provide more, complete and irrefutable physical evidences.

The present trichological work was carried out on guard hair of domestic animal species; Cattle (*Bos indicus*), Buffalo (*Bubalus bubalis*), Goat (*Capra hircus*), Sheep (*Ovis aries*), Horse (*Equus caballus*), Pig (*Sus scrofa domesticus*), Dog (*Canis lupus familiaris*) and Cat (*Felis catus familiaris*) and the wild animal species; Spotted Deer (*Axis axis*), Nilgai / Blue Bull (*Boselaphus tragocamelus*), Sambar (*Cervus unicolour*), Hanuman Langur / Gray Langur (*Semnopithecus hypoleucos*), Tiger (*Panthera tigris*) and Leopard (*Panthera pardus*), to study physical, microscopic, scanning electron microscopic and molecular parameters of guard hair. This study would be useful to provide up-dated and iconographic data, which would be useful for identification of various domestic and wild animal species. Hence the present study is proposed to be undertaken with following objectives.

- To study the scanning electron microscopy to reveal the scale pattern of hair.
- To study the histo-morphology of hair of different species of animals.
- To study the molecular characters of the hair.
- To create markers for rapid identification of species of domestic as well as wild animals from hair.

2

Review of Literature

Hausman (1920a) worked on the microscopic structure of hair from the commercial fur. The hair samples of 27 different species of animals were used for the study. He described in detail the structural features of guard hair. He classified medulla of hair into 4 different types as discontinuous, continuous, fragmental and interrupted on the basis of composition of medullar cells. He also classified the cuticular scale pattern into two types as imbricate interrupted and imbricate coronal. He mentioned in detail the methods for the preparation of slides for the study of microscopic structures of hair. He also studied the cortex and pigment granules.

Hausman (1920b) studied structural characteristics of guard hair from 166 mammalian species. He stated that the hair of mammals could be classified on the basis of cuticular scale pattern, pigment granules and medullary pattern. He classified the medulla into three types as discontinuous, continuous and fragmented. He further classified discontinuous medulla type into two sub groups as simple and compound. The simple discontinuous medulla pattern was further classified into three categories according to the shape of individual medullary cell as ovate, elongate and flattened, while compound discontinuous medulla was subdivided as ovate and flattened. The continuous medulla pattern was also divided into two groups as nodose and homogenous.

He classified the cuticular scale patterns broadly into two groups according to the arrangements of cuticular scales as imbricate and coronal. He further divided the imbricate cuticle pattern into five sub groups on the basis of shape of individual cuticular scales as ovate, acuminate, elongate, crenate and flattened. Similarly, the coronal cuticular pattern was also classified as simple, serrate and dentate.

He further mentioned that the protective or guard hair of mammals mostly presents cuticular scales of flattened or crenate type and medulla of continuous nodose and continuous homogenous type.

Kirk *et al.* (1949) discussed the various techniques of negative cast preparation for the study of cuticular scales of hair. They described the procedure of negative cast preparations with the help of transparent nail enamel. They further studied the new and modified technique for negative cast preparation with the help of "Vinylite', the plastic sheet or "Lucite" the thermoplastic sheet. They prepared negative scale cast of cuticular scale of around 13 different species and observed variations in the cuticular scale pattern, scale margin and scale margin distance. They concluded that the species and individual could be identified by cuticular scale pattern.

Lyne and Mc Mohan (1950) observed the surface structures of hair of Tasmanian Monotremes and Marsupials. They studied dorsal guard hair from 22 species. They worked on various surface structures of hair such as principle dimensions, cross sectional shapes and cuticular scale patterns. They demonstrated various methods used for the preparations of slides for the study of cuticular pattern viz. dry mount, negative impression and positive impression. They concluded that the surface morphology and cross sectional shapes could be used for differentiation of animal species.

Vernall (1961) studied the size and shape of cross sections of hair from men of four different races. The hair samples were collected from 86 males (20 Chinese, 21 from Great Britain, 26 Asiatic Indians, 19 Negroids, 9 Nigerians, 9 from USA and 1 from Jamaica). The Parlodion was used as embedding media for the cross sectioning of hair. The blocks were cross sectioned on the Hardy sectioning device with Gillette microtome blades. He found that the Chinese hair cross sections were largest, while that of Western European were smallest. The shape of cross section in Chinese was observed as rounded, flattened elliptical in Negroid. Asiatic Indian and Western European had similar cross sectional shape of hair. He finally concluded that the difference in size and shape of cross section could be attributed to the racial difference.

Shelley and Ohman (1969) developed new technique for the cross sectioning of human hair samples for histochemical and morphological studies. They took around 10-50 segments of hair, arranged parallel in bundle and wrapped in a tea bag paper. The wrapped hair were pretreated with diethyl ether for removal of lipid and rehydrated with distilled water. The rehydrated hair were embedded in paraffin with routine method. The blocks were cut with parsona super stainless steel double edge razor blade fitted in special holder on a regular rotary microtome. They concluded that the technique was proved useful for sectioning of animal and human hair. Method proved best for study of cross section morphology and melanin content without any staining procedure.

Bruinner and Coman (1974) prepared a photographic key for the identification of hair from mammalian species. They explained in detail the procedures for the preparation of slides for hair profile, whole mount, negative cast of cuticular scales and the cross section of hair. They also mentioned about the types of hair present in the pelage of mammals and the different components of hair viz. cuticle, cortex, medulla, pigment granules and ovoid bodies.

They classified the medulla pattern mainly in four groups as unbroken, broken, ladder and miscellaneous medulla. They classified the cuticle as per the margins of scale, scale pattern and distance between the scale margins.

Koppikar and Sabnis (1976) worked on gross and microscopic morphology of hair from 21 mammalian species of Melghat tiger reserve, Amravati. They considered three basic regions of hair shaft as proximal, middle and distal for the observations. They worked on total length, diameter, colour, cuticular pattern and medullary pattern of hair. They found that Langur (*Presbytis entellus*) hair were 6.2cm long with 60μ diameter and uniformly gray coloured. The cuticular scales of proximal and middle region were imbricate, while plain at distal region. The medulla was absent throughout the length of hair. The tiger (*Panthera tigris*) hair were 4-8 cm long, straight with curved tip and diameter at proximal region 84 μ. The hair had three colour bands as white at proximal region, dark gray at distal region and both bands were separated from each other by yellow band at middle region. They observed that the cuticular scales had spiny borders at proximal region while plain borders at middle and distal part of hair. The medulla was present throughout the length except at distal region.

They found that the hair of panther (*Panthera pardus*) were 3 to 4 cm long with 45 μ diameter. The hair showed four colour bands from proximal to distal as light yellow, black, brown and yellow. The cuticular scales were crenate at proximal and middle region while plain at distal region. The medulla was continuous throughout the length except at distal part where it was fragmented.

They stated that the domestic goat (*Capra hircus*) hair were 3 to 5 cm long with 45 μ diameter. The cuticular scales were imbricate with crenate edges at the proximal region of hair, while smooth edges at middle and distal region of hair. They also commented that the medulla was fragmented in the proximal and distal regions, while in middle region, it was discoid.

The hair of cattle were curved and 1 to 2cm long with 30μ diameter. The imbricate cuticular scales with crenate edges were observed in the proximal and middle region. In the distal region, the cuticular scales were flattened with minute spines. The medulla appeared continuous in the proximal and middle region and fragmented in distal region.

They stated that the sambar hair were tapering at both proximal and distal region and broad at middle region. The hair were 3 to 5 cm long with the 180μ diameter. Three colour bands were there as pure white in proximal, yellowish grey in the middle and black in the distal region. The cuticular scales were smooth in proximal region, imbricate crenate in middle and spiny in the distal region of the hair. They found that the medulla was reticular polygonal in the proximal and middle region and was not visible in the distal region.

They further commented on hair of spotted deer that the hair were 3 to 4 cm long with 84μ diameter and profile was slightly wavy. The hair were with three colour bands as white at proximal, brown at middle and yellowish brown at the distal region. The cuticular scales were imbricate compressed ovate at medial and spiny at distal region. The medulla was continuous in proximal and middle region and fragmented in the distal region.

The hair of nilgai showed 23 to 27 cm length and 140μ diameter. The hairs were white in colour at proximal region, brown in middle and black in the distal region. The cuticular scales were imbricate with crenate edge in proximal and middle regions with fine long bristles at distal region. They found that the medulla was continuous in the proximal and middle region and was not visible in the distal region.

Pepper *et al.* (1977) developed a new technique for cross sectioning of free hair. They used Araldite as an embedding medium for hair. The embedded hair were sectioned with glass knife on routine rotary microtome at 1-2 μm thickness sections. The sections were taken on glass slide and placed on a drop of water, which was allowed to dry on hot plate for 2-3 min. The cross sections were observed under phase contrast microscope for its shape and size. They further mentioned the disadvantage of the method that only one hair could be sectioned at a time and so the method was time consuming for large number of hair samples.

Homan (1978) studied the hair characters of 36 species from Heteromyidae family. He worked on dorsal guard hair collected from dried skin specimen for cuticular scales, medullary cells and cross section shape. The cuticular scale pattern was studied by using scanning electron microscope. The medullary cells and cross section shapes were observed under camera lucida and drawings were made for further evaluations. He finally stated that the species of Heteromyidae family could be differentiated from the dorsal guard hair characteristics features.

Riggott and Wyatt (1980) studied the morphology of guard hair of adult rat by using scanning electron microscope. The hair samples were collected from 25 male and 25 female adult rats from abdomen, flank, back, head and ear regions along with vibrissae. They stated that the number of cuticular scales were more in vibrissae and ear hair as compared to hair from other body regions. The cuticular scale size was found smaller in ear hair and vibrissae. They observed significant difference in mean cuticular scale number and cuticular scale size in ear hair and vibrissae. But no significant difference was noticed in hair of male and female.

Taylor (1985) worked on the hair morphology of forty Tasmanian mammals. He mentioned that the hair could be primarily classified according to the cross-sectional shape in seven groups as circular, oval, oblong, eye or lemon shaped, concavo-convex, dumb-bell shaped and cigar shaped. He demonstrated photographic key for the identification of seven Tasmanian mammals viz. *Sarcophilus harrisii*, *Thylacinus cynocephalus*, *Bettongia gaimardi*, *Carcartetus lepidus* and *Pseudomys higginsi*.

Rajaram and Menon (1986) presented a report on scanning electron microscopic study of hair keratins from the hair of Lion tailed macaque (*Macaca silenus*),

Tiger (*Panthera tigris*), Porcupine (*Hystrix indica*), Cow (*Bos Indica*), Mangoose (*Herpestes edwardsi*), Sloth Bear (*Melurus ursinus*), Black buck (*Antilope cervicapra*), Sambar (*Cervus unicolor*), Chital (*Axis axis*), Hog (*Sus Scrofa*) and Human from Indian subcontinent. The samples were collected from rump part of male animals and were processed as per the standard procedure for surface and cross section study. They found that the porcupine quills were tubular with spongy medulla surrounded by solid cortex and cuticular scales directed towards root. They stated that the medulla was absent in the hair of lion, tailed macaque, buffalo, hog and human hair. The hair of above species showed a small hole in the center while bear hair showed fine cuticular pattern. The porcupine quill, sheep, blackbuck and sambar showed regular medulla pattern. They also stated that the black buck hair showed circular and peanut shaped cross section.

Teerink (1991) worked on the hair of almost all mammals from West Europe. He mentioned types of hair present in the pelage of animals and studied the dorsal guard hair. He classified the cuticular scale pattern, medulla pattern and cross-section shape in details. He considered the cuticular scale position, cuticular scale pattern, scale margin and scale margin distance. He noted that the size and shape of individual cuticular scale vary according to the position on the shaft of hair. He stated that the distal part of each cuticular scale overlap the proximal part of the next scale.

He reported that the medulla was generally composed of clearly visible shrunken dead cells and the air filled intercellular connections was responsible for specific characters of medulla. He classified the medulla on the basis of composition, structure and form of medulla margins.

He mentioned that the shape and diameter of hair cross section plays very important role in species identification, and demonstrated as easy techniques for the preparation of slides to study hair profile, cuticular scale pattern, medulla pattern and cross section shape.

Marinis and Agnelli (1993) prepared identification key on the hair of Italian mammals from Insectivora, Rodentia and Lagomorpha order. They worked on various parameters such as cuticular structure, cortex, medullay features and cross sectional details of the hair samples. They prepared the photographic key of hair from around 21 species.

Agren (1995) studied the hair residue present on Viking age Penannular Brooches from Birka. He found hair on 23 brooches out of 73 brooches present in Birka. He observed that all the hair samples present on brooches were underhair, which were used to prepare livery of Saami. The hair found on brooches were mostly from sheep and beavers, while some hair remained unidentified because of their poor conditions.

Chakraborty and De (1995) worked on the structure and pattern of cuticular scale in marbled cat (*Felis marmorata charltoni*) gray. The hair samples were collected from five individuals and subjected for the study of cuticular scale pattern, number of scales per millimeter of hair length, proximo-distal length and side to side length of an individual scales. They observed that the number, shape and pattern of cuticular scales were different in basal, transitional and apical region of an individual hair. They concluded that the cuticular scale structure and patterns could be used for species identification.

Bahuguna and Mukherjee (2000) worked on the hair of animals from Tibetan region to recognize bending of wool products of Tibetan village using scanning electron microscope. They worked on 20 wool fibers and 10 guard hair from pashmina, angora, shashmina, ibex, blue sheep and shahtoosh. They worked on the cuticular scale pattern of various animals under study. They found variations in the cuticular scale pattern of wool fiber and guard hair of same individual and amongst the species. They stated that Angora hair showed parallel wavy scale pattern with smooth margin and near scale margin distance. The pashmina hair showed irregular wavy distance however, shashmina hair showed broad petal pattern with crenate margin, Ibex hair showed irregular pattern with crenate margin while shahtoosh / Tibetan antelop hair showed a different pattern of regular mosaic with smooth scale margin. The wool fibers of Angora, Pashmina, Shashmina, Ibex and shahtoosh showed parallel wavy with smooth margin, irregular wavy with highly crenate margin, coronal scales with crenate margin, irregular wavy with crenate margin and single scale pattern respectively.

Broeck *et al.* (2001) studied the different types of hair in feral, New Zealand white and Angora rabbits with the help of scanning electron microscope. The skin and hair samples were collected from dorsal, lumbar, lateral abdominal, inguinal and femoral regions of 2 feral rabbits, 2 New Zealand white and 2 Angora rabbits. They worked on the cover / guard hair, wool hair and tactile hair. They found that the cross section profile of guard hair, wool hair and tactile hair was oval, angular oval to rectangular and circular respectively. The guard hair showed deep groove towards the mid part of shaft, which gave dumbbell / bean shaped cross section profile to the hair. The cuticular scale pattern was observed regular to irregular wave in guard and tactile hair, while petal to double chevron in wool hair. The medulla composition was multiserial ladder in guard hair, uniserial ladder in wool hair and hollow canal like in tactile hair.

Chakraborty and De (2002) studied the structure of mid dorsal guard hair from dry skin specimens of hunting leopard (*Acinonyx jubatus venaticus*) and lesser panda (*Ailurus fulgens F. cuvier*). They worked on surface structure, medullary configuration and cross section details of dorsal guard hair. They found that the cuticular scale pattern was diamond petals with smooth scale margins and distant distance between scale margins in lesser panda. They also studied the medullary configuration as unbroken cellular and circular shaped cross section in lesser panda. In hunting leopard, they observed the cuticular scale type as imbricate to crenate, cuticular scale pattern as irregular wave with irregularly ripples scale margins and intermediate distance between scale margins. The medullary configuration was found as simple unbroken amorphous and cross section shape was almost circular in hunting leopard. They finally concluded that the hunting leopard could be differentiated from the species of panther family and lesser panda from the other species of Ursidae and Procyonidae family on the basis of structural characters of dorsal guard hair.

Toth (2002) worked on the dorsal guard hair of Hungarian mustelidae and other small carnivorous of Hungary. The 20 dorsal guard hair from five adult individuals of each species were collected for the quantitative and qualitative characteristics. He prepared the identification key on the basis of various cuticular scale pattern and medullary pattern of hair. He found pectinate, diamond petals and mosaic cuticular scale pattern in *Mustela eversmannni*, *Lutra lutra* and *Mustela erminea* respectively. He also stated that the various medullar patterns observed were unicellular ladder in *Mustela ermine*, cloisonned pattern in *Mustela foina* and multicellular ladder in *Felis silvestris*.

Deedrick and Koch (2004) prepared practical guide and manual on microscopy of animal hair. They stated that the cuticle scales could be classified into three main types as coronal, spinous and imbricate. They mentioned that the medulla might or might not present in the hair, if present, it might be fragmented, discontinuous or continuous. They stated that medulla formed main body of hair. The cortex generally composed of spindle shaped cells, pigment granules and large oval shaped structures, the ovoid bodies. They commented on the group characters of many animals like deer and antelope, commercial fur animals and domestic animals.

They mentioned that the hair of animals from deer family and antelope were difficult to distinguish on gross characters. They stated that coarse diameter was 300 μm, medulla was made up of spherical cells with wine glass shaped root and had constant diameter throughout the shaft of hair with regular wave or crimp cuticle.

They studied the fur of fur animal, rabbit, mink, seal, muskrat and chinchilla and reported that the fur fibers had very fine to medium diameter (20 to 150 μ), serial or vacuolated medullary pattern, variations in diameter of shaft and colour bands.

Koch (2004) studied the structure and composition of human hair. He stated that the cuticular scales were always directed from the proximal or root end to the distal or tip end and cuticular scales were of three types, viz. the coronal, spinous and imbricate.

The cortex or main body of hair composed of elongated and spindle shaped cells along with cortical fusi, pigment granules and large round to oval shaped ovoid bodies.

He stated that the pelage of animals composed of guard, fur and tactile hair, while the human body hair could be considered as combination of guard and fur hair. He further stated that the stage of hair could be determined on the basis of shape of hair root. He found that the human guard hair had double medulla.

Soni *et al.* (2004) discussed importance of hair structure in identification of various species of genus Panthera. Ten hairs were randomly collected from the body of three large cats; *Panthera leo persica* (Asiatic lion), *Panthera tigris* (Tiger) and *Panthera pardus* (Leopard). They studied the total length, diameter, cuticular scale structure, and cross section shape of dorsal guard hair. They found highest length of hair in Asiatic lion (34.89 mm) followed by leopard (21.05 mm) and tiger (20.3 mm) and maximum diameter of hair was 0.000455 mm in Asiatic lion, 0.000537 mm in tiger and 0.00052mm in leopard. The cuticular scale pattern in Asiatic lion was observed as irregular wave with smooth scale margin associated with near scale margin distance and round shaped cross. The hair of tiger showed regular wave cuticular scale pattern, rippled scale margin with near scale margin distance and oval shape cross section. The cuticular scale pattern was diamond, smooth scale margins with distant scale margin distance and cigar shaped cross section shape in leopard. They finally concluded that the combination of morphological characters of guard hair viz., total length, diameter, cuticular scale pattern and cross section shape could be used for identification of three large cats (Asiatic lion, tiger and leopard).

Chakraborty and De (2005) studied the dorsal guard hair of nine Indian species of family Viverridae, scheduled as schedule I and II animals under wildlife (Protection) Act 1972. They summarized that all the species except *Arctictis binturong* showed similar characters as unbroken vacuolated medulla, shielded and spatulated hair profile, oval and circular cross section. From their work, they finally concluded that the identification of species from the hair samples was possible with the help of combination of morphological characters of hair.

Marinis and Asprea (2005) studied the morphological changes occurred in the hair structure during domestication of sheep and goat. The hair samples from five different locations of dorso-lateral aspects of 21 specimens (12 wild and 9 domestic) were collected from wild goat (*Capra aegagrus*) and mouflon (*Ovis orientalis*). They observed same medulla in wild sheep and wild goat, as partially filled lattice with scalloped medulla margin. They found medulla of domestic sheep and goat as multicellular, continuous with scalloped medullar margin, which was different from the wild sheep and goat. They commented that the cuticular scale pattern in domestic sheep and goat were regular wave with rippled scale margin and distant scale margin distance. The wild sheep and goat showed irregular wave pattern with smooth scale margin and distant scale margin distance.

They concluded that the domestic sheep and goat showed different medullary structure from the wild sheep and goat. Simultaneously, they stated that the cuticular scale structure of domestic sheep showed variations from that of wild ancestors, while domestic goat did not show variations from their wild relatives.

Marinis and Asprea (2006) prepared photographic hair identification key of wild and domestic ungulates of Southern Europe. 105 hair samples from ten wild and five domestic ungulates were collected from the dorsolateral region of the body. They observed cuticular scales pattern, margins of scale, scale margin distance, medulla pattern, medulla margins and composition of medulla along with the pigment distribution in the cortex of hair. They stated that the hair of adult wild boar among the ungulates could only be identified without the use of microscope as the hair were spilt at least once. They classified the cuticle as per the position of cuticular scales, structure of scale margin, distance between scale margins and scale pattern as transversal / intermediate, smooth / rippled, distant / near and regular / irregular respectively. They differentiated medulla type on the basis of composition, structure, pattern and form of the margins as unicellular / multicellular, amorphous / uniseriate / multiseriate / filled lattice / partially filled lattice / vacuolated, continuous / fragmental, irregular / straight / scalloped respectively. They also studied the variations in structure of hair with the change in season.

Sato *et al.* (2006) studied the guard hair of dog and cat using numerical morphology. The tuft of five guard hair from 12 body regions (sinciput, buccal, dorsal neck, ventral neck, dorsum, popliteal region, caudal radix, tail, abdomen, lateral brachium, lateral femur and flank) of the five mongrel dogs and cats were collected. They measured length, maximum width, cross sectional maximum diameter, cross sectional minimum diameter, cuticular thickness of cross section and number of scales per unit length of hair at proximal, middle and distal parts of the hair. They calculated the indices viz. hair width index, medulla index, hair index and cuticle index. All the observations were subjected for statistical analysis. They concluded that the overall morphology of various numerical features were extensively overlapped each other between dog and cat hair.

Lungu *et al.* (2007) worked on the morphological features of hair from domestic and wild ruminants of bovidae and cervidae family. They studied the hair morphology by using whole mount and quick nail polish method. They worked on parameters like diameter of the shaft, medullary diameter, cortex thickness, medullary index, medullary vacuolated cells, cuticular scale pattern and scale length. They found that shaft diameter was maximum in fallow deer (238.3 ± 13.1 μm), and minimum in sheep (68.5 ± 6.4 μm). The shaft diameter in other animals was 96.4 ± 14.9 μm in cow, 128.0 ± 16.6 μm in chamois, 194.3 ± 18.3 μm in deer and 110.2 ± 11.0 μm in roe deer. They further stated that the medulla was absent in sheep hair. The medullary patterns were variable in all species under study as goat- continuous lattice, chamois – regular lattice, cow-fragmentory, deer-vacuolated, roe deer- lattice pattern with air spaces and fallow deer-vacuolated medulla. The cuticle of sheep was made up of rhomboidal shaped scales with toothlike borders, which were protruding out. In cow flattened deep imbricated, obliquely placed cuticular scales were present. The cuticular scales in goat were imbricated elongated transversely and placed in oblique manner. They also found that the cuticular scale length was more in wild ruminants as compared to domestic ruminants. They found that the cortex was present in all investigated wild species with a prominent wavy borderline between medulla and cortex. The cortex of cow was full with cortical fusi and lot of ovoid body.

Sahajpal *et al.* (2008) worked on the characteristics features of the hair from four different species of Indian bears viz.; Himalayan black bear, Brown bear, Sloth bear and Malaya sun bear. They stated that the Indian bear species could be differentiated on the basis of microscopic characters of hair up to species level. They stated that the hair was rough in feel and kinky, which was very different from other carnivorous animals, which had soft hair. The hair of all bear species were black except for brown bear, which was brown in colour. They further commented that the cuticular scale pattern was found similar (irregular to regular wave pattern) in all species of bear except sloth bear, in which pattern was broad petal. They also stated that the scale count index was variable in all the four species of bear under study. The bear hair could not be differentiated from other species on the basis of medullary feature, as it was very narrow and amorphous (medullary index 0.15 or less) except for brown bear with a vacuolated medulla (medullary index 0.36). They also commented that the cross section of all the four bear hair showed oval shape with dark black pigmentation in cortex except brown bear, where the pigment granules were brown in colour.

Kshirsagar *et al.* (2009) studied human and animal hair in relation with diameter and medullary index. They stated that the mean diameter of shaft of human hair varied from 30 to 80 μm, while that of animals varied between 25 μm to 160 μm. They recorded the mean shaft diameter in dog as 25 μm, cat as 30 μm, rat as 40 μm, donkey and squirrel as 50 μm, camel as 80 μm, horse and sheep as 90 μm, goat as 100 μm, buffalo as 110 μm and 160 μm in cow. The mean human medulla diameter was 5 μm, while the medulla diameter of animals showed variation from 20 μm to 100 μm. The minimum medulla diameter was found in cat and rat as 20 ± 0.63 μm and 20 ± 2.28 μm respectively. Maximum medullary diameter was found in cow which as 100 ± 5.47 μm. They further calculated the medullary index in human as 0.1 to 0.25 while in animals was 0.44 to 0.70.

Sahajpal *et al.* (2009a) recorded the medullary cuticular and cross section shape of five different species of mongoose. They observed different medulla pattern in all four species as, wide cortical intrusion in Indian grey mangoose and ruddy mongoose, wide with vacuoles in crab eating mongoose, while medium with vacuoles in Bengal mongoose. The cross sections of hair showed variations in shape as oval to oblong (crab eating mongoose) and oval (Bengal mongoose). During their study, they mentioned that the cuticular pattern showed no differences. The medullary index and cuticle scale count index were highest in Indian grey Mongoose (0.792 + 0.0005 and 170 + 1.23) and lowest in Bengal mongoose (0.596 + 0.0004 and 114 + 1.27).

Sahajpal *et al.* (2009b) worked on the microscopic characteristics of hair from the ten bovide species listed under schedule I of wildlife (Protection) Act of 1972 of India and three domestic bovine species. They studied ten hair form each individuals of species; Black buck (*Antilope cervicapra*), Indian gazelle / chinkara (*gazelle bennettii*), Chousingha (*Tetracerus quadricornis*), Nilgiri tahr (*Hemitragus hylocrius*), Sumatran serow / Southern serow (*Capricornis sumatraensis*), Indian Bison /Gaur (*Bos gaurus*), Himalayan blue sheep /bharal (*Pseudois nayur*), Argali / mountain sheep (*Ovis ammon hodgsonii*), Urial /Arkars / Shapo (*Ovis vegnei vignei*), Tibetan antilope / Chiru (*Pantholops hodgsonii*), domestic cattle (*Bos taurus*), domestic goat (*Capra hircus*) and domestic sheep (*Ovis aries*). They used combination of cuticular scale pattern, medulla pattern and cross section for differentiation of bovidae species from each other. They found that the cuticular scale pattern was regular wave, scale margin was smooth and cuticular scale distance was near in Black buck, Indian gazelle, Chousingha and Argali, while irregular wave pattern, crenate margin and near distance was observed in Nilgiri tahr, Sumatran serow and Indian Bison. In domestic animals, the cuticular scale pattern was observed as irregular wave, cuticular scale margin as crenate and scale distance as near. They noted wide cellular lattice medulla in Black buck, Indian gazelle, Himalayan blue sheep, Argali, Urial and Tibetan antilope, while wide medulla lattice type medulla was observed in chousingha, Nilgiri tahr and domestic goat. They reported presence of narrow medulla lattice type of medulla in Sumatran serow, median width amorphous medulla in Indian bison, wide amorphous medulla in domestic cattle and narrow medulla with vacuoles in domestic sheep.

They found oval cross section shape of medulla in Sumatran serow, argali, domestic cattle and domestic sheep, while circular shape in Indian bison, Himalayan blue sheep, urial and Tibetan antelope. The kidney shaped cross section of hair was observed in black buck, oblong to dumb bell shaped in chousingha and Nilgiri tahr and oval to oblong in domestic goat. They concluded that the use of single parameter for species identification might overlap between some bovidae species. So, combination of all the microscopic characteristics of hair could be helpful for the identification of bovidae species.

Zafarina and Pannerchelvam (2009) worked on the hair samples of rare living animal species ie. Jenglots using microscopical and molecular techniques. The hair samples were collected from three Jenglots from Irian Jaya, Indonesia. They calculated medullary index and the diameter of the hair by using whole mount slide, cuticular scale pattern by using negative scale cast and analyzed the mtDNA extracted from hair. They compared the questioned hair samples with the human hair samples and finally concluded that the questioned hair samples were of human origin on the basis of morphological analysis of hair samples and mtDNA sequence based analysis.

Bahuguna (2010) worked on trichotaxonomy of Indian species of genus Ratufa Grey. Approximately 10 hair samples were collected from dorsal, ventral, head and tail regions of Ratufa macroura (Pennant), Ratufa indica (Erxleben) and Ratufa bicolor (Sparman). She noted pelage colour, total length, length index, profile, cuticular scale pattern, medullar type and cross section shape. She stated that the medulla type was wide aeriform lattice in all species except ventral hair of Ratufa macroura which was simple medullar type. The cuticular scale pattern showed similarities in all species under study as regular wave pattern with rippled margin and near scale margin distance except in head and tail of Ratufa bilcolor and tail of Ratufa macroura. She also reported that the cross section shape was oblong in hair of all the regions of all three species of genus Ratufa.

Davis (2010) worked on the technique for rapidly quantifying mammalian hair morphology for zoological research. The hair samples were collected from ventral aspect of neck of six white tailed deer (*Odocoileus virginianus*), three male and three female of 0.5 to 5 years old. He used scanned images of 120 hair for the measurement of surface area, length, width, width / length ratio, mean Hire and mean darkness using FoveaPro image analysis programme. He concluded that the method could be considered over manual microscope measurements.

Sahajpal and Goyal (2010) identified the species of animal from single hair with a small piece of skin by microscopy and nucleotide sequencing (FINS). The hair was grey coloured and showed diamond petal shaped cuticular scales, while medulla was wide with vacuoles. The features were indicative of species Viverridae. They further subjected the same sample for mitochondrial DNA sequencing and on the basis of mitochondrial DNA sequencing, they confirmed that the skin sample along with the single hair strand belongs to small Indian civet (*Viverricula indica*) as it matched at 350 bp on cytochrome b gene, which showed 100% similarity with the 125 rRNA gene of *Viverrcula indica.*

Chattha *et al.* (2011) worked on various methods of hair mounting as techniques for trichological study. They described materials and method for dry mount, wet mount, whole mount method, scale replicate method, Polaroid print method and Crocker method. They finally concluded that the hair mounting technique was the best simplest technique in conservation of small and large carnivores.

Sarkar *et al.* (2011) studied macroscopic structure, surface structure, medullary configurations and medullary index of hair from five species of the family cercopithecidae. They studied dorsal guard hair of *S. entellus* (langur), *T. geci, T. pileatus, T. johnii* and *T. phayrei* for species identification. The hair of all the five species showed slightly wavy profile, but colours were different as brown in langur, Prout's brown in capped langur, raw umber in golden langur, black in nilgiri langur and tawny olive in phayre's leaf monkey. They further noted that the cuticular scale pattern was of regular wave type in all species, except langur, which was irregular wave. The scale margin distance was distant in all the five species of primate. They also observed that the scale margins were crenate in langur and nilgiri langur, while smooth in other three species. The golden langur, nilgiri langur and phayre's leaf monkey showed interrupted medulla, capped langur showed uniserial ladder, while langur showed simple medulla pattern. They further stated that the cross section shape was oval in capped langur, nilgiri langur and pharyre's leaf monkey and round in langur and golden langur.

Anwar *et al.* (2012) prepared photographic key for the identification of mammalian hair particularly of the prey species of snow leopards (*Panthera uncia*) from Pakistan. The guard hair from abdominal region of five domestic animals as cattle (*Bos taurus*), sheep (*Ovis aries*), goat (*Capra hircus domesticus*), yak (*Bos* grunniens) and zo/zomo (cross between cattle and yak) and five wild animals, Himalayan ibex (*Capra ibex sibirica*), Astore markhor (*Capra falconeri falconeri*), Musk deer (*Moschus chrysogaster*), Marmot (*Marmota caudata*) and Pika (*Ochotona roylei*). For the preparation of Photographic key they considered seven different hair characteristics viz. average number of scale across mid shaft region, average scale diameter, diameter of hair at mid-shaft region, medulla diameter, cuticular pattern, medulla pattern and pigmentation. The photographic key was also supported by the cross-sectional shape.

Dharaiya and Soni (2012) studied cross sections of the hair collected from the faeces of lion (*Panthera leo*) and leopard (*Panthera pardus*) to determine the prey species of two large cats found in Gir protected area. They used sharp blade for taking cross sections of hair. After sectioning the imprints were taken on thin film of gelatin or Kores correcting fluid. They reported that the medullary shape was oval in domestic cow, irregular oval in domestic sheep, large oval in chital, sambar, ratel, leopard, large Indian civet and wild boar, large rounded in blue bull and domestic buffalo, small rounded in hare, Asiatic lion and langur, elliptical in chinkara and cigar shaped in Jackal and domestic goat. They observed similarity in black buck and blue bull with respect to structure of cuticular scale margin, cuticular pattern and distance between scale margins. The cow hair showed rippled cuticular scale margin, regular wave pattern with close margin distance between scale margins, sheep hair showed smooth scale margin, irregular wave pattern with close margin distance and dentate scale margin, diamond scale pattern with distant scale margin was found in goat. The sambar hair showed smooth scale margin, regular wave pattern with near scale margin distance. The chital hair showed similar pattern of cuticle with the exception in scale margin distance as close. The langur hair showed crenate cuticular scale margin, regular wave pattern with close distance between scale margins.

Joshi *et al.* (2012) studied comparative trichology of wild hervivores like mouse deer (*Trangulus memina*), spotted deer (*Axis axis*), Sambar (*Cervus unicolor*) and barking deer (*Muntiacus muntjak*) for species identification. They noted the colour variation in hair of different species of animals. They reported that hair colour was brown or yellow brown in mouse deer and spotted deer, while brown or yellow brown or black brown in sambar and it was white brown or brown in barking deer. The cuticular scale margin was smooth proximally and crenate distally in mouse deer and spotted deer, while rippled in sambar and barking deer. The colour of cortex varied from brown to dark brown in mouse deer and spotted deer, while brown with black and brown pigmentations in sambar and barking deer. They also observed black to brown striations in the distal region of the hair of sambar deer. They further stated that the medullary pattern had very little variations amongst the species. The medullary pattern varied from wide lattice in mouse deer and spotted deer to wide lattice and thicker but barely visible in sambar deer and barking deer.

Vinayak *et al.* (2012) compared the partially burnt hair from a burnt women and towel recovered from the crime spot. They studied the hair colour, cuticular scale structure, medullary index, hair shape and structure of ends of hair. They found that the medullary index of both the hair samples was less than 1/3, a characteristic feature of human hair. They also observed fragmented medulla, dark brown pigmented cortex with ends showed signs of burn. They compared the hair recovered from crime spot with the animal hair like dog, cat, rabbit, guinea pig, sheep and rat. Finally they concluded that the hair samples were human origin and not of the animals.

Aparna and Yadav (2013) mentioned the role of hair as evidence in forensic investigations. They discussed the importance of hair in identification of individuals or species on the basis of composition and serologic analysis. They further stated that the age, sex and colour could be determined on basis of hair characters. They mentioned that the heavily pigmented hair are lighter in weight than the grey hair. White hair contains more medulla, the bleached hair are more sensitive to the radiations and possess lower tensile strength than heavily pigmented hair. They mentioned that the trace element content of hair were indicatiors of geographical region. They stated that the time of death could be determined on the basis of fungal growth and changes in proximal end morphology, which showed direct relation with post-mortem interval.

Dahiya and Yadav (2013) used scanning electron microscopic characters and elemental analysis of hair as a tool in identification of wild animals of fellidae species. They collected hair from lion, tiger and leopard. Ten hair from different body parts of 30 individuals of each species were collected. They observed significant difference in scale layers and scale pattern between the three species. They noted the values of sodium, potassium, sulfur and calcium of hair by using SEM-EDS.

Gharu and Trivedi (2013) worked on the dorsal guard hair of Hanuman langur and grey slender Loris for species identification. They observed that the cuticular scales in Hanuman langur were imbricate with crenate scale margin, irregular wave pattern and distant scale margin distance. The cuticular scale pattern in grey slender Loris was imbricate and elongated with comparatively smooth scale margins. The medulla was found fragmented in adult Hanuman langur with scattered light grey pigments in hair cortex, while in slender grey Loris, the medulla was found interrupted uniserial ladder like with the similar scattered light grey pigment granules in cortex.

They finally concluded that the cuticular scale pattern and medulla pattern was different in both the species with the similarity in pigment granules.

Kamalakannan *et al.* (2013) worked on the surface characteristics of dorsal guard hair from three dry specimens of Indian Chevrotain (*Moschiola indica*) gray. They observed the profile of dorsal guard hair, as straight. They measured the length and diameter of dorsal guard hair as 13 – 21 mm and 50 – 70 μ respectively. They found that the hair had two colour bands, claret brown at the tip and white towards the base. They noted cuticular scales arrangement as transversal scale position, smooth scale margin and scale margin distance as distant. Medulla was found filled in the entire width of the hair. The medullary configuration was unbroken lattice, structure of medulla was multicellular and medullary margin was observed as scalloped. There was no distinct margin found between medulla and cortex.

Kitpipit and Thanakiatkrai (2013) studied morphological features of hair from four adult tigers (*Panthera tigris*) of Thailand. They worked on various quantitative and qualitative parameters by observing whole mount, casting and cross section of hair from four body regions; head, dorsum, ventrum and extremities. They found that the tiger hair were mainly of four colours: white, black, yellow and brown. Mostly the proximal and distal parts of hair were dominated by white and black colour, whereas black, yellow and white colours were equally distributed in the middle part of the hair. The average length of hair was found as 14.4 mm with the range of 5.2 to 45.0 mm. The ranges of proximal and maximum hair width were 15.0 to 97.5 μm and 22.5 to 120 μm respectively with the hair width index of 68.3 ± 13.3.

They observed that the medulla of tiger hair was narrow with the medullary index of 32.5 ± 11.3. They observed three types of medulla simple, uniserial ladder and the mixture of both simple and uniserial ladder. They also mentioned that the type of medulla did not show change along the hair length.

They observed five different types of cuticular scale patterns, regular wave, single chevron, irregular wave, streaked and mixture of all four. They observed highest width values of cuticular scale width as 46 ± 12.6 μm at the middle part of hair, while height value at proximal part of hair as 8.6 ± 2.1 μm.

They noted three cross section shape in tiger hair; circular, concavo-convex and oval with medium sized medulla.

Sarkar and De (2013) worked on the dorsal guard hair of Indian species of rodents belonging to subfamily Sciurinae. They studied 5-6 individuals of 15 different species of animals. They worked on various parameters such as colour, bands, diameter, total length, cuticular scale pattern, scale margin, scale margin distance, scale count, side to side cuticular scale length, medulla pattern, medulla configuration and cross section shape. Based on the study of all above characteristics features of hair, they prepared a photographic key for the identification of species belonging to the subfamily Sciurinae.

They observed irregular mosaic cuticular scale pattern with wavy scale margins. They also found that the cortex was filled with the rectangular box like cells. The medulla was broad and amorphous with high medullary index. The pigment granules were equally distributed throughout the length of hair and ovoid bodies were seen towards cortex. They worked on the cortico-medullary index of hair, which was observed as 1.387.

Taru *et al.* (2013) studied morphological hair characteristics of South African Blue wildebeest (*Connochaetes taurinus*), Black wildebeest (*Connochectes gnow*) and Red Rock Hare (*Pronolagus crassicaudatus*) using scanning electron microscope. They collected ten samples of dorsal guard hair from dried skin of each species. They found that the hair of Blue wildebeest had concavo-convex cross section shape, irregular mosaic wave pattern of cuticular scale with smooth margins and near scale margin distance and small to medium medulla. The hair of black wildebeest showed circular shaped cross section, irregular wave pattern of cuticular scales with rippled margins and near scale margin distance with no distinct medulla. They stated that the hair of Red rock hare showed dumbbell shaped cross section, coronal cuticular scale pattern with scale margins and near scale margin distant and medulla with ten large cavities. They stated that hair colour was scale gray to dark brown; light brown to black and light colour in blue wildebeest, black wildebeest and red rock hare respectively.

Baddi *et al.* (2014) studied morphological feature of hair of sloth bear (*Melursus ursinus*) under different microscopes such as light, dark field, phase contrast and polarized microscope. They reported that the cuticular morphology was best observed under light, dark field and phase contrast microscope, cortex under dark field microscope and medulla under polarizing and dark field microscope.

They observed irregular mosaic cuticular scale pattern with wavy scale margins. They found that the cortex was filled with the rectangular box like cells. The medulla was broad and amorphous with high medullary index. The pigment granules were equally distributed throughout the length of hair and ovoid bodies were seen towards cortex. They worked on the cortio-medullary index of hair, which was observed as 1.387.

Bhat *et al.* (2014) worked on the microscopic features of hair of Lion, Stripped Hyena, Sloth Bear, Nilgai and Chinkara for species identification. The hair were collected from four different body parts; neck, back, abdomen and tail. They studied the microscopic features of hair such as medullary configuration, cuticular scale pattern and cross sectional shape. The medulla of all the species under study was multicellular with variations in the medullary pattern. The chinkara hair showed cloisonné and sloth bear hair had dark, filled medulla. The nilgai showed four cell layered medulla without air spaces between the cells, while hyena and lion showed similar medullary pattern as fine grained transparent structures with few circular vacuoles. The vacuoles were more in number in stripped hyena than that of the lion.

They further stated that there was remarkable difference in the cuticular scales of hair among the species under study. The lion hair showed transversal scale position, regular wave pattern with smooth scale margin and near distance between scale margins. The chinkara also showed similar features as that of lion hair with the difference of distant scale margin distance. The stripped hyena, sloth bear and nilgai had transverse irregular wave with rippled scale margin, while distance between scale margins was close in hyena and near in sloth bear and nilgai.

They also mentioned about the importance of cross section shape of hair for species identification. They found that the cross section shape of hair of lion, sloth bear and nilgai had circular outline, ablong outline in nilgai while concavo-convex outline in hair of chinkara. They finally concluded that all the morphological features of hair could be considered for the species identification.

Feilx *et al.* (2014) reviewed the thrichological study for species identification of mammals and its use in research and agriculture. They stated that the trichological study could be used for conservation of native breeds of sheep in Italy.

Khan *et al.* (2014) studied human hair from four different castes such as Awan, Gujjar, Rajput and Butt of the Pakistan. They observed hair for inner cuticle margin, cuticle thickness and ovoid bodies. They found that the inner cuticle margin and cuticle thickness were similar in majority samples as distinct and thick respectively. They further stated that the ovoid bodies were different amongst the caste. The ovoid bodies were highest in the individuals of butt caste while least in the samples of Gujjar caste. They also mentioned that there was non-significant difference in all the three parameters amongst the sex.

Lee *et al.* (2014) prepared photographic identification key of Korean mammalian hair. They worked on 22 endangered wild species and one domestic species of mammals. The dorsal guard hairs from adult animals were studied for cuticular scale pattern and medulla pattern. They noted that the hair profile of most of the species was straight and undulated, with the exception of uniquely wavy profile in lernidae and Moschidae, zigzag in Insectivora. They commented that the hair of *Sus scrofa* could be identified by its unique morphological feature, thick, straight profile with split ends. The cuticular scales of all species showed different pattern viz. broad petal shaped, narrow diamond shaped, broad diamond petal shaped, elongated petal shaped. They also noted the medulla pattern of all species at the shield part of hair.

Tridico *et al.* (2014) discussed about the myths, misconceptions, possibilities and pitfalls of morphological identification of animal hair. They finally concluded that the morphological identification of animal hair could be considered as robust and valid forensic technique.

Farag *et al.* (2015) conducted light microscopic and electron microscopic studies on hair from three body regions; dorsal neck, dorsum and flank of five wild animals viz. American black bear, Blue Nile monkey, Barhary sheep, Bacterian camel and Lama. They studied diameter of hair shaft, medulla index, cuticular index and scale count per unit length of hair for species identification from hair. They found no statistical difference in the diameter at proximal, middle and distal parts of hair shafts of various body regions amongst the individuals of same species, but the statistical difference was found in diameter of hair shaft in different species.

They found that there was small or no significant difference in the medullary index of individuals from the same species, but the significant difference was observed at species level. They commented that the medullary index was comparatively less in human hair and more in animal hair except bacterian camel in which it was more than the human hair. The cuticular index showed great significant difference among different species. The cuticular scale pattern showed variations between different species, but showed similarity in the individual of same species along the length of hair shaft and even in the different parts of the body. The scale count showed great variations in different species, but no significant difference was observed in the individual of same species.

Gharu and Trivedi (2015a) identified species of some wild and domestic carnivores by studying cuticular scale pattern, medulla and pigmentation of mid dorsal guard hair. Domestic cat had transversely placed, imbricate, flattened, regular cuticular scale pattern with distant scale margin distance and smooth scale margins. They also commented on scale margins that the margins were notched or chevron at some place throughout the length of hair. They observed that the domestic dog had flattened irregular and imbricate cuticular scales with smooth scale margins. One or two scales per row were present. Many scales were found with depression in the middle and the distance between scale margins was near to distant. In tiger cub, it was flattened, transverse or obliquely placed, imbricate regular wave pattern with rippled scale margins and scale margin distance as near and one two scales per row. The leopard cub showed regular, imbricate, petal or chevron scale pattern with smooth margins and one to two scales per row.

They further mentioned that the domestic cat had uniserial ladder and continuous medulla, domestic dog had continuous or fragmental, amorphous or broad, tiger cub had amorphous continuous or fragmental, while the leopard cub had amorphous, fragmental and broad.

The pigmentation in the hair of domestic cat was amorphous, but was not found in cuticle and cortex. In the hair of tiger cub, the pigmentation was present only in medulla which was coarse, fine and scattered. Very few brown granules and streaks were found in hair of dog and some pigments granules were present in cortex. The leopard hair showed broader distribution of pigments, which was found in the form of dark brown granules and streaks throughout the width of hair.

Gharu and Trivedi (2015b) compared the cuticular scales, medulla and pigments in the hair in sheep, goat, cow and buffalo. They noted that the hair of goat showed imbricate, mosaic, irregular wave pattern of cuticular scale, the scales were flattened and transversely placed. They also stated that no significant difference was observed in the cuticular scale pattern of male and female. Similar cuticular scale pattern was observed in sheep with the exception of smooth and distant scale margin. They stated that the cow hair showed regular or irregular wave pattern, the scales were transversely placed with one scale in one row and the margins of scales were crenated or rippled and distant and the margins were notched at some places. The buffalo hair showed irregular wave, crenate and mosaic cuticular scale pattern, the scale margins were rippled and near. They further commented that no sexual dimorphism was seen in cuticular scale pattern of both cow and buffalo.

The medulla in goat was simple, continuous broad with wavy to crenate margins, while in sheep hair, medulla was continuous, nodular and at some places narrow lattice with irregular margins. The dorsal guard hair of cow showed continuous simple, straight, irregular medulla with straight margins and some amorphous zones. The medulla of buffalo hair was continuous, simple with straight continuous margin.

The pigment granules were dark brown to light brown, regularly distributed throughout the cortex of goat hair while, in sheep the hair cortex was transparent with few sparely distributed pigment streaks. In cortex of cow hair, the pigments granules were distributed in regular manner and were dark brown to black in colour, while in buffalo, hair pigment granules were densely distributed and dark brown in colour with few streaks.

Gharu and Trivedi (2015c) studied the cuticular scale pattern, medulla pattern and pigments in the hair of animal from equidae family. The dorsal guard hair were collected from five adult individuals of horse and donkeys and two adults of mules. No significant difference was observed in cuticular scale pattern of horse, donkey and mule. The cuticular scale pattern was imbricate irregular in all the three species, while the scale margin distance was uneven in horse and broader in donkey and mule. The medulla of horse, donkey and mule was multicellular, continuous and fragmental. The medulla was found narrow in horse, broad in donkey and broadest in mule. The pigment granules were found dark brown to reddish brown with scattered distribution in horse, donkey and mule. The pigments were the combination of granules, streaks or linear arranged parallel to the long axis of hair. They finally stated that the hair of horse, donkey and mule could be differentiated by using medullar size.

Mukherjee *et al.* (2016) worked on comparative trichological analysis of common domestic herbivores and carnivores of Jabalpur district. They studied the medullary and cuticular scale pattern in hair of cow, buffalo, goat, rabbit, cat and dog. They reported that the cuticular patterns of buffalo was imbricated, irregular and rippled, while that of cow was crenate. The medulla of buffalo was brown in colour, wide occupying almost entire width of hair, thick, solid, while the medulla of cow was dark brown in colour and thin. The cuticular scale pattern in cow was crenated, irregular wave with imbricated margins. The medulla pattern and cuticular scale pattern was similar in both buffalo and cow with only difference in the distance between scale margin. The cuticular pattern in goat was imbricated, mosaic to irregular wavy with transversely, placed flattened scales, while medulla showed central dark core. In rabbit imbricated elongated pattern and triangular spinous and petal like scales, while the medulla was thin. The cuticular scales of cat were densely crenate or imbricated, tooth like and prominent over main point of shaft and the medulla was relatively wide. The hair of dog showed clearly imbricated cuticular pattern and vacuolated medullary pattern. They concluded that the variation in the cuticular and medullary pattern could be useful for the identification of herbivores and carnivores.

Prates *et al.* (2016) studied the hair remains found in the grave of hunter gatherer from Patagonia. They collected the hair samples from the hide pouch of grave. They studied the cuticular scale shape and structure, cuticular scale pattern and medulla pattern by preparing molds of cuticular scales and medulla slides. They concluded that the leather pouch found in grave was made out of skin from *Lagidium viscacia* and hair found inside it were of human.

Sahajpal *et al.* (2016) discussed the role of hair as physical evidence. The dorsal guard hair from *Pantholops hodgsonii* and *Capra hircus* species were collected. The samples were subjected for microscopic, scanning electron microscopy, keratin pattern study using SDS-PAGE and mitochondrial DNA study. They found distinct cuticular pattern in both the species as regular mosaic in *Pantholops hodgsonii* (Tibetan antelope) and regular wave pattern in *Capra hircus* (Pashmina goat). They noticed that the medulla was wide lattice in Tibetan antelope and wide in Pashmina goat. The cross section shape was also different as circular in Tibetan antelope and oblong to dumbbell in Pashmina goat. They concluded that the wool fibers / hair from different species could be identified on the basis of group of microscopic characters, molecular weight of keratins and the DNA mapping.

Verma *et al.* (2016) worked on bones, hair and skin of chital for species identification. The dorsal primary hair samples from 1400 chital were collected. They studied the cuticular scale pattern under scanning electron microscope and found that the pattern was regular with crenate scale margin and near scale margin distance. They used bone pieces for molecular analysis. They concluded that the DNA isolation technique and molecular identification technique by using SEM and EDX were accurate for species identification.

Kamalakannan (2017a) studied the microscopic structure of hairs of the male and female Hoolock gibbon. The hair samples from male and female hoolock gibbons were collected from the dry flat skin. He worked on the structural difference in medulla, cuticle and shape of cross section amongst the sex. He noted that the dorsal guard hair were sexually dimorphic. The coat colour was blackish in male, while in female, it was golden blonde. Other structural characters such as scale position, scale pattern, structure of scale margin, distance between scale margins, composition of medulla, structure of medulla and form of the medulla margin were similar in male and female. He also found that the shape of cross section was oval in male, while circular in female. He further concluded that the male and female Hoolock gibbon could be differentiated from the structural characters of dorsal guard hair. Similarly, the coat colours could be used for differentiating the Hoolock gibbon from the other primates.

Kamalakannan (2017b) studied the microscopic characteristics of hair from Namdapha flying squirrel (*Biswamoyopterus biswasi*) and differentiated with the other species of the squirrel on the structural pattern of dorsal guard hairs. The hair samples were collected from the dry flat skin of the flying squirrel. He specifically observed the cuticular scale characteristics, medullary configuration and cross section shape. He noted that in the flying, squirrel the scale position was transversal, scale pattern was regular wave, structure of scale margin was smooth and the scale margin distance was near. He found that the composition of medulla was multicellular, structure of medulla was multiserial ladder form of medulla margin was scalloped and the cross section shape was biconvex. He further concluded that the Namdhapa flying squirrel had unique microscopic characteristic features of dorsal guard hair as multicellular and multiserial medulla and the biconvex cross section shape. The most important feature for the differentiation of flying squirrel from all other squirrels was its hair coat colour, which was morocco red shaded at tip with white colour.

Kamalakannan (2017c) compared morphological characters of dorsal guard hair of large Indian civet (*Viverra zihetha*) and small Indian civet (*Viverricula indica*). He noted that the coat colour of large civet was dark hoary grey while yellowish or brown tinge fur in small civet. He found that the cuticular scales were different in both the species. The dorsal guard hair of large Indian civet had longitudinal scale position, large diamond petal scale pattern with smooth scale margins and distant scale margin distance, while small Indian civet cat had longitudinal or transversal scale position, regular wave or broad diamond petal scale pattern with smooth scale margin and distant scale margin distance. The medulla of mid dorsal guard hair of both large and small Indian civet showed similar observations as mùlticellular medulla composition, wide lattice medulla pattern with straight margin. He further stated that the cross sectional shape of hair were different in both the species as oval in large Indian civet and oval or oblong in *small Indian civet.* He finally concluded that the large and small Indian Civets could be differentiated on the basis of morphological features of hair.

Kamalakannan (2017d) studied the morphological, medullary, cuticular and cross sectional characteristics of dorsal guard hair of Nilgai (*Boselaphus tragocamelus*) using optical and scanning electron microscope. The hair samples were collected from three dried skins. Under morphological characters, he found that the length of hair was 198.2 ± 18.3 mm and diameter was 31.9 ± 11.7 μm. The hair showed two colour bands with earth yellow colour towards the base and black colour towards the tip. The profile was slightly wavy. The cuticular scales showed transversal scale position, regular wave pattern, smooth scale margins and near scale margin distance. He found that 186.6 ± 39.8 cuticular scales per mm length of hair 94.7 ± 2.2 μm length and 10.1 ± 1.1 μm width of the cuticular scale. He worked on the medullary features of hair and noted medullary composition as unicellular regular, simple medullary structure, straight medulla margin with 93.3 ± 4.7 μm medullar diameter. He noted oval cross sectional shape of dorsal guard hair and concluded that the combination of microscopic hair characters could be used for identification of species.

Kamalakannan (2017e) studied dorsal guard hair of *Ratufa bicolor* (black giant squirrel), *Ratufa indica* (Indian giant squirrel) and *Ratufa macroura* (Grizzled giant squirrel). The hair samples were collected from flat dried skin specimen present in the museum of zoological survey of India, Kolkata. The medullary structure of all three species was found same as multicellular in row, multiserial ladder medullary pattern and scalloped margin. He observed that the cuticular scale pattern was same in *R indica* and *R. macroura* as transversal scale position, irregular wave cuticular scale pattern, rippled scale margins and near scale margin distance while, different in *R. bicolor* as transversal scale position, regular wave cuticular scale pattern, smooth cuticular scale margin with near scale margin distance. The oval cross sectional shape was noticed in *R. indica* and *R. macrousa,* while oblong shaped cross section in *R. bicolour*. He concluded that the photographic representation of the three giant squirrel species could be used for differentiation of the endangered giant squirrel species.

Kamallakannan and De (2017a) compared the hair morphology of three species of Indian otter viz. *Aonyx cinerea*, *Lutra lutra* and *Lutrogale perspicillata*. The hair samples were collected from the mid dorsal region of the dried skin. They worked on the cuticular characters like scale position, scale pattern, structures of scale margins, distance between scale margins and medullary characters such as width, composition, structure and form of margins of the medulla and cross section shape of the hair. They reported that hair of *Aonyx cinerea* had transversal scale position, irregular wave pattern, rippled scale margins and close distance between scale margins. The medulla showed multicellular composition, wide aeriform lattice structure and scalloped margin and biconvex cross section shape. Similar structural characters were observed by them in hair of *Lutra lutra* with the exception in cross section shape as oblong. The hair of *Lutrogale perspicillata* showed transversal scales position, regular wave scale pattern, the smooth scale margin and close scale margin distance. The medulla was unicellular irregular lattice, scalloped medullar margin and oblong shape in cross section. Finally, they concluded that the hair characters of *Aonyx cinerea* and *Lutra lutra* were similar. However, the shape of hair cross section was different, while the shape of *Lutrogale perspicillata* were totally different from other two species of otter with the similarity of cross section shape, which was same as that of *Lutra lutra*.

Kamalakannan and De (2017b) microscopically examined morphological characters of the dorsal guard hair of Striped Hyena. They noted transversal scale position, scale pattern regular wave, structure of scale margins as smooth and the distance between the scale margin near in the cuticle. The medulla showed unicellular regular composition, structure of medulla as simple or narrow medulla lattice and form of medulla margins straight. They reported that the shape of cross section of hair was oblong, which was unique in hyena.

Negi *et al.* (2017) compared medullary index of various domestic animals. The hair samples were collected from young and adult animals of different breeds of cow, buffalo, dog and cat. They found that the medullary index of cow was 0.52 - 0.78, buffalo was 0.532 - 0.78, dog was 0.542 - 0.75 and cat was 0.52 - 0.78. They further mentioned that there might be similarities in the medullary index of different species and concluded that the medullary index could be the one of the parameters of hair considered for species identification.

Rana *et al.* (2017) compared the morphology of tactile hair of tiger and leopard. The shaded tactile hair were collected from one individual of each species. They observed two colour bands with white at base and brown at tip in both the species. The cuticular scale pattern was observed as streak with crenate margin and near scale margin distance in tiger, while in leopard, it was regular wave with smooth scale margin and near scale margin distance. They found similar medullary pattern in the tactile hair of both tiger and leopard. The medulla pattern observed was uniserial ladder, but slight difference was observed in the length of individual medullar segment. They further concluded that the cuticular scale pattern could be used as a tool for identification of tiger and leopard species.

3

Material and Methods

The present study on trichology of domestic and wild animals was conducted in the Postgraduate laboratory of the Department of Veterinary Anatomy and Histology, Nagpur Veterinary College, Nagpur. The study was undertaken with an aim of species differentiation from the guard hair of domestic as well as wild animals. The hair samples were collected from a total of eight domestic animal species and six wild animal species. Six animals from each species were studied and only three animals studied from the Sambar species as only three animals from this species was reported for post-mortem to the Nagpur Veterinary College, Nagpur within the period of research.

The domestic animal species considered for the present study were Cattle (*Bos indicus*), Buffalo (*Bubalus buballis*), Sheep (*Ovis aries*), Goat (*Capra hircus*), Horse (*Equus caballus*), Pig (*Sus scrofa domesticus*), Cat (*Felis catus domesticus*) and Dog (*Canis lupas familiaris*) while, the wild animal species were Spotted Deer (*Axis axis*), Sambar Deer (*Cervus unicolour*), Nilgai / Blue Bull (*Boselaphus tragocamelus*), Hanuman Langure / Grey Langure (*Semnopithecus entellus*), Tiger (*Panthera tigris*) and Leopard (*Panthera pardus*). The hair samples were collected from the animals, which were brought to the Department of Veterinary Pathology, Nagpur Veterinary College, Nagpur for Post mortem examination after accidental and natural death. Some samples were collected from Transit Centre, Seminary Hills, Nagpur and Gorewada Research Centre, Gorewada, Nagpur during their treatment. The samples were collected from six different body regions viz. head, neck, back, lateral abdomen, lateral thigh and tip of tail. The tufts of 20-25 hair were plugged with the help of tweezers from all the six body regions. The hair samples thus collected were stored in zip locked pouches without adding any preservatives till further processing.

The samples were then processed for physical, microscopic and scanning electron microscopic observations. The hair were also processed for isolation of DNA to study the species specific bands using PCR technique.

3.1. Physical characteristics

The hair samples were first washed with regular soap solution for physical observations. The hair samples were dried on the blotting paper inside the air tight chamber. The hair was placed on microscope glass slide along with the longitudinal axis of the slide, so that profile stands out clearly. It was then covered with the cover glass with a small drop of glue on each corner of the cover glass to restrict the movement of hair (Teerink, 1991). The slides were observed under light microscope for colour, number of colour bands and hair profile. Nomenclature of colour of hair was followed after Ridgway (1886) and the nomenclature of hair profile was followed after Brunner and Coman (1974).

Total length of hair was measured by straightening the hair on plain paper and then measured with the help of a ruler scale (Sato *et al.* 2006). The hair diameter was measured at proximal (part of the hair shaft towards the hair root), middle and distal (part of the hair shaft towards the tip of hair) parts of the hair shaft by making use of Aerospace 150 mm Digimatic Vernier Calliper.

3.2. Cuticular characteristics_

The cuticular scales were studied by using negative scale cast of hair. The negative scale cast was prepared after modifying the method given by Deedrick and Koch (2004). The hair samples were cleaned thoroughly with soap solution for 10 minutes to remove debris from the hair surface. The hair were then rinsed with distilled water to remove the excess soap from it and tap dried with blotting paper. The dried hair then subjected to the solution containing equal amount of petroleum ether and absolute alcohol (1:1) for 10 minutes to remove greasy material from the hair surface. A thin film of transparent nail enamel was prepared on glass slide with the help of brush provided with the enamel bottle. The hair was placed on semidried nail enamel film with its one end free from the surface of nail enamel. The nail enamel was allowed to dry for 20-30 minutes in closed container to avoid dust particle to get stick to nail enamel and then the hair was plucked off with the free end by using forceps. The negative cast thus obtained was observed under light microscope at magnification 100x, 200x and 400x for cuticular scale position, cuticular scale pattern, cuticular scale margins and cuticular scale margin distance. The structural nomenclature of cuticular configuration is followed after Brunner and Coman (1994), Teerink (1994) and Marinis and Asprea (2005).

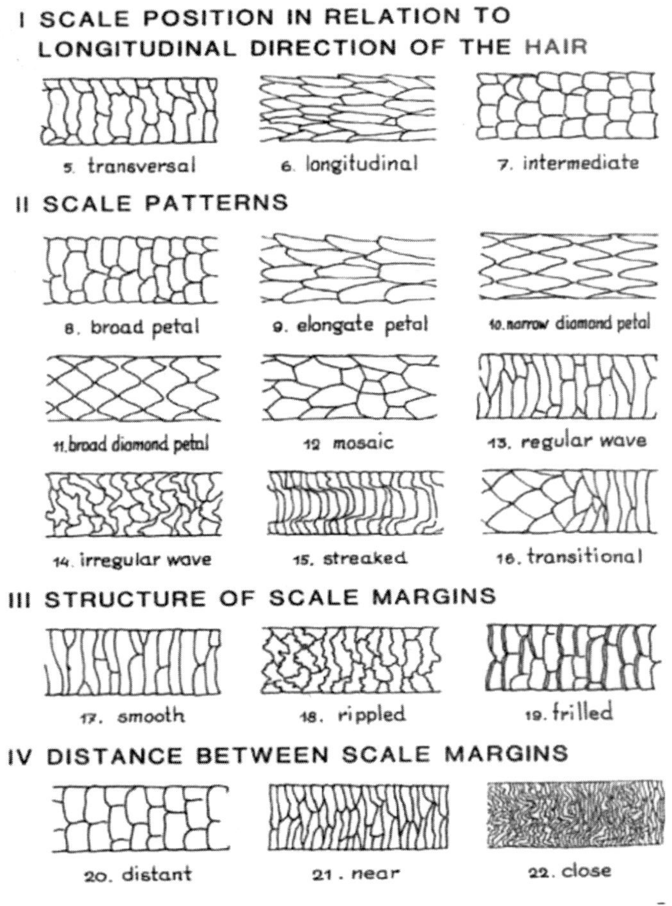

Fig. 3.1. Structural nomenclature of cuticular configuration (Teerink, 1994)

The cuticular scale width, cuticular scale length and number of cuticular scale per mm length of hair were noted using Zen software. The slides were micro-photographed using Axiocam ERs 5s camera fitted to the Zeiss photomicroscope. The numerical observations were subjected for statistical analysis as per the standard methods given by Snedecor and Cochran (1994).

The cuticular scale width, cuticular scale length and number of cuticular scale per mm length of hair / The numerical observations were noted from the negative scale-cast at three different positions of hair viz; proximal (part of the hair shaft towards the hair root), middle and distal (part of the hair shaft towards the tip of hair) as follows:

1. Scale width (µm) was measured from five randomly selected cuticle scales.
2. Scale height (µm) was measured from five randomly selected cuticle scales.
3. Number of cuticular scales per mm length of hair was observed at three randomly selected regions of the hair shaft.

3.3. Characteristics of medulla and cortex of hair

The medulla and cortex characters were studied by using the whole mount of hair, which was prepared by the method given by Kshirsagar *et al.*, (2009). The hair were cleaned with soap solution followed by distilled water 2-3 times to remove dust and debris if any. The hair were tap dried with blotting paper and then transferred to the solution containing petroleum ether and absolute alcohol in equal proportion (1:1) for 2-3 minutes. The ether alcohol was used to remove fatty or greasy material present on the hair surface. The hair were again blot dried with blotting paper. The dried hair were then transferred to the hydrogen peroxide for 2 hours, to make the line of demarcation between cortex and medulla lucid / transparent. Then the hair were tap dried with blotting paper and shifted to the clearing agent, the xylene for minimum of 48 hrs. depending upon thickness of hair. The cleared hair were mounted on microscope glass slide with Natural Canada Balsam. The whole mounted hair slides were observed under transmitted light microscope at 100X, 200X, 400X and 1000X. The slides were photographed with the help of Axiocam ERs 5s camera fitted to the Zeiss photomicroscope. The whole mounts were used to note the composition of medulla, structure of medulla, medulla margin, colour of medulla, colour of cortex and pigment granules.

The structural nomenclature of medullary configuration is followed after Brunner and Coman (1994), Teerink (1994) and Marinis and Asprea (2005).

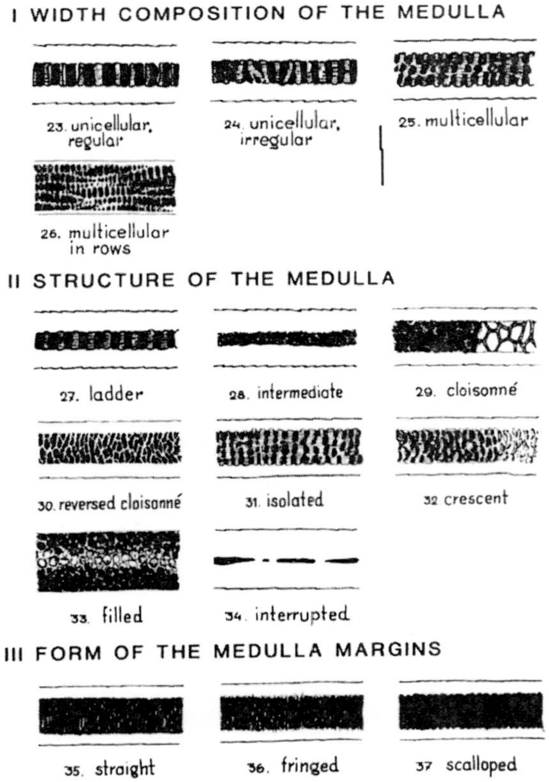

I WIDTH COMPOSITION OF THE MEDULLA

23. unicellular, regular
24. unicellular, irregular
25. multicellular
26. multicellular in rows

II STRUCTURE OF THE MEDULLA

27. ladder
28. intermediate
29. cloisonné
30. reversed cloisonné
31. isolated
32 crescent
33. filled
34. interrupted

III FORM OF THE MEDULLA MARGINS

35. straight
36. fringed
37 scalloped

Fig. 3.2. Structural nomenclature of medullar configuration (Teerink, 1994)

The whole mount was also used for determination of medulla thickness / width, hair width and medullary index. The whole mount of guard hair were photographed with the help of Axiocam ERs 5s camera fitted with the Zeiss microscope and micrometry was done by using Zen software. The observations thus noted were statistically analysed as per the standard methods given by Snedecor and Cochran (1994).

The numerical features were noted as per Kitpipit and Thanakiatkrai (2013) as follows:

1. The Hair width (μm) was measured at three positions of hair viz; proximal (part of the hair shaft towards the hair root), middle and distal (part of the hair shaft towards the tip of hair).
2. The medulla width (μm) was measured at three positions of hair viz; proximal, middle and distal as above.
3. Medullary Index was calculated as per the formula given by Sahajpal *et al.*, (2008).

$$\text{Medullary Index (MI)} = \frac{\text{Medulla width (MW)}}{\text{Hair width (MW)}} \times 100$$

3.4. Characteristics of cross section of hair

The cross sections of the hair were obtained after modifications in the method given by Shelly and Ohman (1969). Ten to twenty small segments of hair were wrapped in a small piece of tea bag paper. The wrapped hair were kept in diethyl ether for 10 min at room temperature.

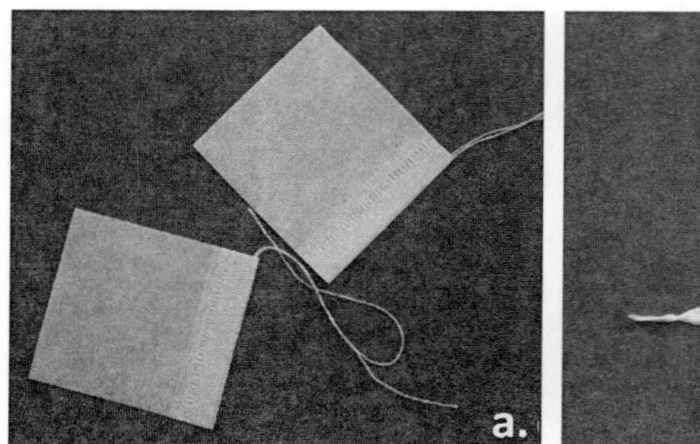

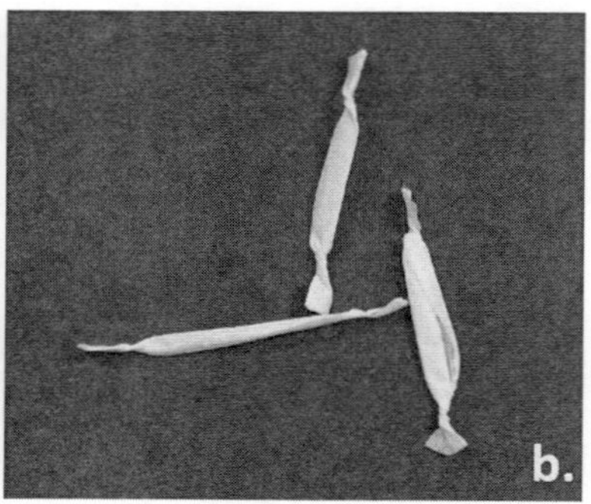

Fig.3.3. Photograph showing (a) tea bag and (b) wrapped hair for cross section of hair

The hair were then rehydrated in distilled water at 60°C for 10-15 min. The wrapped hair were removed from distilled water and embedded in paraffin at 60°C for 30 min. The blocks were prepared by using 'L' blocks. The prepared blocks were trimmed and mounted on block holder and sectioned with the help of fresh surgical blade No. 22, which were fitted in the specially designed blade holding knife for rotary microtome. The adhesive used for the adhesion of cross section was Mayer's Egg Albumin. Thus prepared slides were deparaffinised and mounted with the help of mountant, the Natural Canada Balsam. The unstained cross sections of hair were observed under transmitted light microscope at 50X, 100X, 200X, 400X and 1000X magnification. The cross sections were observed to note cross section shape, medulla shape and medulla margin. The structural nomenclature of medullary configuration is followed after Brunner and Coman (1994), Teerink (1994) and Marinis and Asprea (2005).

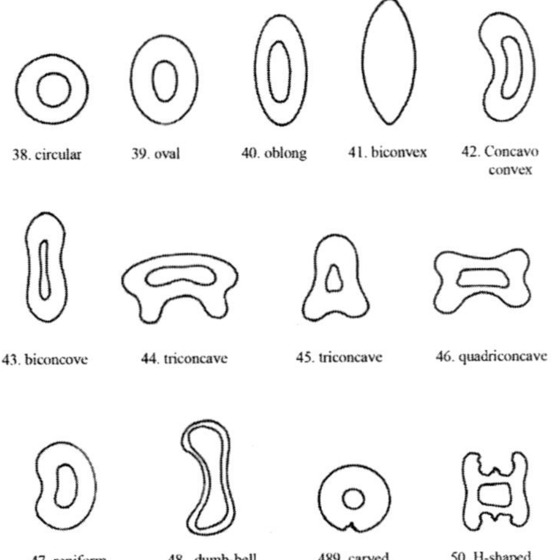

Fig. 3.4. Structural nomenclature of cross sectional configuration (Teerink, 1994)

The cross sections were micro-photographed with the help of Axiocam ERs 5s camera fitted with the Zeiss microscope and micrometry was done by using Zen software. The cross sections were also observed to note the hair diameter, diameter of medulla, cortex width and cuticular width. The observations thus noted were statistically analysed as per the standard methods given by Snedecor and Cochran (1994).

The cross sections were used to note the following numerical feature at three randomly selected cross sections as follows :

1. Cuticle width (μm)
2. Minimum hair diameter (μm) was measured at the minor axis at the widest point of the hair.
3. Maximum hair diameter (μm) was measured at the major axis at the widest point of the hair
4. Medulla diameter (μm)
5. Cortex width (μm)
6. Hair index

$$\text{Hair index} = \frac{\text{Maximum diameter of cross section}}{\text{Minimum diameter of cross section}} \times 100$$

7. Cuticle index

$$\text{Cuticle index} = \frac{\text{Maximum diameter of cross section}}{\text{Cuticular thickness of cross section}} \times 100$$

3.5. Scanning electron microscopy

The hair samples were washed with absolute alcohol at room temperature for about 24 hrs. followed by packaging in zip-lock bags and stored at 4°C till further analysis. For scanning electron microscopy, the hair samples were dissected in pieces of 5mm size and placed on sample holder. Finally, they were coated with gold by using JEOL JFC-1600 auto fine coater and observed under a JEOL JSM-6610LV scanning electron microscope at different magnifications at Department of Veterinary Anatomy, College of Veterinary Science and Animal Husbandry, Pantnagar, Uttarakhand and were photographed. The photoelectronmicrographs were studied for differential features of the surface pattern.

The structural nomenclature of cuticular, medullary and cross sectional configuration is followed after Brunner and Coman (1994), Teerink (1994) and Marinis and Asprea (2005).

3.6. Molecular characterization of hair samples

Identification of animal species on the basis of biological samples like hair plays very important role in legal cases. Species identification can be done on the basis of macroscopic as well as microscopic studies of hair, when the hair samples are available in good amount. Many times, because of unavailability of required quantity of sample, the macroscopic and microscopic analysis can restrict the identification to the level of group of species. The molecular characterisation of hair plays very important role in narrowing the identification from group of species upto species levels.

3.6.1. DNA isolation

Hair samples from the back region of all the species under study were considered for the isolation of DNA. The hair were chopped finely to make the hair powder. The chopped hair were washed with absolute ethanol to remove the dirt and greasy contents from the hair.

DNA extraction was carried out by using Qiagen's Q1Aamp DNA Investigator Kit (Cat. No. 56504) as per the manufactures protocol as depicted below.

The washed hair samples were dried properly and then 300 µl buffer ATL, 20 µl Proteinase k and 20 µl 1M Dethiothreitol (DTT) were added in microcentrifuge tube. The mixture was vortexed for 10 seconds and kept in dry bath minimum for

1 hr or maximum upto complete lysis of hair at 56⁰C.

After incubation following steps were followed for the isolation of DNA.

Microcentrifuge tubes were centrifuged at 8000 rpm for 10 min.

500 µl buffer AL was added to tube and mixed by vortexing for 10 sec

↓

The tubes were incubated in dry bath for 10 min at 10⁰C

↓

The microcentrifuge tubes were centrifuged at 8000 rpm for 10 min

↓

150 µl absolute ethanol was added and vortexed for 15 sec.

↓

The tubes were centrifuged at 8000 rpm for 10 min

↓

Supernatant was transferred from microcentrifuge tube to Q1Aamp Mini Elute column

↓

The columns were centrifuged at 8000 rpm for 1 min

↓

Q1Aamp MiniElute column was placed in a clear 2 ml collection tube and discarded the collection tube containing the flow through.

↓

In the Q1Aamp MiniElute column 700 µl buffer AW2 was added and centrifuged at 8000 rpm for 1 min.

↓

After centrifugation the flow through was discarded and in the pallet 700 µl absolute Ethanol was added

↓

The column were centrifuged at 8000 rpm for 1 min

↓

The flow through was discarded and columns were centrifuged at 14000 rpm for 3 min.

↓

The Q1Aamp Mini Elute column was placed in 1.5 ml microcentrifuge tube and discarded flow through along with collection tube.

↓

The tubes were incubated at room temperature for 10 min with open lid.

↓

30 µl buffer ATE was added to the centre of the Q1 Aamp Mini Elute membrane.

↓

The microcentrifuge tubes were incubated at 20°C for 1 min

↓

The tubes were then centrifuged at 14000 rpm for 1 min.

All the centrifugation of the above protocol was carried out at room temperature ie. between 15-20°C.

The quantification of extracted DNA was done by using Nanodrop spectrophotometer (Thermo scientific, USA).

3.6.1.a. Molecular characterization of hair sample using *cytochrome b* and *12srRNA* gene

Molecular characterization for identification of species from hair was performed using conventional polymerase chain reaction. The product was amplified by using following universal primer and amplified *cytochrome b* (398bp) and *12s rRNA* (400bp) gene segments as per the details given by Sahajpal and Goyal 2010.

The primer sequence used for *cytochrome b* & *12srRNA* is mentioned table below.

Details of primer sequence

Target Gene	Sequence (5' to 3')	Base Pair (bp)	Source
Cytochrom b	TACCATGAGGACAAATATCATTCTG	398	Sahajpal and Goyal 2010
	CCAAGAAGAGCTGCAACAGATA	398	
12s rRNA	CAAACTGGGATTAGATACCCCACTAT	400	
	GAGGGTGACGGGCGGTGTGT	400	

3.6.1.b. Conditions for optimization of *Cytochrome b*

The PCR reaction was carried out with 25µl reaction mixture using following ingredients mentioned in table below.

Details of ingredients used for amplification of *cytochrome b* gene:

Ingredients	**Volume (µl)**
2X Go Taq green master	12.5
Nuclease Free Water (NFW)	8.5
Forward Primer (10pmol/ µl)	1.0
Reverse Primer (10pmol/ µl)	1.0
DNA Template	2.0
Total	**25 µl**

In order to optimize the amplification following denaturation, annealing and extension conditions applied are mentioned in table below as follows

3.6.1.c. Conditions for optimization *cytochromeb*

Stages	**PCR conditions**	**Temp. (^{0}C)**	**Time**
Stage 1	Initial Denaturation	95	10 min
Stage 2	Denaturation	95	45 sec
	Annealing	51	45 sec
	Extension	72	2 min
Stage 3	Final Extension	72	10 min
Stage 4	Holding	4	∞

3.6.1.d. Agarose Gel Electrophoresis

Agarose gel electrophoresis was performed to visualise the amplified PCR product for *cytochrome b* primers using horizontal gel electrophoresis assembly (GENAXY GENECO – H- MAXI). The gel tray was adjusted on plain surface and gel comb was placed in casting tray, 1mm above the base of platform. Molecular grade agarose (Invitrogen) was boiled in 1X Tris Acetic Acid EDTA (TAE) to prepare ---- agarose gel. The gel was cooled about 45-50^{0}C and ethidium bromide was added @ 0.5 µg/ml in the gel. The agarose gel was poured in gel caster and kept for solidification. After complete solidification the comb was removed and tray was submerged in electrophoresis buffer (1X TAE) tank and 10µl PCR product was loaded in wells. Similarly in one well 5µl of 100bp DNA marker (Promega, USA) was loaded and electrophoresis was carried out at 100V till complete migration of PCR product. After complete electrophoresis the gel was visualized under UV rays in Gel Documentation system (Biorad) for determination of amplicon size of 398bp.

3.6.1.e. Conditions for optimization of *12srRNA*:

The PCR reaction was carried out with 25µl reaction mixture using following ingredients mentioned in table below.

Details of ingredients used for amplification of *12srRNA* gene:

Ingredients	Volume (μl)
2X Go Taq green master	12.5
Nuclease Free Water (NFW)	8.5
Forward Primer (10pmol/ μl)	1.0
Reverse Primer (10pmol/ μl)	1.0
DNA Template	2.0
Total	**25 μl**

For amplification of DNA following optimization conditions were followed as detailed in table below.

Conditions for optimization of *12srRNA*:

Stages	PCR conditions	Temp. (^{0}C)	Time
Stage 1	Initial Denaturation	94	5 min
Stage2	Denaturation	95	45 sec
	Annealing	60	45 sec
	Extension	72	1 min
Stage3	Final Extension	72	1 min
Stage4	Holding	4	∞

After completion of PCR the gel electrophoresis was followed as detailed above in-----

3.6.2. Confirmation of species

After confirmation of amplicon size using each gene primers the PCR products (100 μl each) were further forwarded for sequencing to Eurofins Bengaluru and then BLAST at NCBI website. After BLAST the confirmed sequences were submitted to NCBI and phylogenetic tree was generated using the data available on NCBI.

4

Result and Discussion

4.1. Physical characteristics

4.1.1. Colour, number of bands and hair profile

The physical characteristics of hair like colour, number of bands and hair profile showed huge variations in the same individual from region to region particularly in the domestic animal species. So, these features of hair cannot be considered as an important tool for species identification from hair. These findings are in accordance with the observations of Sarkar et al. (2011). Who mentioned that the coat of mammalian species varies with season, habitat and sex of the individuals.

In cattle (*Bos indicus*) the colour of hair varied from mummy brown to Prout's brown colour in various body regions of the same individual. No colour bands were noted and had curved hair profile at all the body regions except tail (Fig. 4.1.). The hair from tail region, were slightly wavy, black coloured and no colour bands were found. In agreement with the findings of present study Koppikar and Sabnis (1976) also reported curved hair profile and high variation in hair colour in Cattle.

The colour of hair in Buffalo (*Bubalus bubalis*) was found black throughout the length of shaft of hair in all the body regions. No colour bands were observed in the hair of any of the body regions. The profile was noted as straight (Fig. 4.2.). Similar findings regarding hair colour was noted by Dharaiya and Soni (2012). They observed black colour at the proximal as well as distal part of hair in Buffalo.

The hair from Sheep (*Ovis aries*) showed uniform colouration through the length of hair shaft whereas, variation in colour was noted amongst the body regions. Mostly the hair had straight profile at the proximal part of hair while, wavy profile was noted at the middle and distal parts of hair in almost all six body regions (Fig. 4.3.). In contrast with the findings of the present study Dharaita and Soni (2012) noted white colour at the proximal part of hair, while brownish colour at the distal part of hair in Sheep. These variations in the color of hair may be attributed to the variation in breeds. However, the present findings regarding hair profile are in agreement with the findings of Marinis and Asprea (2005). Who mentioned that the Sheep hair had undulating profile.

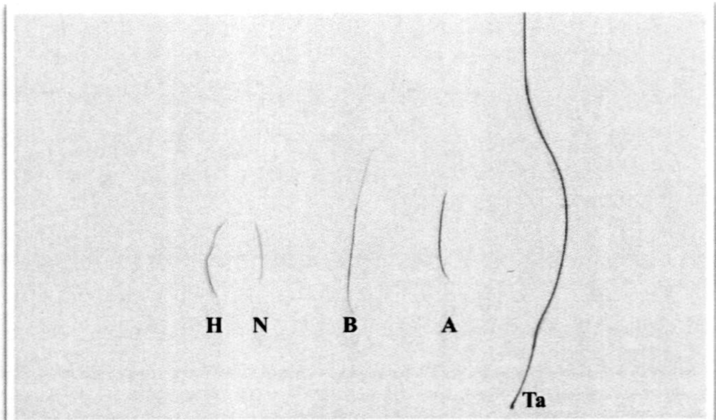

Fig. 4.1: Photograph showing hair profile in Cattle (*Bos indicus*)

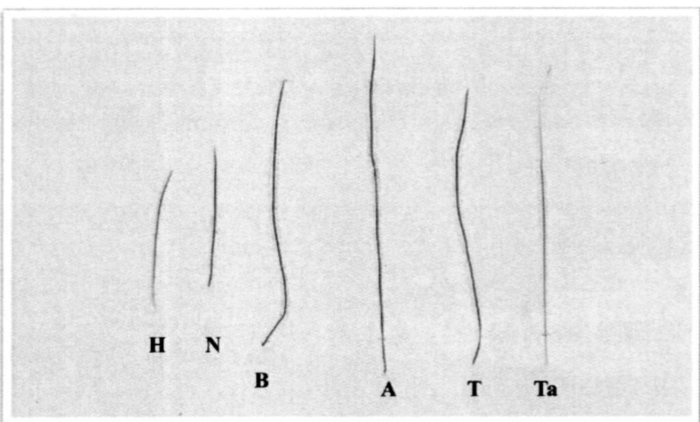

Fig. 4.2: Photograph showing hair profile in Buffalo (*Bubalus bubalis*)

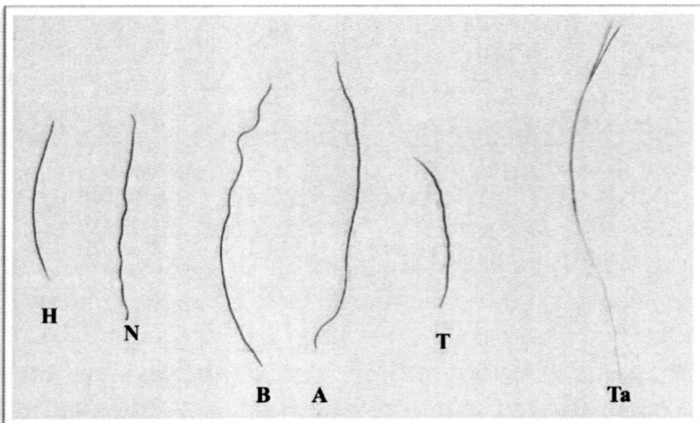

Fig. 4.3: Photograph showing hair profile in Sheep (*Ovis aries*)

Goat (*Capra hircus*) hair from all the six body regions showed uniform black colour through the length of hair shaft. The hair from head region had curved profile whereas hair from other five body regions showed straight profile with slight undulance at the tip of hair (Fig. 4.4.). Similar, hair profile was observed by Koppikar and Sabnis (1976) in Goat hair. However, Marinis and Asprea (2005) noted straight hair profile in Goat, which was in partial accordance with the present findings. The findings of the Dharaiya and Soni (2012) observed darker colour at the proximal part of hair and lighter colour at the distal part of hair. These findings are not in agreement with the observations of the present study.

The hair from head, neck, back, abdomen and thigh regions of Horse (*Equus caballus*) had straight to slightly wavy profile (Fig. 4.5.) with darker colour shade at the proximal part of hair and light colour shade at the distal part of hair, while the hair from tail region had straight profile. The colour of hair from tail was not uniform throughout the length of hair shaft.

The hair of Domestic Pig (*Sus scrofa domesticus*) showed splitting at the tip part. The hair from all the regions were splitted into two to three parts (Fig. 4.6.). Similar findings were recorded in Wild Boar by Marinis and Asprea (2006). They reported that the hair of Wild Boar could be easily identified with the help of gross observations on the basis of splits in the hair. Domestic Pig hair had uniform black colour throughout the length of hair, with straight profile at all the six body regions (Fig. 4.6.). These findings regarding hair profile were in concurrence with the findings reported by Marinis and Asprea (2006).

The hair of Cat (*Felis catus familiaris*) showed variations in the number of bands and colour of hair among all the body regions. However, profile was straight and was uniform in all six body regions (Fig. 4.7.).

In Dog (*Canis lupus domesticus*), variation in colour was noted amongst all the body regions. The hair profile was straight at all the six body regions (Fig. 4.8.).

The hair of Spotted Deer (*Axis axis*) mostly showed tow colour bands at all the six body regions. The raw brown colour band was observed at the distal half of hair, while white coloured band was seen at the proximal half of the hair. The hair from head, neck, back, abdomen and thigh regions had straight profile while, tail region showed straight profile at the proximal half and slightly wavy profile at the distal half of hair (Fig. 4.9.). Dharaiya and Soni (2012) observed while colour at the proximal part of hair and brown to yellow colour at the distal part of hair, which was akin with the observations of the present work. Koppikar and Sabnis (1976) reported slightly wavy profile in the hair of Spotted Deer and white colour at the proximal part, brown at the middle part and yellow at the distal part of the hair. These findings are in accordance with the findings recorded during the present study.

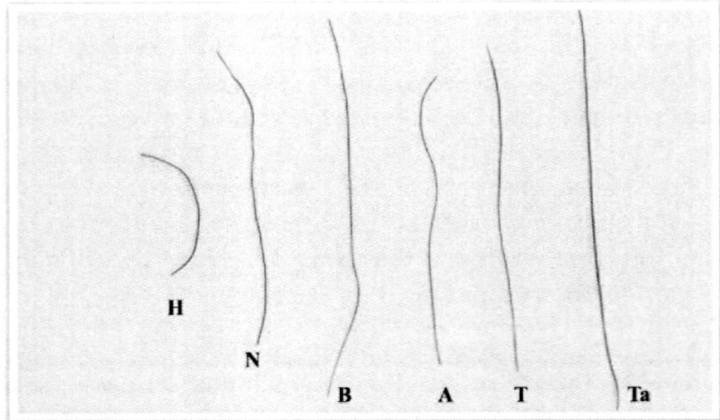

Fig. 4.4: Photograph showing hair profile in Goat (*Capra hircus*)

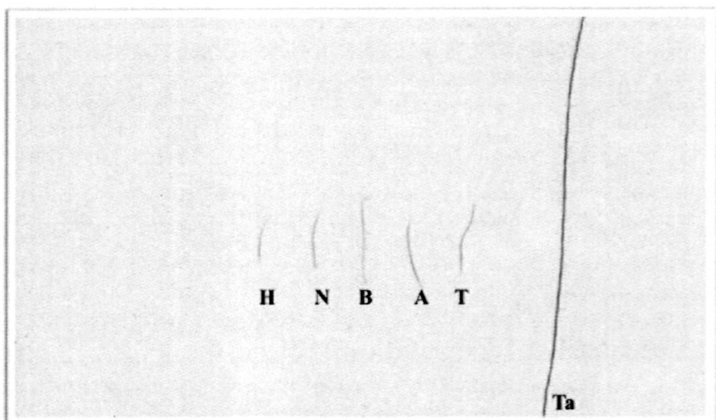

Fig. 4.5: Photograph showing hair profile in Horse (*Equus caballus*)

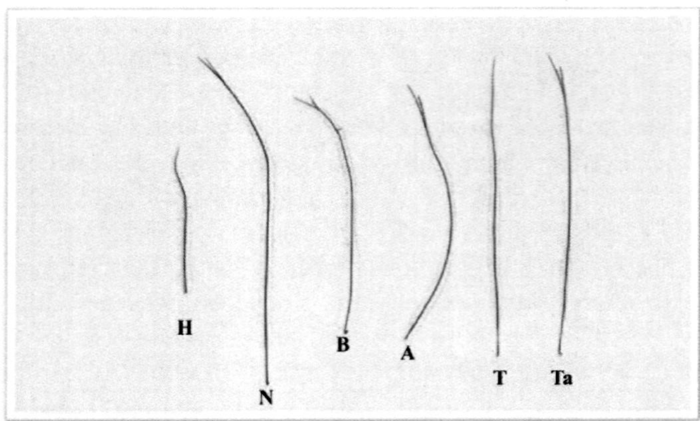

Fig. 4.6: Photograph showing hair profile in Domestic Pig (*Sus scrofa domesticus*)

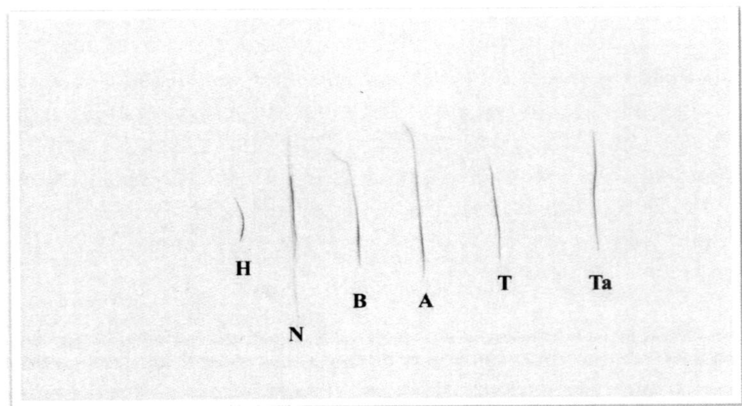

Fig. 4.7: Photograph showing hair profile in Cat (*Felis catus familiaris*)

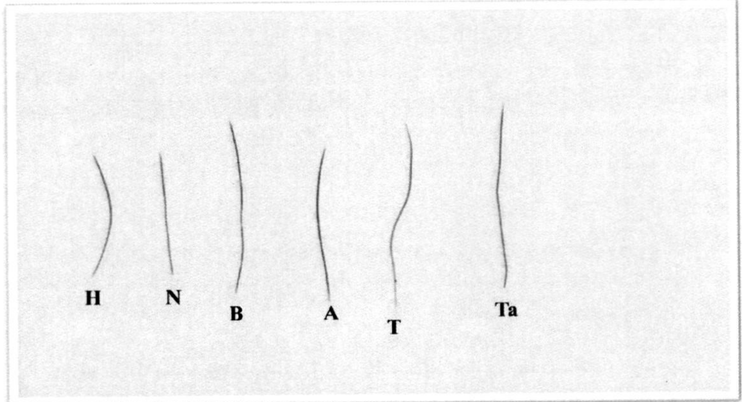

Fig. 4.8: Photograph showing hair profile in Dog (*Canis lupus domesticus*)

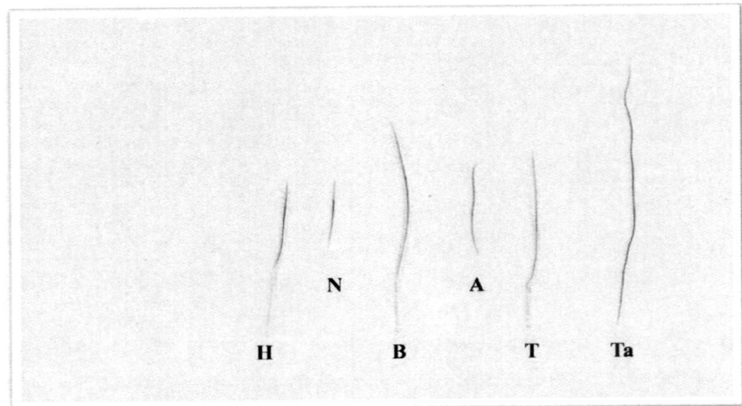

Fig. 4.9: Photograph showing hair profile in Spotted Deer (*Axis axis*)

Sambar (*Rusa unicolor*) hair had small white coloured band at the base of hair while, other part of hair shaft had cinnamon brown colour at all the regions of body. The profile of hair was straight at the head, neck, abdomen and back region. The hair from back and tail regions showed straight hair profile at the proximal half of hair and wavy profile at the distal part of hair (Fig. 4.10.). Dharaiya and Soni (2012) observed white colour at proximal hair part and yellowish colour at the distal hair part. These observations are in partial agreement with the findings of the present study. Similar observations were reported by Koppikar and Sabnis (1976) in Sambar.

Nilgai (*Boselaphus tragocamelus*) hair had two colour bands at all the body regions. The band towards the tip of hair was black coloured and the band towards proximal part was tawny olive brown coloured. The band at the distal part was nearly one third and proximal band was two third of the total length of hair. The hair of Nilgai showed straight profile in the hair of all the regions of body (Fig. 4.11.). Koppikar and Sabnis (1976) reported white colour at proximal part, brown at middle part and black at the distal part of hair, these findings are in partial agreement with the present findings. De and Chakraborty (2012) reported straight profile of hair, and uniformly black coloured hair in Nilgai. The observations regarding hair profile are in partial accordance with the findings of the findings of the present study.

The hair from wild herbivores considered for the present study could be distinguished from that of the domestic herbivore on the basis of its gross examination because of its metallic appearance.

The hair of Hanuman Langur from all the six body regions had slightly wavy profile (Fig. 4.12.) and was uniformly hair brown coloured. These observations are in full accordance with the findings reported by Sarkar *et al.* (2011). They reported slightly wavy profile, hair brown coloured and number of colour bands as none in the hair of Hanuman Langur. The observations of Koppikar and Sabnis (1976) and Soni and Dharaiya (2012) were not in agreement with the present findings. Who reported uniform gray coloured hair.

In the hair of Tiger (*Panthera tigris*) huge colour variations were observed at all the body regions. Some hair were found with uniform white, tawny olive or black colouration. While some are with the five bands of different colours. Mostly the proximal part had white colour and distal part had black colour and in the middle tawny olive and hair brown coloured bands were separated by a thin buffy white band. These findings of the present work regarding hair colour are in agreement with the findings reported with the findings reported by Kitpipit and Thenakiatkrai (2013). The straight hair profile was observed in the hair of all the six body regions (Fig. 4.13).

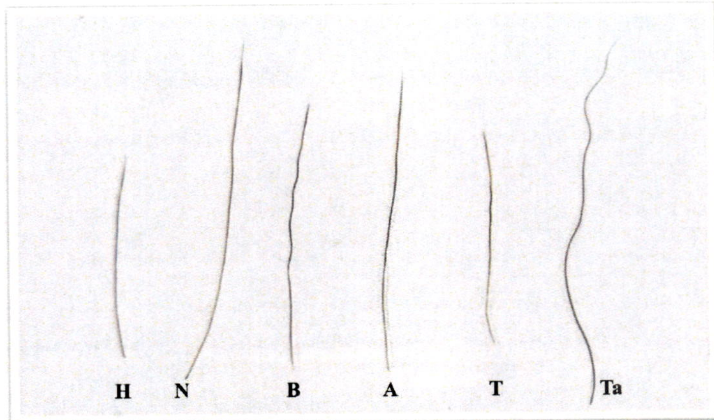

Fig. 4.10: Photograph showing hair profile in Sambar (*Rusa unicolor*)

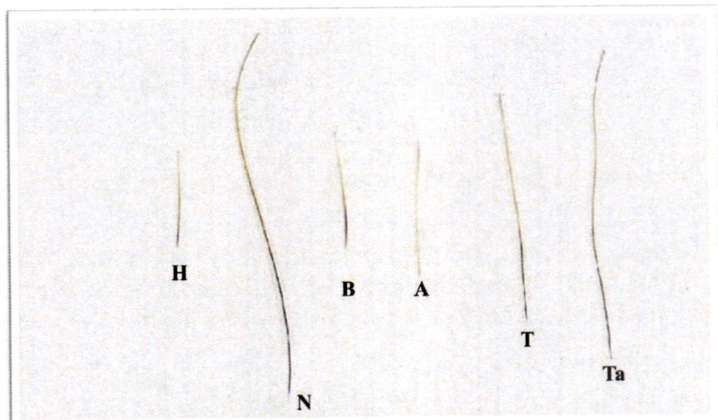

Fig. 4.11: Photograph showing hair profile in Nilgai (*Boselaphus tragocamelus*)

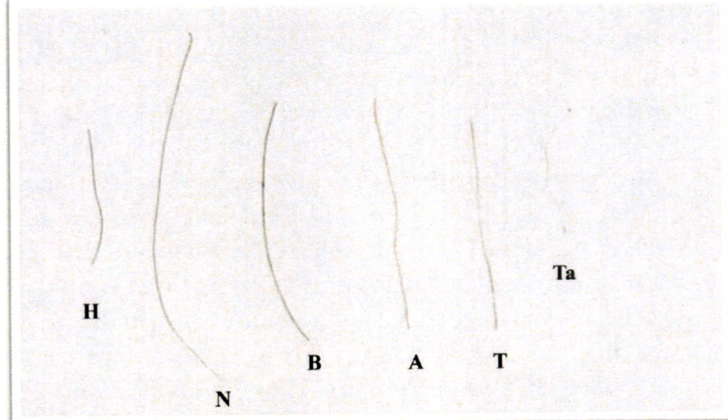

Fig. 4.12: Photograph showing hair profile in Hanuman Langur (*Samnopithecus entellus*)

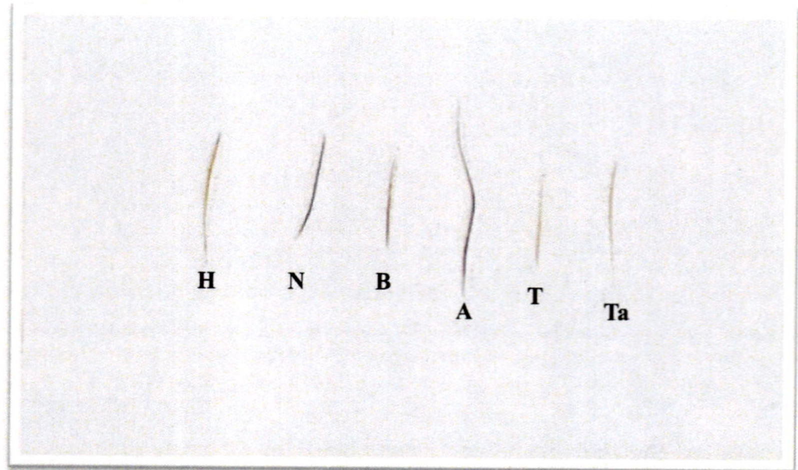

Fig. 4.13: Photograph showing hair profile in Tiger (*Panthera tigris*)

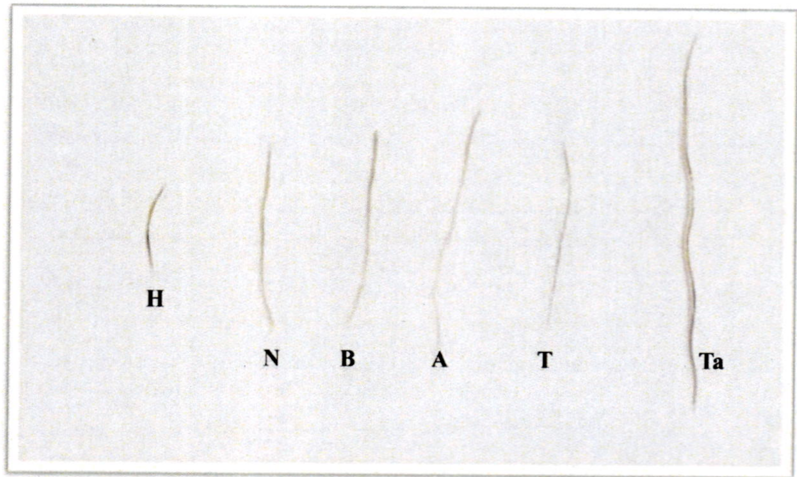

Fig. 4.14: Photograph showing hair profile in Leopard (*Panthera pardus*)

Slightly wavy profile was observed in the hair of Leopard (*Panthera pardus*) from all the six body regions (Fig. 4.14). Some hair of Leopard were raw umber coloured or white coloured and some were with three colour bands. The band at the base of hair was white in colour and band towards distal part was burnt umber coloured and these two bands were separated by a band of raw umber colour. The above findings regarding colour of hair was in partial concurrence with the findings reported by Koppikar and Sabnis (1976). They observed light yellow band at proximal part, brown at middle and black at the distal part of hair.

4.1.2. Total length

The findings of the present study revealed that the shortest hair were found in the head region of most of the species under study except Cattle, Horse, Pig and Tiger (Table 4.1). The shortest hair were present in the thigh region of Cattle (15.25 ± 0.79 mm), Pig (49.00 ± 5.48 mm) and Tiger (14.83 ± 0.69 mm). In Horse (10.92 ± 0.94 mm), shortest hair were found in back region. The shortest length of hair in the back region of Horse recorded during present study may be attributed to the fact that Horses used for the present study were riding Horses from the military school. In riding Horses saddles are kept on the back which may not allow hair to grow longer like other body regions. The longest hair were observed in the tail region of Cattle, Buffalo, Goat, Horse, Spotted Deer, Nilgai, Sambar, Tiger and Leopard (Table 4.1). This presence of long hair in the tail region of these species could be probably due to the reason that the tail is a part in all these animals, which is not exposed for any kind of friction like other parts of body. The maximum length of hair in back region was found in Pig (82.42 ± 10.12 mm), Dog (35.18 ± 4.76 mm) and Cat (33.67 ± 1.71 mm). The maximum length of hair was recorded in neck region of Sheep (132.83 ± 9.05 mm), while it was maximum in abdomen region of Hanuman Langur (81.42 ± 8.37 mm).

The maximum total length irrespective of body region amongst all the species was observed in Sheep and minimum in Tiger (Table 4.1). This observation of the present study is in consonance with the findings reported by Dharaiya and Soni (2012). Amongst all the fourteen species and six body regions, shortest hair length was noted in Tiger (18.35 ± 0.83 mm) followed by Cat (23.51 ± 1.28 mm), while longest hair length was observed in Sheep (83.57 ± 24.21 mm) followed by Horse (80.57 ± 5.83 mm). Amongst all species with respect to various body regions, the shortest hair length was reported in head region of Cat (13.83 ± 1.74 mm), while longest hair length was observed in tail region of Horse (395.25 ± 26.80 mm).

Table 4.1. Means and SE for total length of hair (mm) from various regions of body in different species.

Species→ Region ↓	Cattle	Buffalo	Sheep	Goat	Horse	Domestic Pig	Cat	Dog	Spotted Deer	Sambar	Nilgai	Hanuman Langur	Tiger	Leopard	Pooled Mean ± SE
Head	24.08 ± 2.62^{c}	43.42 ± 7.64^{a}	34.83 ± 2.93^{a}	24.50 ± 2.18^{a}	14.58 ± 1.19^{a}	65.75 ± 5.61^{c}	13.83 ± 1.47^{a}	21.00 ± 2.83^{a}	15.58 ± 1.48^{a}	38.17 ± 1.20^{a}	17.33 ± 1.71^{a}	31.67 ± 4.63^{a}	15.00 ± 1.02^{a}	14.50 ± 0.76^{a}	26.73 ± 3.83
Neck	20.58 ± 1.26^{bc}	54.08 ± 5.23^{b}	132.83 ± 9.05^{f}	54.58 ± 7.31^{d}	34.58 ± 6.82^{b}	77.83 ± 12.83^{d}	20.83 ± 1.64^{b}	32.17 ± 5.64^{bc}	23.42 ± 2.31^{b}	65.17 ± 8.50^{d}	60.33 ± 7.61^{d}	57.92 ± 6.26^{b}	16.83 ± 1.33^{ab}	23.83 ± 2.09^{b}	48.21 ± 8.08
Back	15.42 ± 1.73^{ab}	52.42 ± 4.21^{b}	98.42 ± 7.05^{e}	26.25 ± 5.45^{a}	10.92 ± 0.94^{a}	82.42 ± 10.12^{e}	33.67 ± 1.71^{c}	35.18 ± 4.76^{b}	30.33 ± 3.92^{c}	55.17 ± 1.74^{c}	44.58 ± 7.13^{c}	73.33 ± 11.79^{c}	21.33 ± 3.21^{b}	20.83 ± 0.95^{b}	42.88 ± 6.82
Abdomen	24.75 ± 2.95^{c}	56.33 ± 6.31^{b}	66.08 ± 5.79^{b}	36.92 ± 2.45^{b}	14.08 ± 1.56^{a}	56.00 ± 5.46^{b}	21.42 ± 2.04^{b}	28.92 ± 3.05^{ab}	36.92 ± 5.35^{d}	63.17 ± 1.98^{d}	24.50 ± 6.15^{b}	81.42 ± 8.37^{e}	20.42 ± 1.37^{b}	31.67 ± 1.31^{c}	40.18 ± 5.32
Thigh	15.25 ± 0.79^{a}	58.17 ± 8.46^{c}	91.17 ± 11.79^{d}	45.58 ± 4.65^{d}	14.00 ± 1.04^{a}	49.00 ± 5.48^{a}	21.75 ± 2.19^{b}	24.75 ± 3.95^{a}	30.17 ± 3.76^{c}	46.50 ± 2.67^{b}	28.92 ± 3.19^{b}	68.17 ± 11.89^{d}	14.83 ± 0.69^{a}	24.17 ± 2.32^{b}	38.03 ± 5.89
Tail	73.58 ± 17.70^{d}	69.17 ± 13.98^{d}	75.90 ± 4.95^{c}	66.67 ± 5.62^{e}	395.25 ± 26.80^{c}	59.58 ± 6.36^{b}	29.58 ± 1.56^{c}	34.42 ± 4.68^{c}	51.50 ± 8.23^{e}	90.50 ± 6.69^{e}	83.00 ± 14.75^{e}	36.08 ± 6.36^{a}	21.67 ± 1.90^{b}	54.50 ± 3.06^{d}	81.53 ± 23.87
Pooled Mean ± SE	28.94 ± 4.44^{BC}	55.60 ± 3.38^{E}	83.21 ± 5.83^{G}	42.42 ± 3.16^{D}	80.57 ± 24.21^{G}	65.10 ± 3.67^{F}	23.51 ± 1.28^{AB}	29.41 ± 1.83^{BC}	31.32 ± 2.60^{C}	59.78 ± 4.62^{EF}	43.11 ± 4.87^{D}	58.10 ± 4.53^{EF}	18.35 ± 0.83^{A}	28.25 ± 2.28^{BC}	

Means bearing same superscript for in a column do not differ significantly

These observations regarding length of hair in different body regions of different species of animals, during present study indicated that the effect of species, region and species verses region interaction could be used as a significant source for identification of animal species.

4.1.3. Hair diameter

The mean values for diameter at various parts of the hair from six different body regions of fourteen mammalian species are given in Table 4.5.

The mean diameter of proximal part of hair was found significantly lowest in cat (0.024 ± 0.002 mm) followed by Leopard (0.036 ± 0.002 mm) and Dog (0.036 ± 0.003 mm), while highest diameter was observed in Pig (0.170 ± 0.007 mm) followed by Nilgai (0.104 ± 0.007mm).

Table 4.2. Means and SE for hair diameter (mm) at proximal part of hair from various regions of body in different species.

Species→ Region ↓	Cattle	Buffalo	Sheep	Goat	Horse	Domestic Pig	Cat	Dog	Spotted Deer	Sambar	Nilgai	Hanu-man Langur	Tiger	Leop-ard	Pooled Mean± SE
Head	0.025± 0.002^{a}	0.078± 0.011^{bc}	0.053± 0.006^{c}	0.055± 0.008^{b}	0.027± 0.005^{b}	0.185± 0.010^{d}	0.025± 0.002^{bc}	0.035± 0.008^{a}	0.030± 0.007^{a}	0.050± 0.011^{c}	0.073± 0.007^{a}	0.047± 0.010^{a}	0.038± 0.005^{a}	0.043± 0.005^{c}	**0.055 ± 0.011**
Neck	0.033± 0.003^{b}	0.063± 0.011^{a}	0.038± 0.007^{b}	0.068± 0.006^{c}	0.065± 0.009^{c}	0.178± 0.020^{c}	0.020± 0.003^{ab}	0.037± 0.009^{ab}	0.035± 0.004^{ab}	0.033± 0.002^{ab}	0.163± 0.007^{e}	0.067± 0.009^{b}	0.045± 0.002^{b}	0.043± 0.006^{c}	**0.064 ± 0.012**
Back	0.033± 0.004^{b}	0.085± 0.004^{d}	0.027± 0.005^{a}	0.065± 0.004^{c}	0.030± 0.006^{b}	0.197± 0.016^{e}	0.032± 0.005^{d}	0.035± 0.007^{a}	0.033± 0.005^{ab}	0.027± 0.002^{a}	0.097± 0.010^{c}	0.068± 0.010^{b}	0.058± 0.007^{c}	0.028± 0.002^{a}	**0.058 ± 0.012**
Abdomen	0.030± 0.004^{ab}	0.082± 0.006^{cd}	0.023± 0.003^{a}	0.047± 0.003^{a}	0.018± 0.004^{a}	0.167± 0.011^{b}	0.017± 0.002^{a}	0.035± 0.007^{a}	0.037± 0.013^{b}	0.037± 0.002^{b}	0.068± 0.006^{a}	0.053± 0.006^{a}	0.043± 0.007^{a}	0.030± 0.004^{a}	**0.049 ± 0.010**
Thigh	0.025± 0.003^{a}	0.073± 0.005^{b}	0.027± 0.004^{a}	0.047± 0.005^{a}	0.028± 0.003^{b}	0.173± 0.021^{bc}	0.020± 0.003^{ab}	0.032± 0.006^{a}	0.030± 0.007^{a}	0.033± 0.002^{ab}	0.080± 0.003^{b}	0.078± 0.010^{c}	0.053± 0.004^{c}	0.033± 0.004^{ab}	**0.052 ± 0.010**
Tail	0.083± 0.005^{c}	0.125± 0.003^{e}	0.080± 0.006^{d}	0.090± 0.10^{d}	0.160± 0.004^{d}	0.120± 0.012^{a}	0.028± 0.004^{cd}	0.043± 0.007^{b}	0.067± 0.007^{c}	0.147± 0.006^{d}	0.145± 0.012^{d}	0.078± 0.005^{c}	0.040± 0.005^{ab}	0.038± 0.004^{bc}	**0.089 ± 0.011**
Pooled Mean±SE	**0.038± 0.004^{BC}**	**0.084± 0.004^{F}**	**0.041± 0.004^{BC}**	**0.062± 0.004^{DE}**	**0.055± 0.009^{D}**	**0.170± 0.007^{H}**	**0.024± 0.002^{A}**	**0.036± 0.003^{B}**	**0.039± 0.004^{BC}**	**0.054± 0.010^{D}**	**0.104± 0.007^{G}**	**0.065± 0.004^{F}**	**0.046± 0.002^{C}**	**0.036± 0.002^{B}**	

Means bearing same superscript for particular effect in a column do not differ significantly

CD for Species is 0.009 and CD for Region is 0.006

Amongst the body regions and animal species, the minimum hair diameter at proximal part of the hair was observed in abdomen region of Cat (0.017 ± 0.002 mm) followed by Horse (0.018 ± 0.004 mm), while maximum hair diameter was noted in the back region of Pig (0.197 ± 0.016 mm). Means and SE for hair diameter of proximal part of hair from various regions of the body in different species are given in Table 4.2.

The diameter at the middle part of the hair was lowest in Cat and Sambar (0.022 ± 0.002 mm) followed by Leopard (0.028 ± 0.002 mm) and cattle (0.028 ± 0.003 mm), while highest diameter was noted in Pig (0.138 ± 0.007 mm) followed by Nilgai (0.092 ± 0.006 mm).

Table 4.3. Means and SE for hair diameter (mm) at middle part of hair from various regions of body in different species.

Species→ Region ↓	Cattle	Buf-falo	Sheep	Goat	Horse	Domes-tic Pig	Cat	Dog	Spot-ted Deer	Sam-bar	Nilgai	Hanu-man Langur	Tiger	Leop-ard	Pooled Mean± SE
Head	0.020± 0.003^{a}	0.062 ± 0.011^{b}	0.045± 0.007^{d} .00 0.006	0.042± 0.005^{ab}	0.023± 0.004^{ab}	0.155± 0.012^{c}	0.015± 0.002^{a}	0.033± 0.008^{ab}	0.033 ± 0.005^{b}	0.020± 0.004^{ab}	0.072± 0.004^{b}	0.038 ± 0.006^{a}	0.048± 0.010^{b}	0.030± 0.004^{bc}	**0.045 ± 0.009**
Neck	0.022± 0.002^{b}	0.047 ± 0.006^{a}	0.037± 0.006^{c}	0.047± 0.003^{b}	0.048± 0.004^{c}	0.142± 0.018^{b}	0.017± 0.005^{ab}	0.028± 0.006^{a}	0.037 ± 0.009^{bc}	0.023± 0.002^{bc}	0.148± 0.007^{d}	0.055± 0.006^{b}	0.040± 0.004^{a}	0.033± 0.004^{c}	**0.052 ± 0.011**
Back	0.022± 0.002^{a}	0.070 ± 0.004^{c}	0.022± 0.003^{a}	0.053± 0.003^{c}	0.023± 0.003^{ab}	0.155 ± 0.014^{c}	0.022± 0.003^{b}	0.042± 0.009^{c}	0.033 ± 0.007^{b}	0.017± 0.002^{a}	0.077± 0.011^{b}	0.058± 0.007^{b}	0.058± 0.004^{c}	0.020± 0.003^{a}	**0.048 ± 0.010**
Abdomen	0.022± 0.002^{a}	0.058 ± 0.005^{b}	0.030± 0.006^{b}	0.037± 0.004^{a}	0.018± 0.002^{a}	0.142± 0.015^{b}	0.022± 0.006^{b}	0.028± 0.006^{a}	0.040 ± 0.012^{c}	0.027± 0.006^{c}	0.063± 0.009^{a}	0.058 ± 0.005^{b}	0.047± 0.010^{ab}	0.027± 0.004^{b}	**0.044 ± 0.008**
Thigh	0.020± 0.000^{a}	0.063 ± 0.002^{b}	0.022± 0.003^{a}	0.033± 0.002^{a}	0.027± 0.003^{b}	0.147± 0.016^{b}	0.023± 0.004^{b}	0.033± 0.008^{ab}	0.027 ± 0.003^{a}	0.017± 0.002^{a}	0.072± 0.003^{b}	0.065± 0.007^{c}	0.048± 0.004^{b}	0.028± 0.005^{bc}	**0.045 ± 0.009**
Tail	0.065± 0.004^{b}	0.105 ± 0.004^{d}	0.068± 0.007^{e}	0.078± 0.008^{d}	0.143± 0.006^{d}	0.087± 0.009^{a}	0.032± 0.003^{c}	0.038± 0.008^{bc}	0.057 ± 0.006^{d}	0.027± 0.006^{c}	0.118± 0.015^{c}	0.060± 0.006^{bc}	0.042± 0.007^{a}	0.032± 0.003^{bc}	**0.068 ± 0.009**
Pooled Mean±SE	**0.028± 0.003^{AB}**	**0.068 ± 0.004^{F}**	**0.037± 0.003^{C}**	**0.048± 0.003^{DE}**	**0.047± 0.008^{D}**	**0.138± 0.007^{H}**	**0.022± 0.002^{A}**	**0.034 ± 0.003^{BC}**	**0.038 ± 0.003^{C}**	**0.022± 0.002^{A}**	**0.092± 0.006^{G}**	**0.056± 0.003^{E}**	**0.047± 0.003^{D}**	**0.028± 0.002^{AB}**	

Means bearing same superscript for particular effect in a column do not differ significantly
CD for Species is 0.008 and CD for Region is 0.005

Amongst all the species and body regions, the minimum hair diameter at middle part was noted in head region of Cat (0.015 ± 0.002 mm) followed by back and thigh region in Sambar (0.017 ± 0.002 mm). The maximum diameter of hair at middle part was observed in back region of Pig (0.155 ± 0.014 mm). Highly significant difference was noted in diameter at middle part of hair in all the species as well as different region of body under study. Means and SE for hair diameter at middle part of hair from various regions of the body in different species are presented in Table 4.3.

Table 4.4. Means and SE for hair diameter at distal part of hair from various regions of body in different species.

Species→ Region ↓	Cattle	Buffalo	Sheep	Goat	Horse	Domestic Pig	Cat	Dog	Spotted Deer	Sambar	Nilgai	Hanuman Langur	Tiger	Leopard	Pooled Mean ± SE
Head	0.017 ± 0.002^{a}	0.038 ± 0.007^{a}	0.033 ± 0.006^{c}	0.033 ± 0.006^{b}	0.018 ± 0.002^{bc}	0.072 ± 0.008^{c}	0.013 ± 0.002^{a}	0.030 ± 0.007^{b}	0.028 ± 0.005^{b}	0.013 ± 0.002^{a}	0.057 ± 0.008^{a}	0.027 ± 0.006^{a}	0.038 ± 0.008^{cd}	0.027 ± 0.003^{b}	**0.032 ± 0.004**
Neck	0.015 ± 0.002^{a}	0.037 ± 0.006^{a}	0.033 ± 0.009^{c}	0.032 ± 0.005^{ab}	0.028 ± 0.003^{d}	0.060 ± 0.008^{a}	0.013 ± 0.003^{a}	0.025 ± 0.002^{a}	0.033 ± 0.007^{c}	0.023 ± 0.002^{c}	0.110 ± 0.008^{c}	0.042 ± 0.004^{c}	0.037 ± 0.008^{bd}	0.027 ± 0.003^{b}	**0.037 ± 0.006**
Back	0.015 ± 0.002^{a}	0.047 ± 0.004^{b}	0.028 ± 0.003^{b}	0.032 ± 0.003^{ab}	0.020 ± 0.003^{c}	0.058 ± 0.008^{a}	0.023 ± 0.006^{bc}	0.045 ± 0.011^{d}	0.033 ± 0.003^{c}	0.013 ± 0.002^{c}	0.058 ± 0.008^{a}	0.033 ± 0.008^{b}	0.040 ± 0.004^{d}	0.032 ± 0.007^{c}	**0.034 ± 0.004**
Abdomen	0.015 ± 0.002^{a}	0.048 ± 0.005^{b}	0.032 ± 0.003^{bc}	0.028 ± 0.004^{a}	0.010 ± 0.000^{a}	0.065 ± 0.008^{b}	0.020 ± 0.003^{b}	0.030 ± 0.009^{b}	0.040 ± 0.006^{d}	0.017 ± 0.002^{ab}	0.055 ± 0.006^{a}	0.033 ± 0.003^{b}	0.035 ± 0.008^{bc}	0.018 ± 0.002^{a}	**0.032 ± 0.004**
Thigh	0.013 ± 0.002^{a}	0.050 ± 0.004^{b}	0.022 ± 0.003^{a}	0.027 ± 0.003^{a}	0.015 ± 0.002^{b}	0.065 ± 0.004^{b}	0.018 ± 0.003^{ab}	0.023 ± 0.005^{a}	0.023 ± 0.004^{a}	0.017 ± 0.002^{ab}	0.057 ± 0.004^{a}	0.040 ± 0.004^{c}	0.033 ± 0.004^{b}	0.020 ± 0.000^{a}	**0.030 ± 0.004**
Tail	0.050 ± 0.004^{b}	0.068 ± 0.005^{c}	0.072 ± 0.005^{d}	0.057 ± 0.006^{c}	0.103 ± 0.008^{e}	0.060 ± 0.007^{a}	0.023 ± 0.002^{bc}	0.037 ± 0.008^{c}	0.038 ± 0.004^{d}	0.020 ± 0.004^{b}	0.085 ± 0.012^{b}	0.040 ± 0.005^{c}	0.028 ± 0.004^{a}	0.025 ± 0.003^{b}	**0.050 ± 0.006**
Pooled Mean±SE	$\mathbf{0.021 \pm 0.002^{AB}}$	$\mathbf{0.048 \pm 0.003^{D}}$	$\mathbf{0.037 \pm 0.003^{C}}$	$\mathbf{0.035 \pm 0.002^{C}}$	$\mathbf{0.033 \pm 0.006^{C}}$	$\mathbf{0.063 \pm 0.003^{E}}$	$\mathbf{0.019 \pm 0.001^{AB}}$	$\mathbf{0.032 \pm 0.003^{C}}$	$\mathbf{0.033 \pm 0.002^{C}}$	$\mathbf{0.017 \pm 0.002^{A}}$	$\mathbf{0.070 \pm 0.005^{F}}$	$\mathbf{0.036 \pm 0.002^{C}}$	$\mathbf{0.035 \pm 0.002^{C}}$	$\mathbf{0.025 \pm 0.002^{B}}$	

Means bearing same superscript for particular effect in a column do not differ significantly
CD for Species is 0.006 and CD for Region is 0.004

The lowest mean hair diameter at the distal part of hair was recorded in Sambar (0.017 ± 0.002 mm) followed by Cat (0.019 ± 0.001 mm), while highest diameter was found in Nilgai (0.070 ± 0.005 mm) followed by Pig (0.063 ± 0.003 mm) when compare with all the species under study. Means and SE for hair diameter at distal part of hair from various regions of body in different species are depicted in Table 4.4.

The minimum hair diameter amongst the species and region was noted in the head region of Cat (0.013 ± 0.002 mm) and head and back region of Sambar (0.013 ± 0.002 mm). The maximum diameter at distal part of hair in the neck region of Nilgai was (0.110 ± 0.008 mm). During the present study, it was noted that the species, region and species verses region was significant source of variation for hair diameter at proximal, middle and distal parts of the hair.

Table 4.5. Means and SE for hair diameter (mm) at different parts of hair from various regions of body in different species.

Species→ Region ↓	Hair Parts	Cattle	Buffalo	Sheep	Goat	Horse	Domestic Pig	Cat	Dog	Spotted Deer	Sambar	Nilgai	Hanuman Langur	Tiger	Leopard	Pooled Mean± SE
Head	**Proximal**	0.025 ± 0.002^{b}	0.078 ± 0.011^{c}	0.053 ± 0.006^{c}	0.055 ± 0.008^{c}	0.027± 0.005^{b}	0.185± 0.010^{c}	0.025± 0.002^{b}	0.035 ± 0.008	0.030 ± 0.007	0.050 ± 0.011^{c}	0.073 ± 0.007^{b}	0.047 ± 0.010^{c}	0.038 ± 0.005^{a}	0.043 ± 0.005^{b}	**0.055± 0.011**
	Middle	0.020 ± 0.003^{ab}	0.062 ± 0.011^{b}	0.045± 0.007^{b}	0.042 ± 0.005^{b}	0.023 ± 0.004^{ab}	0.155 ± 0.012^{b}	0.015 ± 0.002^{a}	0.033 ± 0.008	0.033± 0.005	0.020 ± 0.004^{b}	0.072 ± 0.004^{b}	0.038 ± 0.006^{b}	0.048 ± 0.010^{b}	0.030 ± 0.004^{a}	**0.045± 0.009**
	Distal	0.017 ± 0.002^{a}	0.038 ± 0.007^{a}	0.033 ± 0.006^{a}	0.033 ± 0.006^{a}	0.018± 0.002^{a}	0.072 ± 0.008^{a}	0.013 ± 0.002^{a}	0.030 ± 0.007	0.028 ± 0.005	0.013 ± 0.002^{a}	0.057 ± 0.008^{a}	0.027 ± 0.006^{a}	0.038 ± 0.008^{a}	0.027± 0.003^{a}	**0.032± 0.004**
	Mean ± SE	**0.021± 0.001^{AB}**	**0.059 ± 0.005^{E}**	**0.044 ± 0.003^{D}**	**0.043 ± 0.003^{D}**	**0.023± 0.002^{AB}**	**0.137 ± 0.009^{G}**	**0.018 ± 0.001^{A}**	**0.033 ± 0.003^{CD}**	**0.031 ± 0.002^{BC}**	**0.028± 0.005^{ABC}**	**0.067± 0.003^{F}**	**0.037 ± 0.003^{CD}**	**0.042± 0.003^{D}**	**0.033 ± 0.003^{CD}**	
Neck	**Proximal**	0.033 ± 0.003^{c}	0.063 ± 0.011^{c}	0.038 ± 0.007	0.068 ± 0.006^{c}	0.065 ± 0.009^{c}	0.178 ± 0.020^{c}	0.020 ± 0.003^{b}	0.037 ± 0.009^{b}	0.035 ± 0.004	0.033 ± 0.002^{b}	0.163 ± 0.007^{c}	0.067 ± 0.009^{c}	0.045 ± 0.002^{b}	0.043 ± 0.006^{c}	**0.064± 0.012**
	Middle	0.022 ± 0.002^{b}	0.047 ± 0.006^{c}	0.037 ± 0.006	0.047 ± 0.003^{b}	0.048 ± 0.004^{b}	0.142 ± 0.018^{b}	0.017± 0.005^{ab}	0.028 ± 0.006^{a}	0.037 ± 0.009	0.023± 0.002^{a}	0.148 ± 0.007^{b}	0.055 ± 0.006^{b}	0.040 ± 0.004^{ab}	0.033 ± 0.004^{b}	**0.052± 0.011**
	Distal	0.015 ± 0.002^{a}	0.037 ± 0.006^{a}	0.033 ± 0.009	0.032± 0.005^{a}	0.028± 0.003^{a}	0.060 ± 0.008^{a}	0.013 ± 0.003^{a}	0.025± 0.002^{a}	0.033 ± 0.007	0.023 ± 0.002^{a}	0.110 ± 0.008^{a}	0.042± 0.004^{a}	0.037± 0.008^{a}	0.027 ± 0.003^{a}	**0.037± 0.006**
	Mean ± SE	**0.023 ± 0.002^{AB}**	**0.049 ± 0.004^{EF}**	**0.036 ± 0.003^{CD}**	**0.049 ± 0.003^{EF}**	**0.047 ± 0.003^{EF}**	**0.127 ± 0.011^{G}**	**0.017 ± 0.002^{A}**	**0.030 ± 0.003^{BCD}**	**0.035 ± 0.003^{CD}**	**0.027 ± 0.002^{ABC}**	**0.141 ± 0.005^{H}**	**0.054± 0.003^{F}**	**0.041 ± 0.002^{DE}**	**0.034 ± 0.002^{BCD}**	
Back	**Proximal**	0.033 ± 0.004^{c}	0.085± 0.004^{c}	0.027± 0.005^{ab}	0.065± 0.004^{c}	0.030± 0.006^{b}	0.197± 0.016^{c}	0.032± 0.005^{b}	0.035± 0.007^{a}	0.033± 0.005	0.027± 0.002^{b}	0.097± 0.010^{c}	0.068± 0.010^{b}	0.058± 0.007^{b}	0.028± 0.002^{b}	**0.058± 0.012**
	Middle	0.022 ± 0.002^{c}	0.070 ± 0.004^{b}	0.022 ± 0.003^{a}	0.053 ± 0.003^{b}	0.023± 0.003^{ab}	0.155 ± 0.014^{b}	0.022 ± 0.003^{a}	0.042 ± 0.009^{b}	0.033 ± 0.007	0.017 ± 0.002^{a}	0.077 ± 0.011^{b}	0.058 ± 0.007^{b}	0.058 ± 0.004^{b}	0.020 ± 0.003^{a}	**0.048± 0.010**
	Distal	0.015 ± 0.002^{a}	0.047 ± 0.004^{a}	0.028 ± 0.003^{b}	0.032 ± 0.003^{a}	0.020 ± 0.003^{c}	0.058 ± 0.008^{a}	0.023 ± 0.006^{a}	0.045 ± 0.011^{b}	0.033 ± 0.003	0.013 ± 0.002^{a}	0.058 ± 0.008^{a}	0.033 ± 0.008^{a}	0.040 ± 0.004^{a}	0.032 ± 0.007^{b}	**0.034± 0.004**
	Mean ±SE	**0.023 ± 0.002^{AB}**	**0.067 ± 0.003^{F}**	**0.026 ± 0.002^{AB}**	**0.050 ± 0.003^{DE}**	**0.024 ± 0.002^{AB}**	**0.137 ± 0.011^{G}**	**0.026 ± 0.002^{AB}**	**0.041 ± 0.004^{CD}**	**0.033 ± 0.002^{BC}**	**0.019 ± 0.002^{A}**	**0.077 ± 0.005^{F}**	**0.053 ± 0.004^{E}**	**0.052 ± 0.002^{DE}**	**0.027 ± 0.002^{AB}**	

Species→ Region ↓	Hair Parts	Cattle	Buffalo	Sheep	Goat	Horse	Domestic Pig	Cat	Dog	Spotted Deer	Sambar	Nilgai	Hanuman Langur	Tiger	Leopard	Pooled Mean± SE
Abdomen	**Proximal**	0.030± 0.004^{c}	0.082± 0.006^{c}	0.023± 0.003^{a}	0.047± 0.003^{c}	0.018± 0.004^{b}	0.167± 0.011^{c}	0.017± 0.002	0.035± 0.007^{b}	0.037± 0.013	0.037± 0.002^{c}	0.068± 0.006^{b}	0.053± 0.006^{b}	0.043± 0.007^{b}	0.030± 0.004^{b}	**0.049± 0.010**
	Middle	0.022 ± 0.002^{b}	0.058 ± 0.005^{b}	0.030 ± 0.006^{b}	0.037 ± 0.004^{b}	0.018 ± 0.002^{b}	0.142 ± 0.015^{b}	0.022 ± 0.006	0.028 ± 0.006^{a}	0.040± 0.012	0.027± 0.006^{b}	0.063 ± 0.009^{b}	0.058 ± 0.005^{b}	0.047 ±0.010^{b}	0.027 ± 0.004^{b}	**0.044± 0.008**
	Distal	0.015 ± 0.002^{a}	0.048 ± 0.005^{a}	0.032 ± 0.003^{b}	0.028 ± 0.004^{a}	0.010± 0.000^{a}	0.065 ± 0.008^{a}	0.020 ± 0.003	0.030 ± 0.009^{ab}	0.040 ± 0.006	0.017 ± 0.002^{a}	0.055 ± 0.006^{a}	0.033 ± 0.003^{a}	0.035 ± 0.008^{a}	0.018 ± 0.002^{a}	**0.032± 0.004**
	Mean = SE	**0.022 ± 0.001^{ABC}**	**0.063± 0.003^{H}**	**0.028 ± 0.002^{BCDE}**	**0.037± 0.002^{DEFG}**	**0.016 ± 0.001^{A}**	**0.124 ± 0.009^{I}**	**0.019 ± 0.002^{AB}**	**0.031 ± 0.003^{CDEF}**	**0.039 ± 0.004^{EFG}**	**0.027 ± 0.003^{ABCD}**	**0.062 ± 0.003^{H}**	**0.048 ± 0.003^{G}**	**0.042 ± 0.003^{FG}**	**0.025 ± 0.002^{ABC}**	
Thigh	**Proximal**	0.025± 0.003^{c}	0.073± 0.005^{c}	0.027± 0.004^{b}	0.047± 0.005^{c}	0.028± 0.003^{b}	0.173± 0.021^{c}	0.020± 0.003^{ab}	0.032± 0.006^{b}	0.030± 0.007^{b}	0.033± 0.002^{b}	0.080± 0.003^{c}	0.078± 0.010^{c}	0.053± 0.004^{c}	0.033± 0.004^{c}	**0.052± 0.010**
	Middle	0.020 ± 0.000^{b}	0.063 ± 0.002^{b}	0.022 ± 0.003^{a}	0.033 ± 0.002^{b}	0.027 ± 0.003^{b}	0.147 ± 0.016^{b}	0.023 ± 0.004^{b}	0.033 ± 0.008^{b}	0.027± 0.003^{ab}	0.017 ± 0.002^{a}	0.072 ± 0.003^{b}	0.065± 0.007^{b}	0.048 ± 0.004^{b}	0.028 ± 0.005^{b}	**0.045± 0.009**
	Distal	0.013± 0.002^{a}	0.050 ± 0.004^{a}	0.022 ± 0.003^{a}	0.027 ± 0.003^{a}	0.015 ± 0.002^{a}	0.065 ± 0.004^{a}	0.018 ± 0.003^{a}	0.023± 0.005^{a}	0.023 ± 0.004^{a}	0.017 ± 0.002^{a}	0.057 ± 0.004^{a}	0.040 ± 0.004^{a}	0.033 ± 0.004^{a}	0.020 ± 0.000^{a}	**0.030± 0.004**
	Mean = SE	**0.019 ± 0.001^{A}**	**0.062 ± 0.002^{D}**	**0.023 ± 0.001^{A}**	**0.036 ± 0.002^{BC}**	**0.023 ± 0.002^{A}**	**0.128 ± 0.010^{E}**	**0.021 ± 0.001^{A}**	**0.029 ± 0.003^{AB}**	**0.027 ± 0.002^{AB}**	**0.022± 0.002^{A}**	**0.069 ± 0.002^{D}**	**0.061 ± 0.002^{D}**	**0.045 ± 0.002^{C}**	**0.027 ± 0.002^{AB}**	
Tail	**Proximal**	0.083 ± 0.005^{c}	0.125 ± 0.003^{c}	0.080 ± 0.006^{b}	0.090 ± 0.10^{c}	0.160 ± 0.004^{c}	0.120 ± 0.012^{c}	0.028 ± 0.004^{ab}	0.043 ± 0.007^{b}	0.067 ± 0.007^{c}	0.147± 0.006^{c}	0.145 ± 0.012^{c}	0.078± 0.005^{c}	0.040± 0.005^{b}	0.038± 0.004^{c}	**0.089± 0.011**
	Middle	0.065 ± 0.004^{b}	0.105± 0.004^{b}	0.068 ± 0.007^{a}	0.078 ± 0.008^{b}	0.143 ± 0.006^{b}	0.087 ± 0.009	0.032 ± 0.003^{b}	0.038 ± 0.008^{ab}	0.057± 0.006^{b}	0.027 ± 0.006^{b}	0.118 ± 0.015^{b}	0.060 ± 0.006^{b}	0.042 ± 0.007^{b}	0.032 ± 0.003^{b}	**0.050± 0.068**
	Distal	0.050 ± 0.004^{a}	0.068± 0.005^{a}	0.072± 0.005^{a}	0.057 ± 0.006^{a}	0.103± 0.008^{a}	0.060± 0.007^{a}	0.023 ± 0.002^{a}	0.037± 0.008^{a}	0.038 ± 0.004^{a}	0.020± 0.004^{a}	0.085± 0.012^{a}	0.040 ± 0.005^{a}	0.028 ± 0.004^{a}	0.025 ± 0.003^{a}	**0.050± 0.006**
	Mean = SE	**0.066 ± 0.003^{CD}**	**0.099 ± 0.004^{F}**	**0.073 ± 0.002^{D}**	**0.075 ± 0.004^{D}**	**0.136 ± 0.005^{H}**	**0.089 ± 0.006^{E}**	**0.028 ± 0.001^{A}**	**0.039 ± 0.003^{A}**	**0.054 ± 0.003^{B}**	**0.064 ± 0.015^{BCD}**	**0.116 ± 0.007^{G}**	**0.059 ± 0.003^{BC}**	**0.037 ± 0.002^{A}**	**0.032 ± 0.002^{A}**	

Means bearing same superscript for different region (for hair part) in a column do not differ significantly/Means bearing same superscript for different species (hair part of different region) in a row do not differ significantly/ Head Region-CD for Species is 0.011 and CD for hair part is 0.005/Neck Region-CD for Species is 0.012 and CD for hair part is 0.006/ Back Region-CD for Species is 0.011 and CD for hair part is 0.005/Abdomen Region-CD for Species is 0.011 and CD for hair part is 0.00/Thigh Region-CD for Species is 0.010 and CD for hair part is 0.004/Tail Region-CD for Species is 0.011 and CD for hair part is 0.005

The present study revealed that the hair diameter in all the regions reduces towards the distal part of the hair in all the species. It was also observed that the hair from the back region of Spotted Deer had nearly uniform diameter throughout the length of hair shaft (Table 4.5)

The findings of the present study are in agreement with the findings of Lyne and McMohan (1950) who mentioned that the hair shaft diameter, showed variations throughout the length of hair shaft. Kshirsagar *et al.* (2009) reported that the hair diameter of the animal varies from 25 to 160 μm and in human it varies from 30 to 80 μm. These findings are in concurrence with the observations recorded during present study in different species of animal.

It was also noted that the lowest diameter was noted at the distal part of hair from head, neck and thigh regions amongst all the species.

4.2. Cuticular characteristics

4.2.1. Negative Scale Cast

The study of structure of cuticular scales requires special techniques and higher magnification. So, during the present study, the characteristic features of the cuticular scales were observed by preparing negative scale cast.

The present study showed transversal cuticular scale position in relation to the longitudinal direction of hair in all the species under study. It was also observed that the cuticular scale height was less than the cuticular scale width in all the parts of hair in all body regions of all species, except Dog, where the scale width was less than scale height at the middle part of hair of tail region. The observations of the present study indicated that the cuticular structure varies along the length of hair shaft amongst all the species and body regions. In agreement with the findings of the present study, Bhat *et al.* (2014) stated that the position of cuticular scale was found perpendicular to the long axis of the hair shaft in Lion, Stripped Hyena, Sloth Bear, Nilgai and Chinkara. Similar findings regarding position of cuticular scale were noted by Mukherjee *et al.* (2016) in Buffalo, Goat, Dog, Cow, Rabbit and Cat. The observations reported by Marinis and Asprea (2005) in wild and domestic Sheep and Goat are also in concurrence with findings of the present study.

In Cattle (*Bos indicus*), the cuticular scales were arranged in irregular wave pattern with smooth scale margin and distant scale margin distance at the proximal part of hair from all the six body regions. The cuticular scale pattern in middle part of hair from head region showed irregular wave with crenate scale margin and distant scale margin distance (Fig. 4.15). While, in neck, back, abdomen, thigh and tail regions it was arranged in irregular wave pattern with smooth scale margins and distant scale margin distance. The irregular wave pattern with crenate scale margins and near scale margin distance was noted at the distal part of hair from all the body regions of Cattle (Fig 4.15. to 4.20.). These findings are in agreement with the findings reported by Mukherjee *et al.* (2016) and Sahajpal *et al.* (2009) in Cattle. They noted irregular wave pattern of cuticular scales with crenate scale margin and distant to close scale margin distance. While, the findings of the present study regarding scale margin distance are in contradiction with Gharu and Trivedi (2015) mentioned that regular to irregular wave pattern with crenate or rippled or smooth scale margins was observed in Cattle. These findings are in partial agreement with the findings reported during present study.

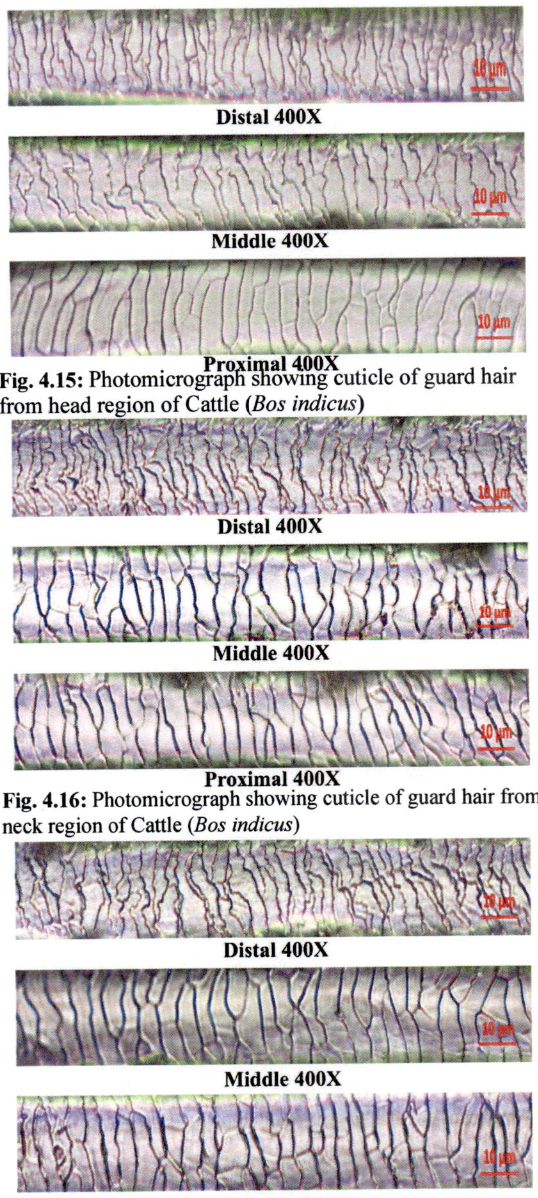

Fig. 4.15: Photomicrograph showing cuticle of guard hair from head region of Cattle (*Bos indicus*)

Fig. 4.16: Photomicrograph showing cuticle of guard hair from neck region of Cattle (*Bos indicus*)

Fig. 4.17: Photomicrograph showing cuticle of guard hair from back region of *Cattle (Bos indicus*)

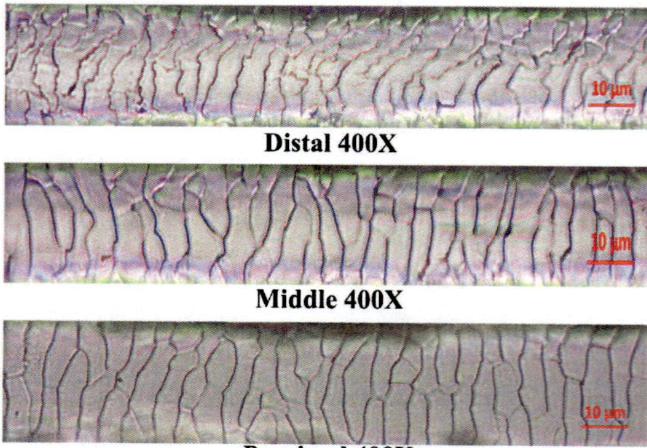

Fig. 4.18: Photomicrograph showing cuticle of guard hair from abdomen region of Cattle (*Bos indicus*)

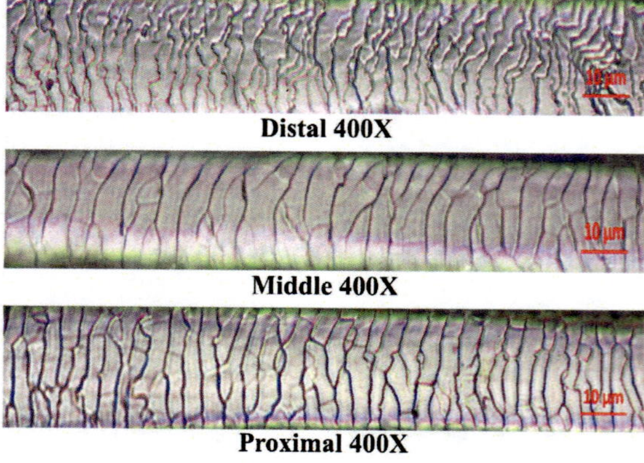

Fig. 4.19: Photomicrograph showing cuticle of guard hair from thigh region of Cattle (*Bos indicus*)

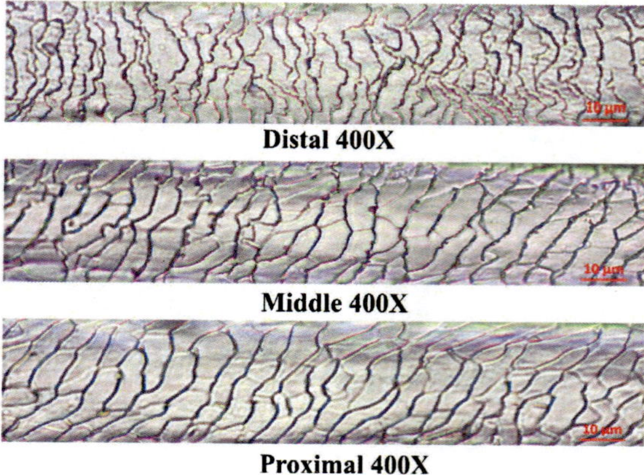

Fig. 4.20: Photomicrograph showing cuticle of guard hair from tail region of Cattle (*Bos indicus*)

In Buffalo (*Bubalus bubalis*), the cuticular scale pattern at proximal part of hair in all the body regions was regular wave type. The middle part of hair however showed regular wave in neck, back, abdomen and thigh region whereas irregular wave pattern was observed in head and tail regions (Fig. 4.22. to 4.25.). The cuticular scales margin was smooth to heavily rippled throughout the length of hair in almost all the regions of the body in Buffalo. The scale margin distance in proximal part of hair in all the body regions in Buffalo was distant. The distal part of hair showed near scale margin distance in all the body regions. The middle part of hair however, showed region wise variations in scale margin distance. It was near in head, back, thigh and tail regions and distant in neck and abdomen regions in Buffalo. These findings of the present study regarding cuticular scale margin distance are in agreement with the observations reported by Gharu and Trivedi (2015b) in Buffalo. However, in contrast with the present findings Mukherjee *et at.* (2016) reported that the cuticle was imbricate, irregular and crenate structure in buffalo.

The hair from the head region of Sheep (Ovis aries) had irregular wave pattern of cuticular scales at all the three parts of hair (Fig. 4.27). The smooth cuticular scale margin was noted at the proximal and middle part of hair, while crenate margin was observed at the distal part of hair and distant scale margin distance at all the three parts of hair. The hair from neck and back regions showed uniform cuticular structure throughout the hair shaft. The mosaic cuticular scale pattern with smooth scale margin distance was found in neck and back region (Fig. 4.28 and 4.29). The proximal and middle part of hair from abdomen and thigh region showed mosaic pattern with smooth scale margins. The scale margin distance was reported distant at proximal part of hair and near at middle part of hair, while, irregular wave pattern with crenate scale margin and near scale margin distance was observed at the distal part of hair of Sheep (Fig. 4.30 to 4.31). The hair from tail region showed altogether different cuticular scale pattern. The tail region showed irregular wave pattern with crenate scale margins and near scale margin distance in all the three parts of hair (Fig. 4.32). Similar findings were reported by Gharu and Trivedi (2015b) in Sheep. However, the present observation are partially in agreement with the findings reported by Dharaiya and Soni (2012) and Sahajpal et al. (2009b), who reported irregular wave pattern with crenate margin and close scale margin distance

During the present work, it was recorded that the guard hair of Goat (*Capra hircus*) had transversely placed cuticular scales. The scales were arranged in regular wave pattern at proximal part of hair from head, neck, back, abdomen and thigh region, while irregular scale arrangement was noted at proximal part of hair from tail region (Fig. 4.33. to 4.38.). The cuticular scales were found arranged in irregular wave pattern at the distal part of hair from head, neck, back and tail regions, while regular wave pattern was noted in abdomen and thigh regions. The middle part of hair from head, back and thigh region showed regular wave pattern of cuticular scales and hair from neck, abdomen and tail regions showed irregular wave pattern of cuticular scales. The cuticular scale margin was noted smooth at proximal part, slightly rippled at middle part and highly rippled at the distal part of hair from all the six body regions in Goat.

Similarly the distance between scale margins was found distant at the proximal part and near at the distal part of hair from all the body regions. While, at the middle part of the hair the distance was observed near at back and thigh region, where as it was noted distant at head, neck, abdomen and tail region.

Marinis and Asprea (2005) reported transversely placed cuticular scales, which were arranged in regular wave pattern with smooth to rippled margin in Domestic Goat (*Capra hircus*) with distant scale margin distance. Similar observations were noted during the present study in Goat.

In contrast with the findings of the present study, Koppikar and Sabnis (1976) stated that the cuticular scales were arranged in imbricate manner with smooth scale margin in Goat hair. This variation may be attributed to the fact that these workers noted the findings by using Camera Lucida and not the microscope.

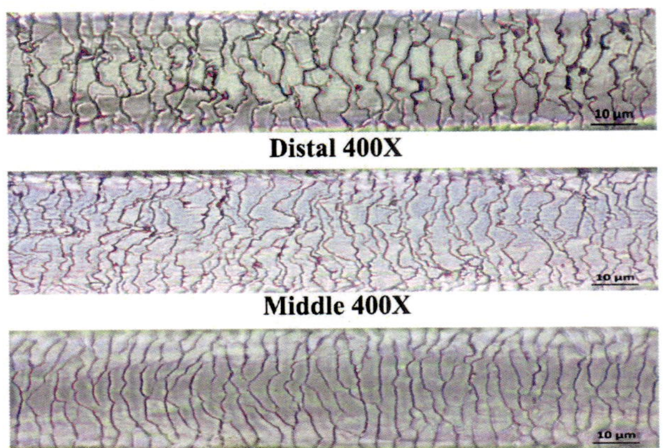

Fig. 4.21: Photomicrograph showing cuticle of guard hair from head region of Buffalo (Bubalus bubalis)

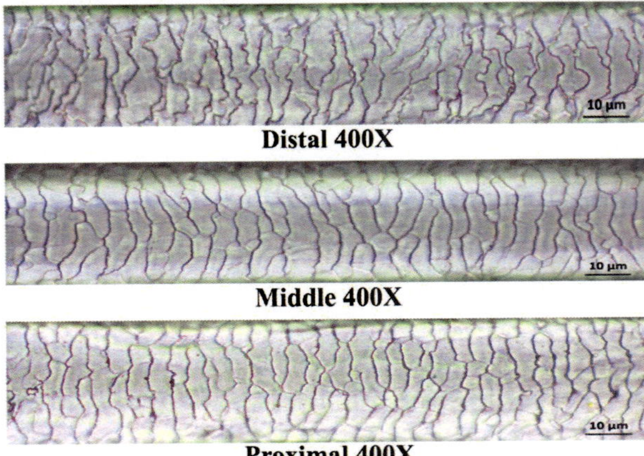

Fig. 4.22: Photomicrograph showing cuticle of guard hair from neck region of Buffalo (Bubalus bubalis)

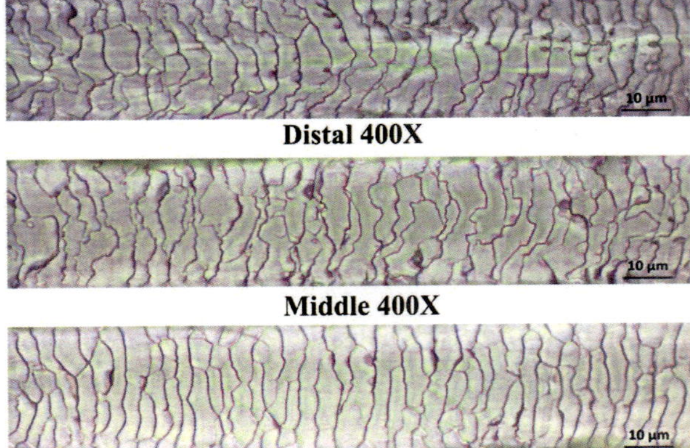

Fig. 4.23: Photomicrograph showing cuticle of guard hair from back region of Buffalo (*Bubalus bubalis*)

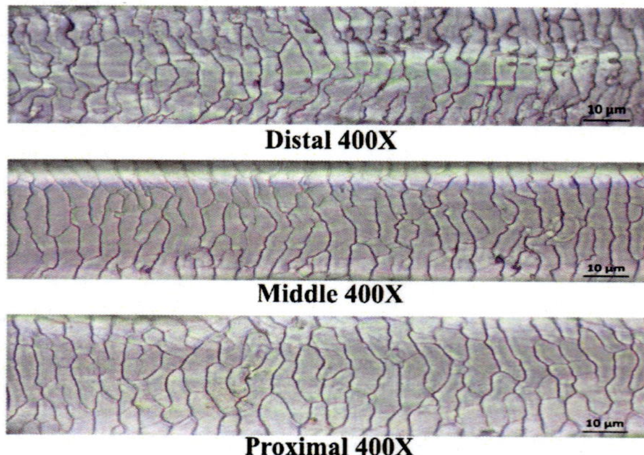

Fig. 4.24: Photomicrograph showing cuticle of guard hair from abdomen region of Buffalo (Bubalus bubalis)

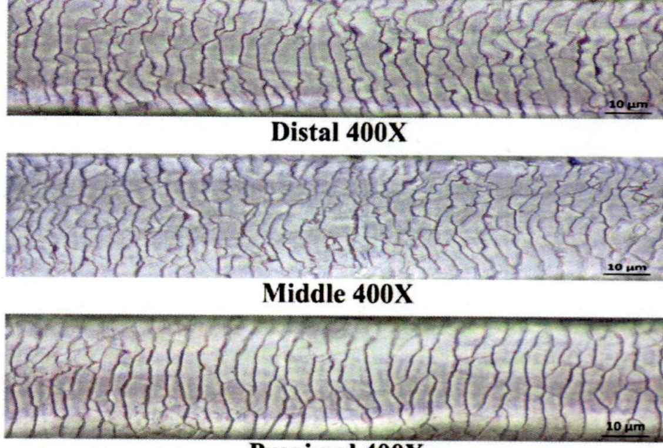

Fig. 4.25: Photomicrograph showing cuticle of guard hair from thigh region of Buffalo (Bubalus bubalis)

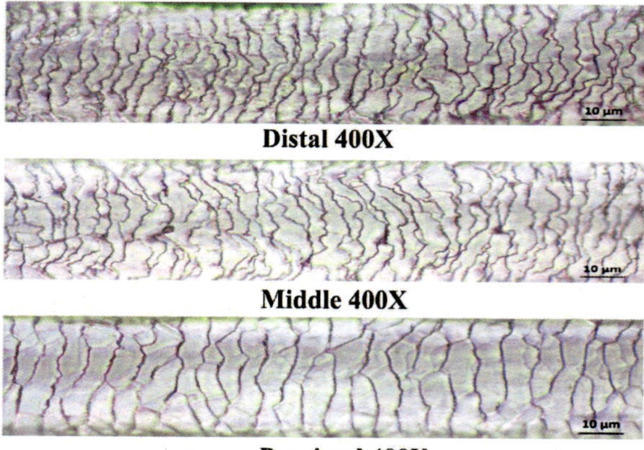

Fig. 4.26: Photomicrograph showing cuticle of guard hair from tail region of Buffalo (Bubalus bubalis)

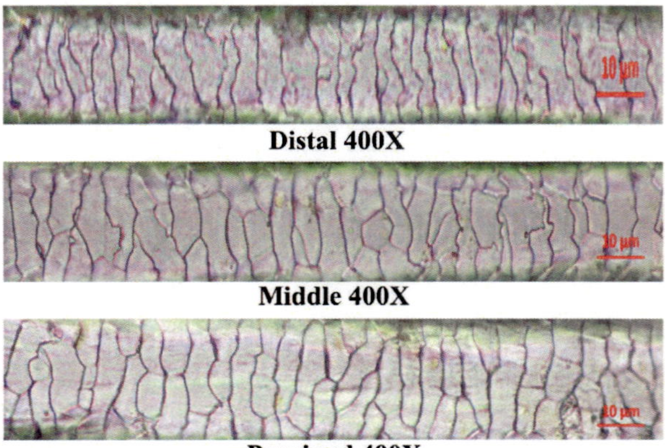

Fig. 4.27: Photomicrograph showing cuticle of guard hair from head region of Sheep (*Oris aries*)

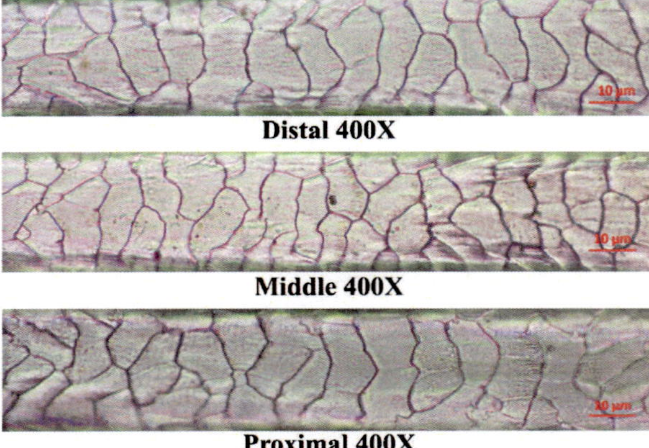

Fig. 4.28: Photomicrograph showing cuticle of guard hair from neck region of Sheep (Oris aries)

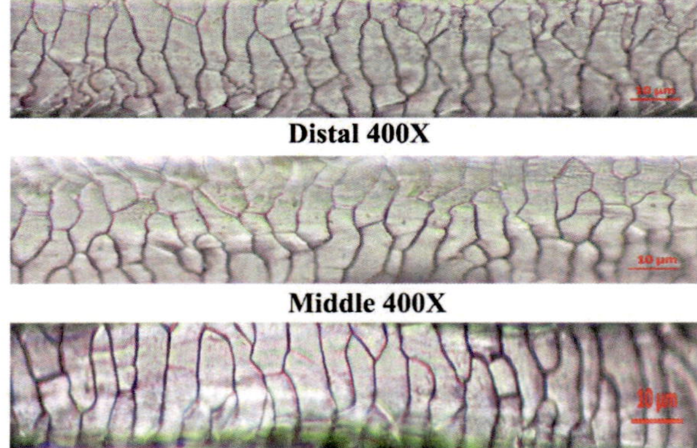

Fig. 4.29: Photomicrograph showing cuticle of guard hair from back region of Sheep (Oris aries)

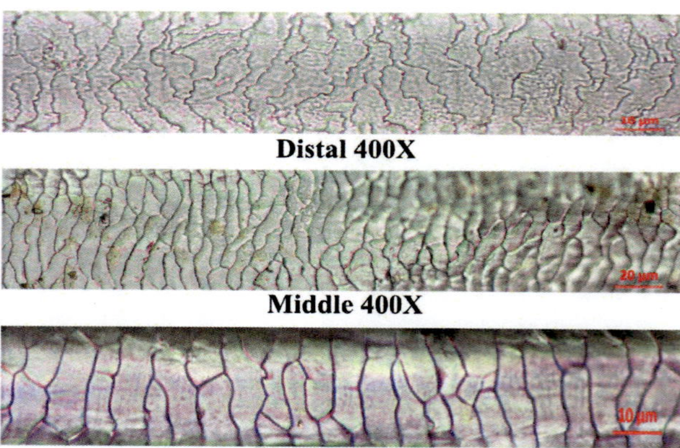

Distal 400X

Middle 400X

Proximal 400X

Fig. 4.30: Photomicrograph showing cuticle of guard hair from abdomen region of Sheep (Oris aries)

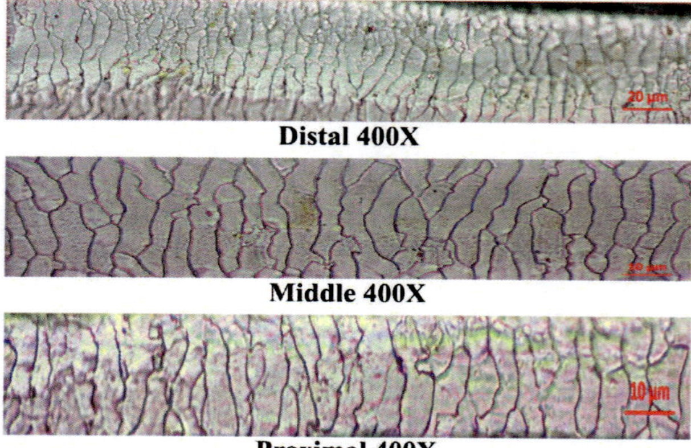

Distal 400X

Middle 400X

Proximal 400X

Fig. 4.31: Photomicrograph showing cuticle of guard hair from thigh region of Sheep (Oris aries)

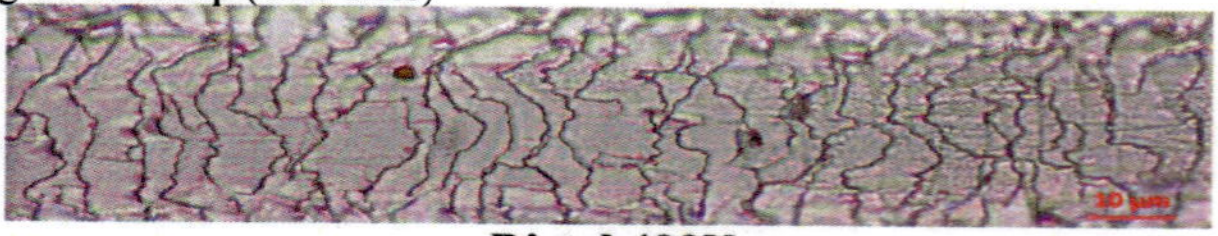

Distal 400X

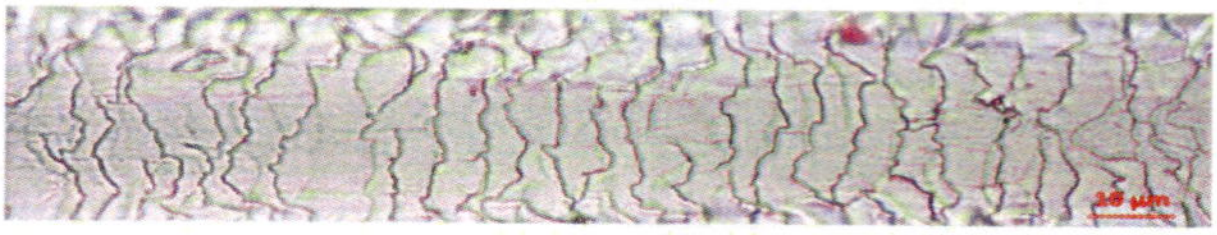

Middle 400X

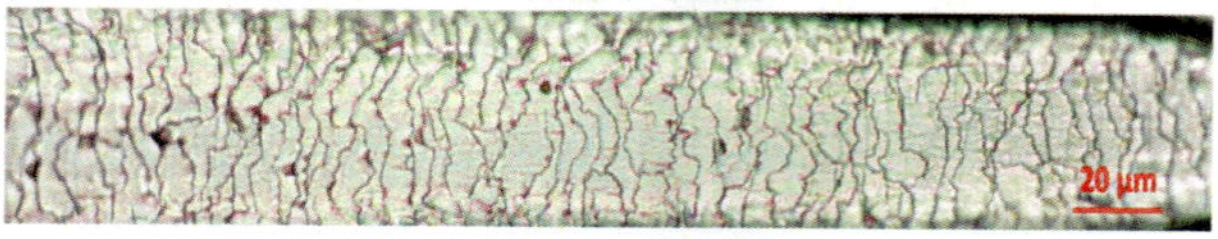

Proximal 400X

Fig. 4.32: Photomicrograph showing cuticle of guard hair from tail region of Sheep (*Oris aries*)

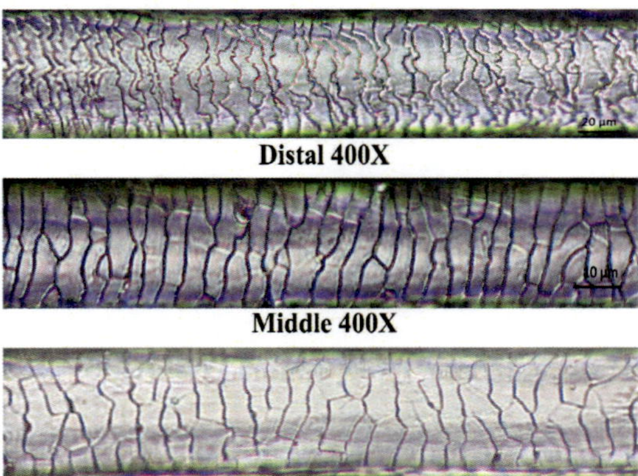

Distal 400X

Middle 400X

Proximal 400X

Fig. 4.33: Photomicrograph showing cuticle of guard hair from head region of Goat (Capra hircus)

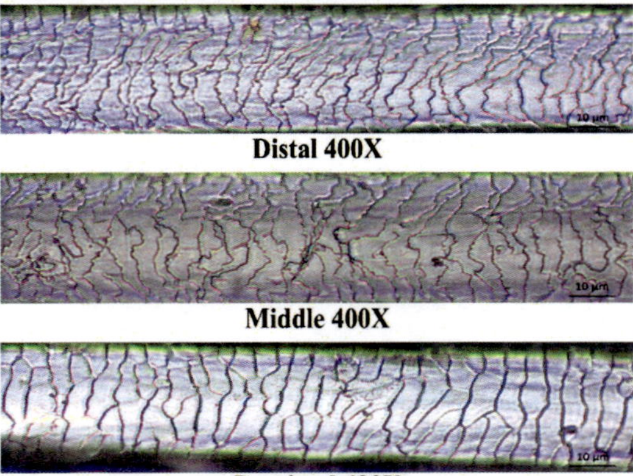

Distal 400X

Middle 400X

Proximal 400X

Fig. 4.34: Photomicrograph showing cuticle of guard hair from neck region of Goat (*Capra hircus*)

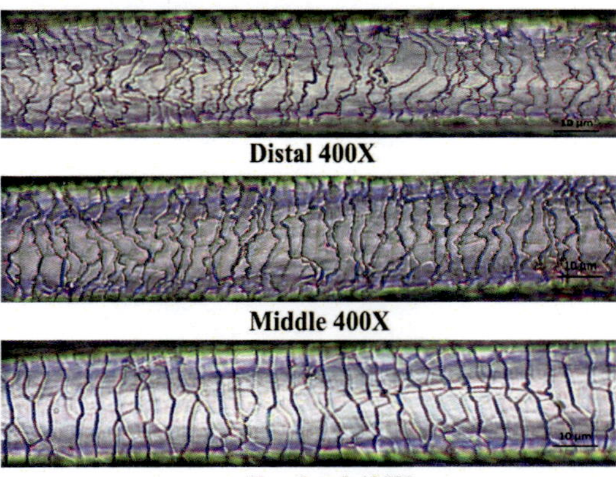

Distal 400X

Middle 400X

Proximal 400X

Fig. 4.35: Photomicrograph showing cuticle of guard hair from back region of Goat (*Capra hircuss*)

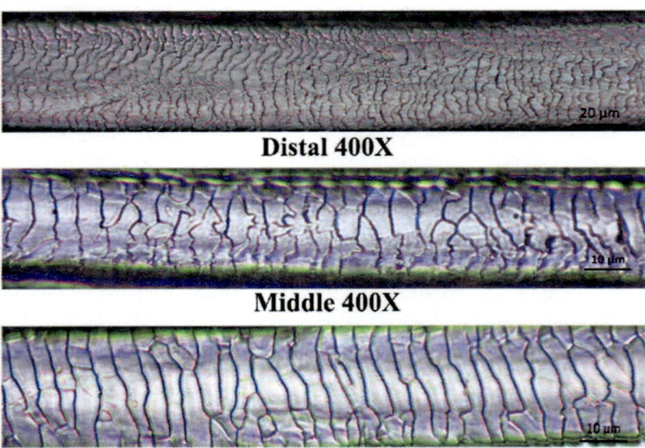

Fig. 4.36: Photomicrograph showing cuticle of guard hair from abdomen region of Goat (*Capra hircuss*)

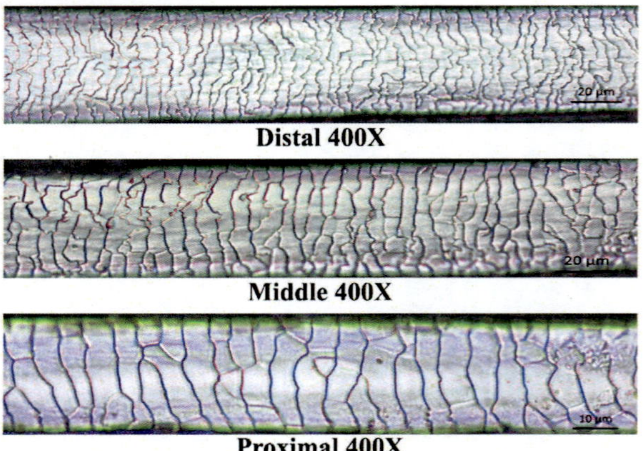

Fig. 4.37: Photomicrograph showing cuticle of guard hair from thigh region of Goat (*Capra hircus*)

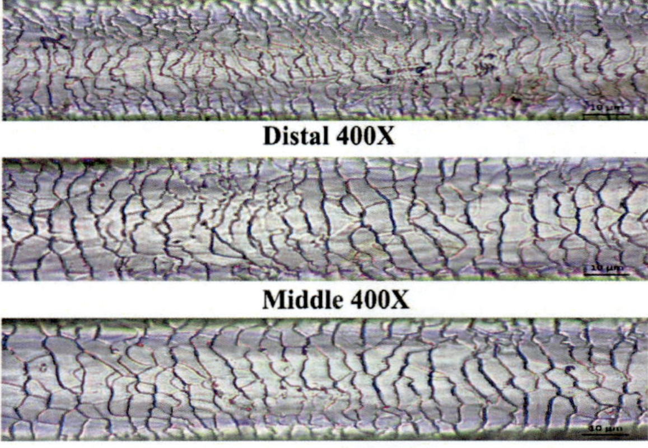

Fig. 4.38: Photomicrograph showing cuticle of guard hair from tail region of Goat (*Capra hircus*)

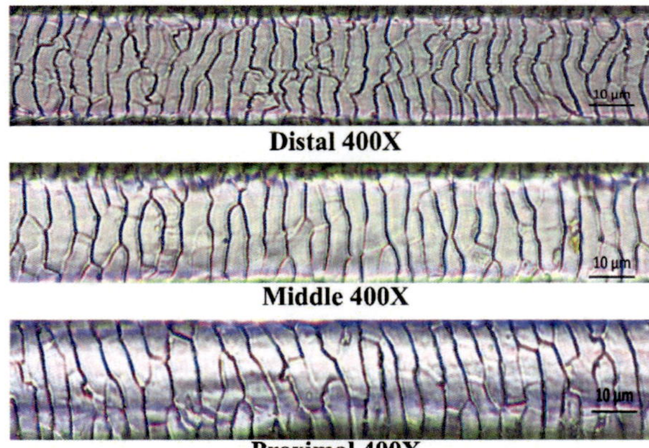

Fig. 4.39 Photomicrograph showing cuticle of guard hair from head region of Horse (*Equus caballus*)

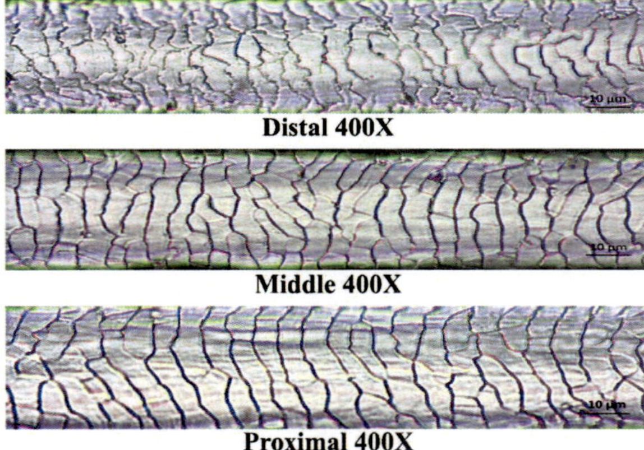

Fig. 4.40: Photomicrograph showing cuticle of guard hair from neck region of Horse (*Equus caballus*)

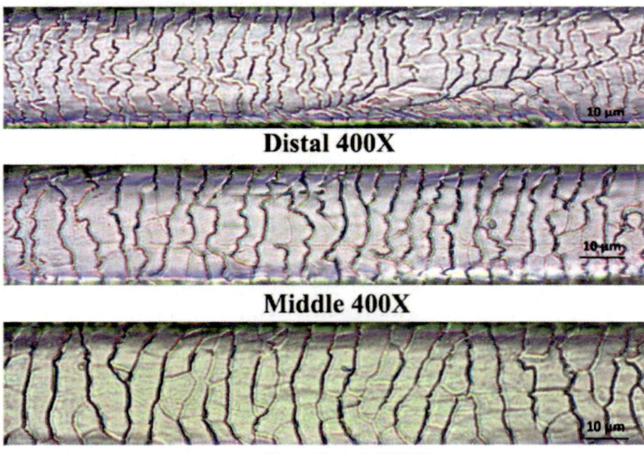

Proximal 400X

Fig. 4.41: Photomicrograph showing cuticle of guard hair from back region of Horse (*Equus caballus*)

The cuticular scale characters studied from the negative scale cast in Horse (*Equus caballus*) showed regular wave pattern with smooth scale margin and distant scale margin distance at proximal and middle part of hair from all the six body regions except at middle part of hair from back and abdomen. The hair from back and abdomen regions showed slightly rippled scale margin. Irregular wave pattern with strongly rippled scale margin and distant scale margin distance was noted (Fig. 4.39 to 4.44.) in distal part of hair in all the six body regions.

The findings of the present study are in concurrence with the findings reported by Marinis and Asprea (2006) and Gharu and Trivedi (2015), who reported, transversal scale position with regular wave pattern, smooth scale margin and distant scale margin distance in Horses at the lower shaft. They also noted irregular wave pattern with rippled scale margin at upper shaft of hair.

During the present study, it was observed that the characteristic features of cuticular scales such as scale pattern, type of scale margin and scale margin distance was similar in the hair of different regions in Domestic Pig (*Sus scrofa domesticus*). These cuticular scale characteristics were found similar in respective part of hair in almost all the regions of body.

The cuticular scales in distal and middle parts of hair in all the body regions were transversely placed, arranged in irregular wave pattern with strongly rippled scale margin and had close scale margins distance. The proximal part of hair in all the body regions however showed irregular wave pattern with rippled scale margin and with near scale margins distance (Fig. 4.45. to 4.50.). These observations of the present study corroborates with the findings reported by Marinis and Asprea (2006) in Wild Boar (*Sus scrofa*).

The cuticular scales in Domestic Cat (*Felis catus domesticus*) were found transversely arranged at all the three parts of hair and body regions. Further, it was noted that the cuticular scales were arranged in regular wave pattern with smooth scale margin and had distant scale margin distance at the proximal part of the hair among all the six body regions. Broad petal cuticular scale pattern was observed at the middle part of hair at all the regions of body with smooth scale margin and distant scale margin distance (Fig. 4.51. to 4.56.) in Cat. The distal part of the hair of Domestic Cat among all body regions showed streaked cuticular scale pattern, rippled margin and near scale margin distance.

In agreement with the findings of the present study Gharu and Trivedi (2015) and Mukherjee *et al.* (2016) stated that the cuticular scales in Domestic Cat were transversely placed and arranged in imbricate pattern with smooth to rippled margins and had distant scale margin distance.

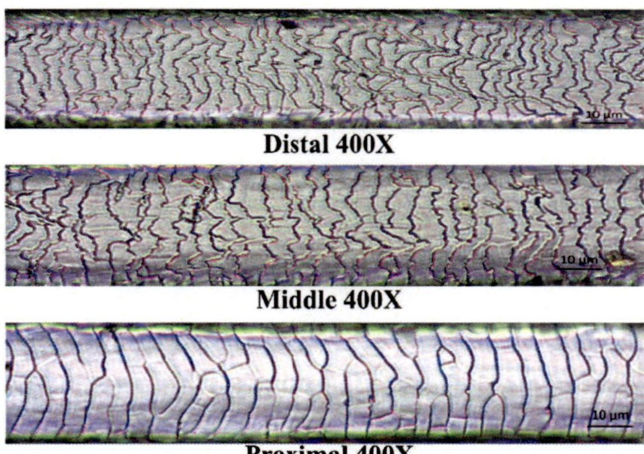

Fig. 4.42: Photomicrograph showing cuticle of guard hair from abdomen region of Horse (*Equus caballus*)

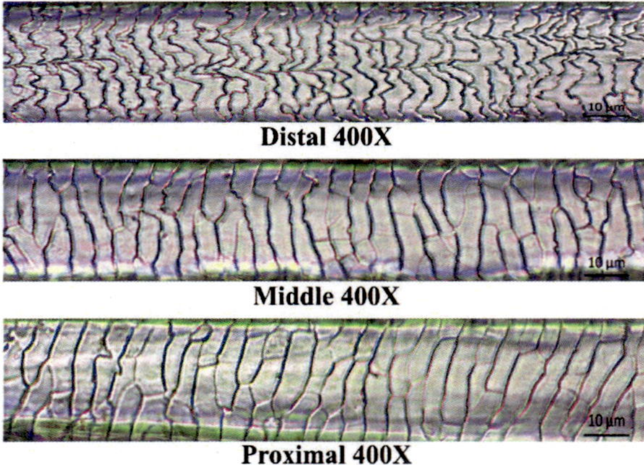

Fig. 4.43: Photomicrograph showing cuticle of guard hair from thigh region of Horse (Equus caballus)

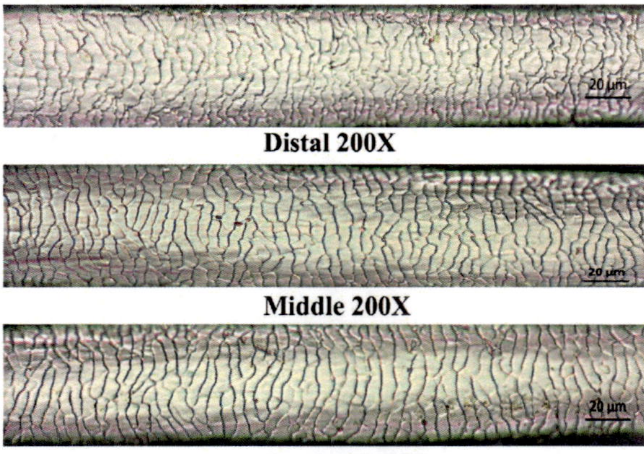

Fig. 4.44: Photomicrograph showing cuticle of guard hair from tail region of Horse (Equus cballus)

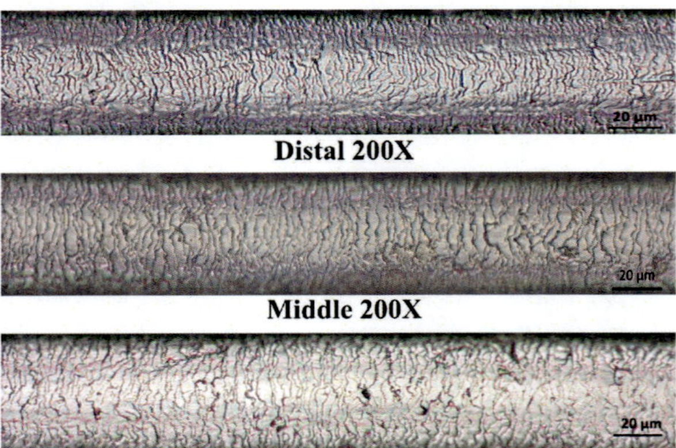

Fig. 4.45: Photomicrograph showing cuticle of guard hair from head region of Domestic pig (Sus scrofa domesticus)

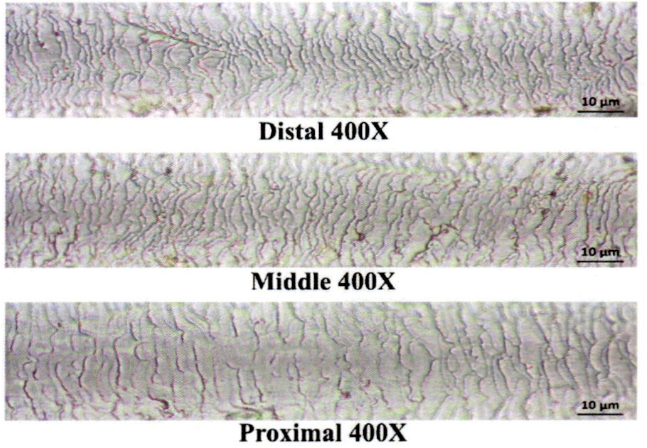

Fig. 4.46: Photomicrograph showing cuticle of guard hair from neck region of Domestic pig (Sus scrofa domestica)

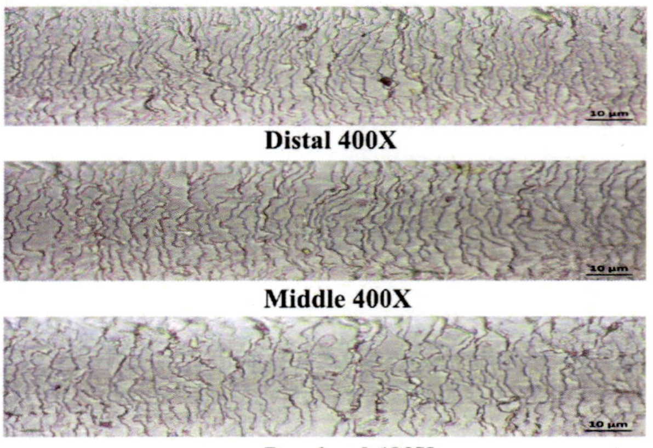

Fig. 4.47: Photomicrograph showing cuticle of guard hair from back region of Domestic pig (*Sus scrofa domestica)*

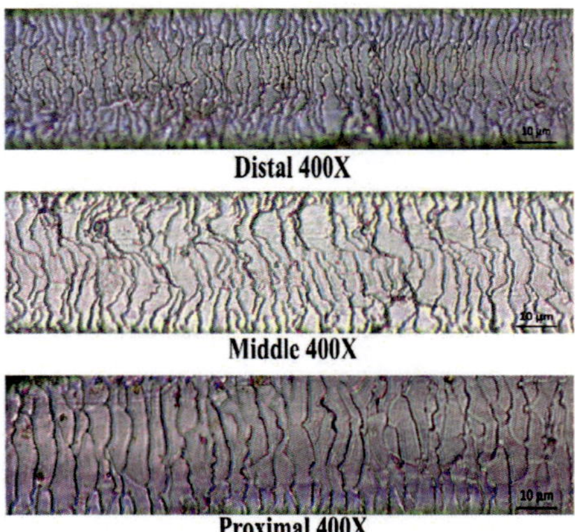

Fig. 4.48: Photomicrograph showing cuticle of guard hair from abdomen region of Domestic pig (*Sus scrofa domestica*)

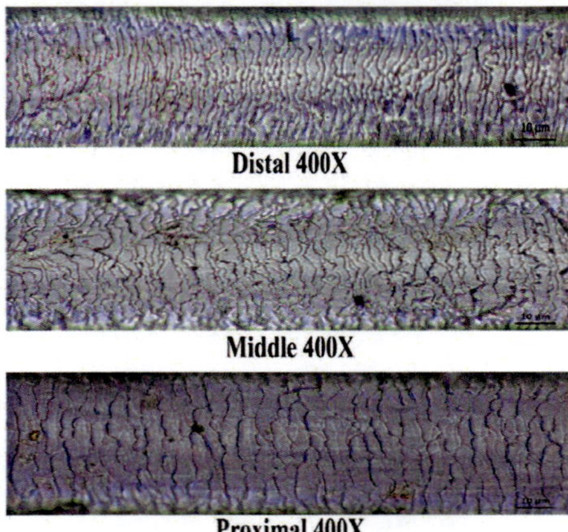

Fig. 4.49: Photomicrograph showing cuticle of guard hair from thigh region of Domestic pig (*Sus scrofa domestica*)

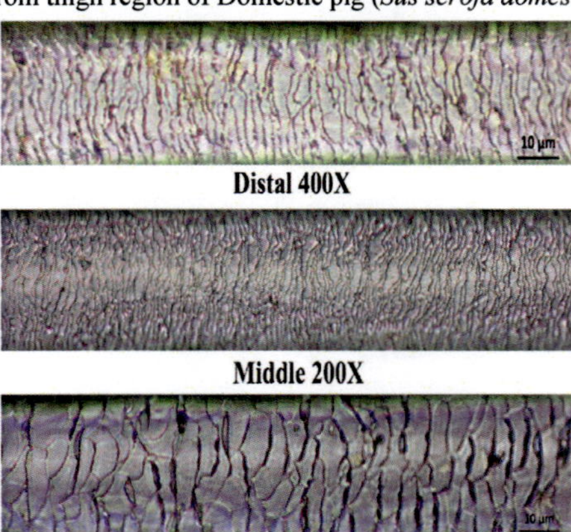

Fig. 4.50: Photomicrograph showing cuticle of guard hair from tail region of Domestic pig (*Sus scrofa domestica*)

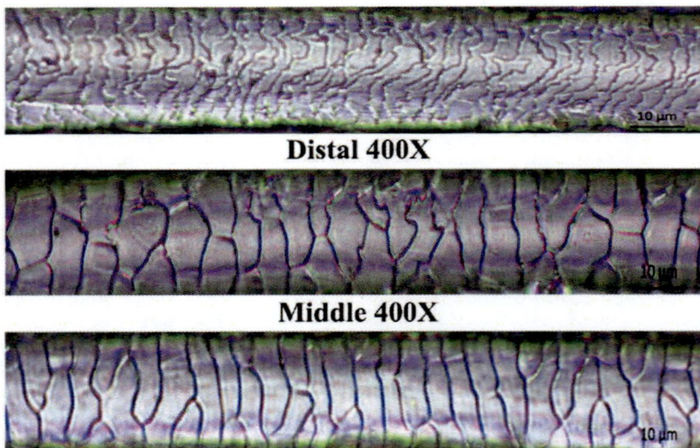

Fig. 4.51: Photomicrograph showing cuticle of guard hair of head region in Cat (*Felis catus domesticus*)

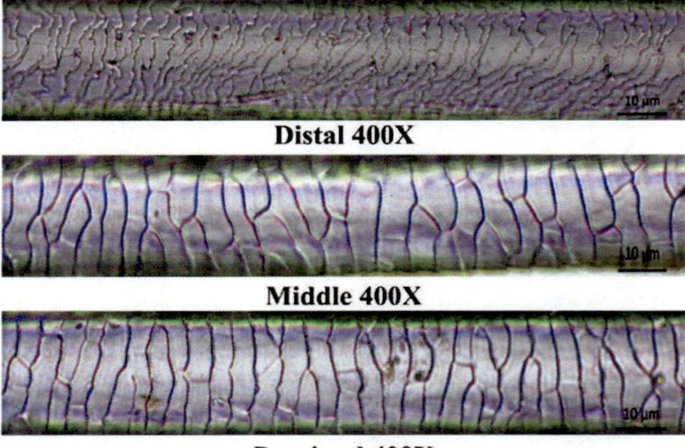

Fig. 4.52: Photomicrograph showing cuticle of guard hair of neck region in Cat (*Felis catus domesticus*)

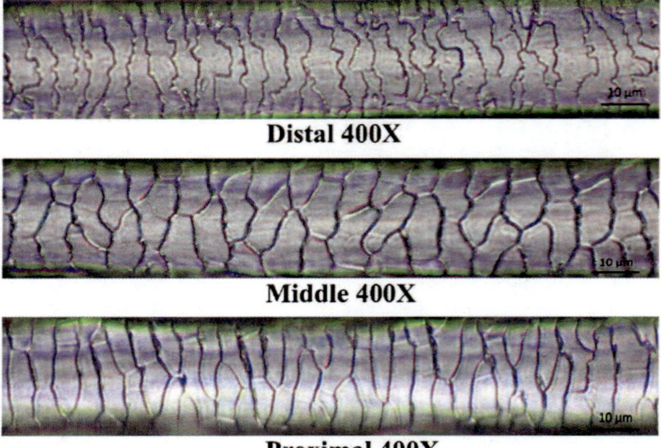

Fig. 4.53: Photomicrograph showing cuticle of guard hair of back region in Cat (*Felis catus domesticus*)

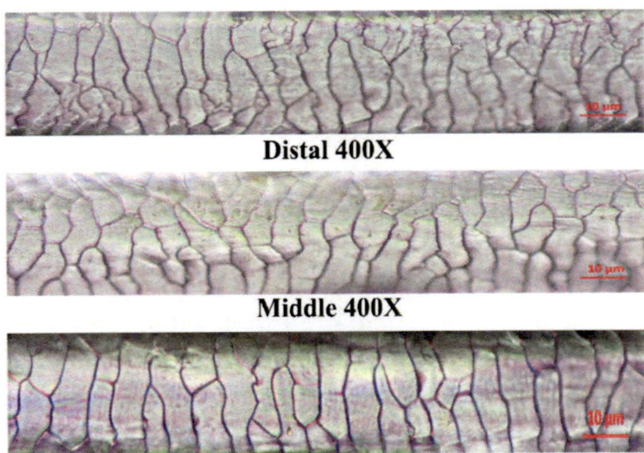

Fig. 4.54: Photomicrograph showing cuticle of guard hair of abdomen region in Cat (Felis catus domesticus)

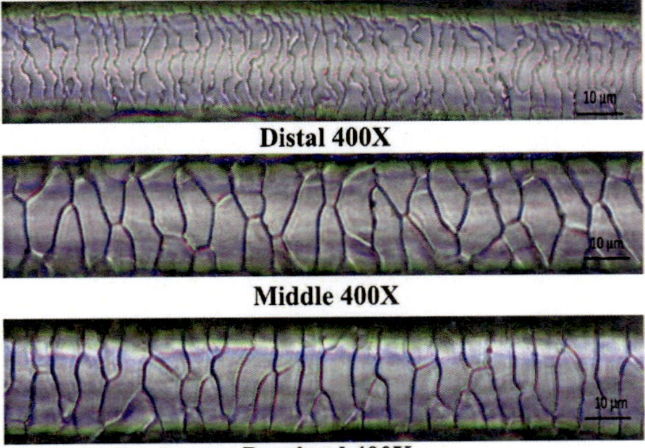

Fig. 4.55: Photomicrograph showing cuticle of guard hair of thigh region in Cat (Felis catus domesticus)

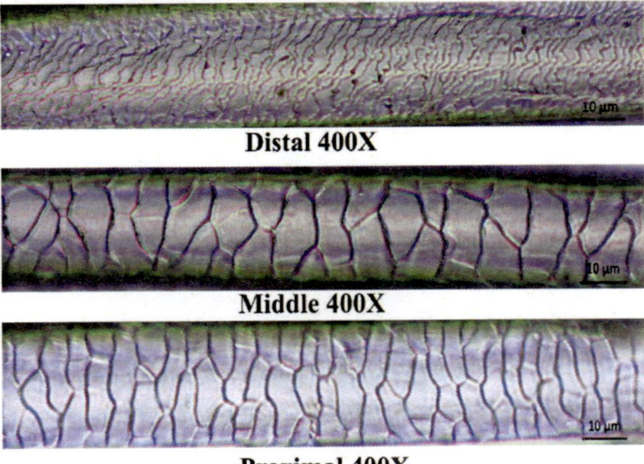

Fig. 4.56: Photomicrograph showing cuticle of guard hair of tail region in Cat (*Felis catus domesticus*)

Domestic Dog (*Canis lupus familiaris*) regular wave pattern and transversal cuticular scale position among all the hair parts of all body regions, except at middle part of hair from tail region. The cuticular scales at middle part of tail region of Dog were found longitudinally placed and arranged in regular mosaic pattern (Fig. 4.61.). The cuticular scale margins were found smooth at the proximal part of hair and scalloped at the distal part of hair among all body regions. At the middle part of hair from head, back, abdomen and thigh region it showed scalloped margin, while neck and tail region showed smooth cuticular margin. It was also observed that the distance between scale margins was distant at proximal and middle part of hair and near at the distal part of hair among all the body regions of Domestic Dog (Fig. 4.57. to 4.62). These observations of the present work are contradictory to the findings reported by Mukherjee *et al.* (2016) in Domestic Dog who mentioned that the cuticular scales were arranged in imbricate pattern. However, they did not mention the part of hair. Furthermore, Gharu and Trivedi (2015) observed coronal cuticular scale arrangement of spines in Dog which do not corroborate with the findings of the present study.

The cuticular scales of Spotted Deer (*Axis axis*) were transversal in position and were arranged in regular wave pattern among all the parts of hair and body regions. It was also noted that the cuticular scale margin was smooth at proximal part, slightly scalloped at middle part and densely scalloped at the distal part of the hair. The scale margin distance was found distant at all three parts of hair and six body region in Spotted Deer (Fig. 4.63 to 4.68.).

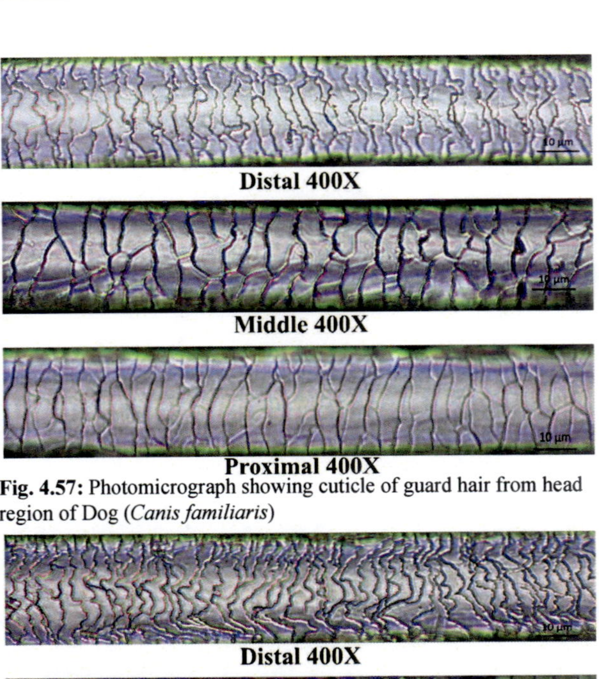

Distal 400X

Middle 400X

Proximal 400X

Fig. 4.57: Photomicrograph showing cuticle of guard hair from head region of Dog (*Canis familiaris*)

Distal 400X

Middle 400X

Proximal 400X

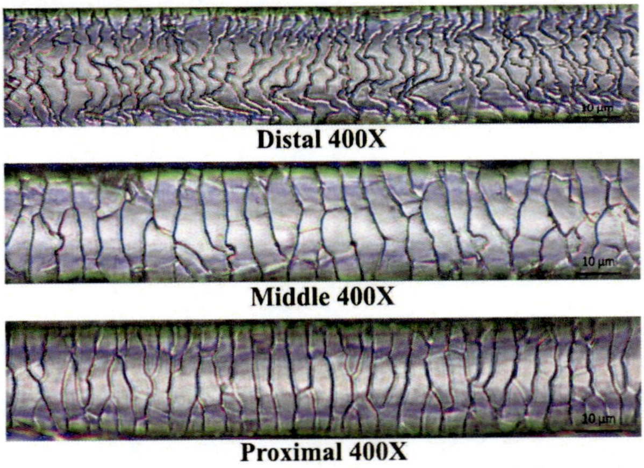

Fig. 4.58: Photomicrograph showing cuticle of guard hair from neck region of Dog (*Canis familiaris*)

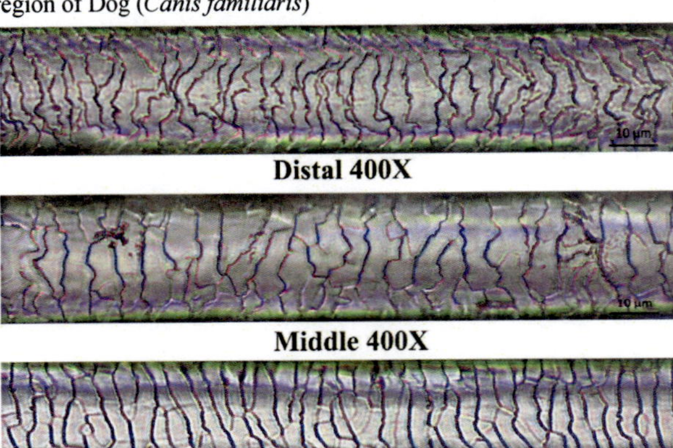

Distal 400X

Middle 400X

Proximal 400X

Fig. 4.59: Photomicrograph showing cuticle of guard hair from back region of Dog (Canis familiaris)

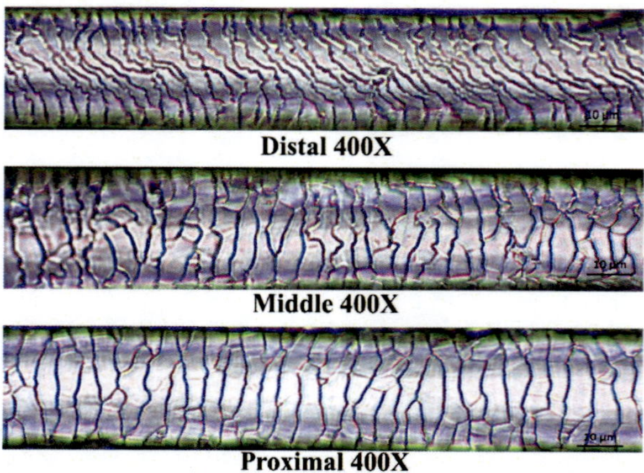

Distal 400X

Middle 400X

Proximal 400X

Fig. 4.60: Photomicrograph showing cuticle of guard hair from abdomen region of Dog (Canis familiaris)

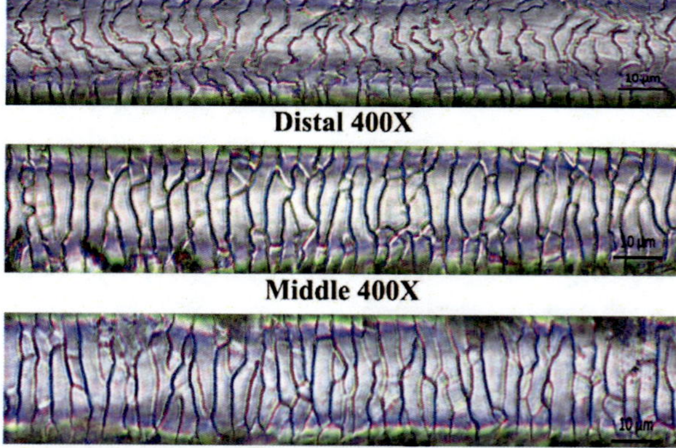

Distal 400X

Middle 400X

Proximal 400X

Fig. 4.61: Photomicrograph showing cuticle of guard hair from thigh region of Dog (Canis familiaris)

Distal 400X

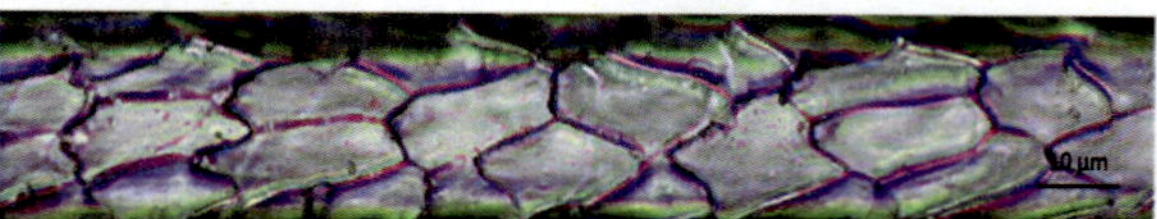

Middle 400X

Proximal 400X

Fig. 4.62: Photomicrograph showing cuticle of guard hair from tail region of Dog (*Canis fmiliaris*)

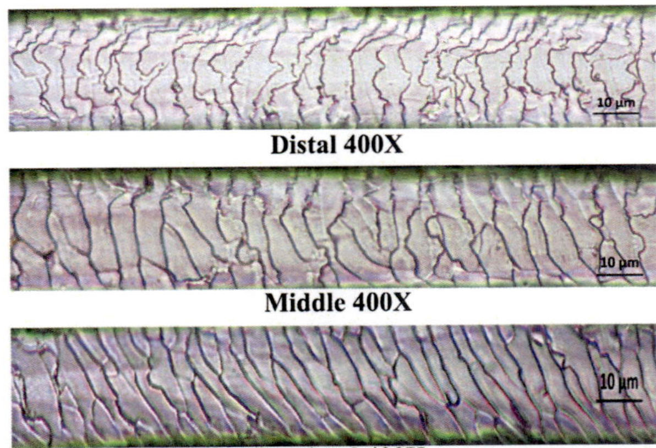

Proximal 400X

Fig. 4.63: Photomicrograph showing cuticle of guard hair from head region of Spotted deer (*Axis axis*)

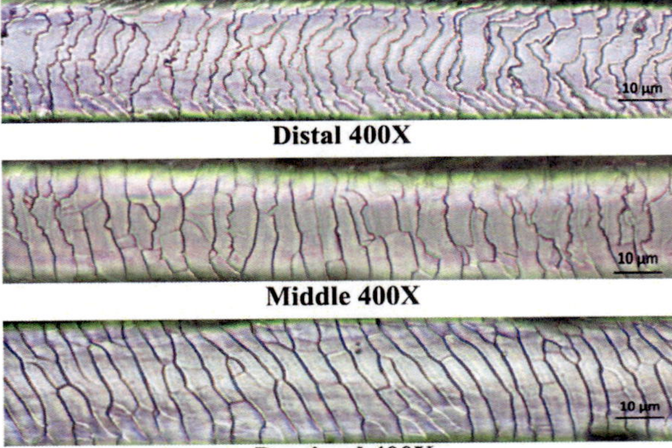

Proximal 400X

Fig. 4.64: Photomicrograph showing cuticle of guard hair from neck region of Spotted deer (*Axis axis*)

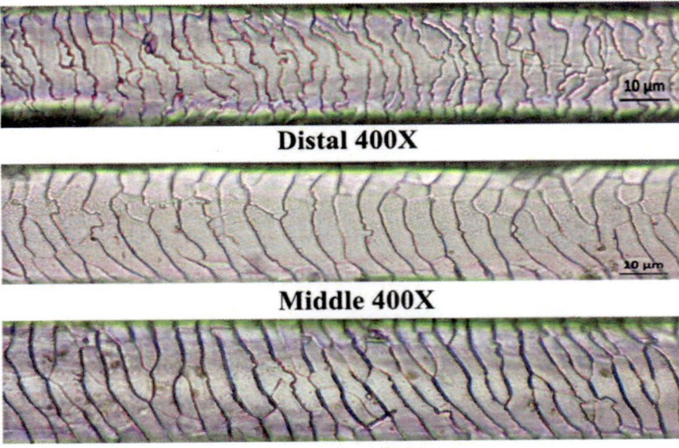

Proximal 400X

Fig. 4.65: Photomicrograph showing cuticle of guard hair from back region of Spotted deer (*Axis axis*)

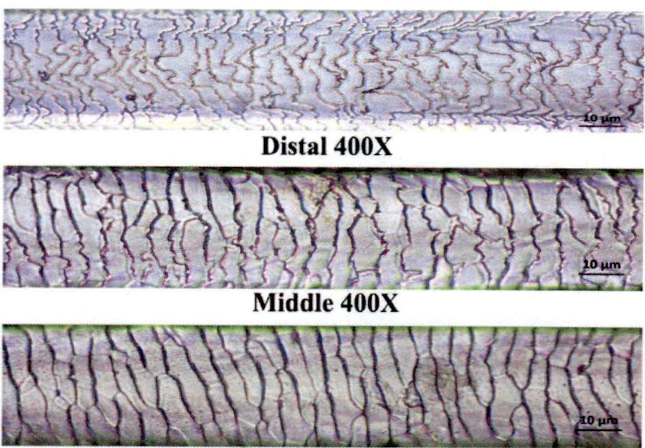

Fig. 4.66: Photomicrograph showing cuticle of guard hair from abdomen region of Spotted deer (*Axis axis*)

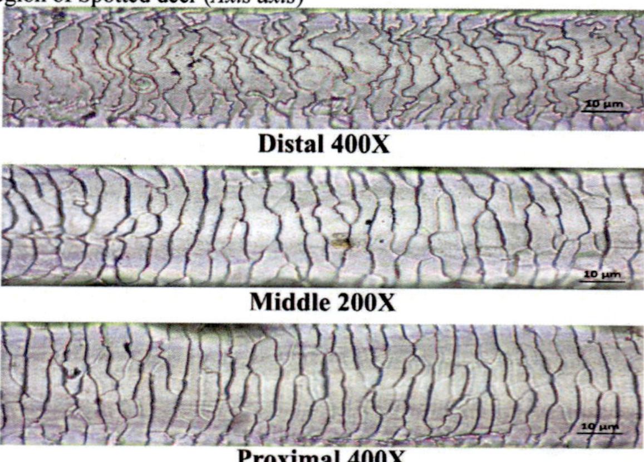

Fig. 4.67: Photomicrograph showing cuticle of guard hair from thigh region of Spotted deer (*Axis axis*)

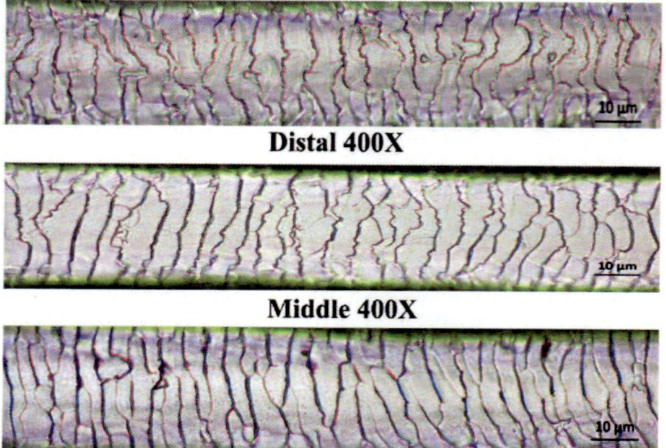

Fig. 4.68: Photomicrograph showing cuticle of guard hair from tail region of Spotted deer (*Axis axis*)

These observations are in partial agreement with the observations of Dharaiya and Soni (2012) who reported that the cuticular scales were arranged in regular wave pattern with smooth scale margin and clear distance between scale margins. Joshi *et al.* (2012) interpreted that cuticular scale had smooth margins except at distal part of hair from thigh region, where it showed crenate margin in Spotted Deer. These findings are also in partial agreement with the findings recorded in the present study. The findings of present study were not in agreement with the observations of Koppikar and Sabnis (1976), who reported imbricate, compressed, ovate type scales at middle part of hair in Spotted Deer.

In Sambar (*Rusa unicolor*), the cuticular scale pattern, among all the body regions and parts of hair was found arranged in regular wave pattern with smooth scale margin at proximal part of hair, smooth to slightly rippled at middle and rippled at the distal part of hair. The hair from tail region had regular wave pattern with smooth scale margin and smooth borders of the cuticular scales through the entire length of hair shaft (Fig.4.69. to 4.74.). The near distance between scale margins was observed through the entire length of hair shaft from all body regions. The finding of the present research work are in line with Koppikar and Sabnis (1976) who reported that the smooth, crenate and spiny or rippled margins at proximal, middle and distal parts of hair and Joshi *et al.* (2012) reported rippled scale margin in Sambar. The findings of the present study are in partial agreement with Dharaiya and Soni (2012) who noted smooth scale margin, regular wave pattern and near scale margin distance.

Nilgai (*Bosellaphus tragocamellus*) showed transversal position of cuticular scale, which were arranged in regular wave pattern with smooth scale margins at proximal part of hair and in irregular wave pattern with highly rippled scale margin at the distal part of hair among all the body regions. The middle part of hair from head, neck and abdomen had cuticular scales arranged in regular wave pattern with smooth scale margin and hair from neck, thigh and tail had irregular wave pattern of cuticular scale width rippled scale margin (Fig. 4.75. to 4.80). It was also noted that the distance between scale margins was near among all the parts of hair parts and body regions in Nilgai.

The results of present work are in accordance with the findings of Kamalakannan (2017d) who reported regular wave pattern with smooth scale margin and near scale margin distance and transversal scale position in mid-dorsal guard hair of Nilgai.

The findings of the present study are in partial agreement with De and Chakraborty (2012) who reported regular wave pattern with smooth scale margin and distant scale margins distance in the dorsal guard hair of Nilgai. Bhatt *et al.* (2004) mentioned that the cuticular scale were transversely positioned and arranged in irregular wave pattern with rippled scale margin and near scale margin distance in the hair from neck, back, abdomen and tail regions of Nilgai. These variations may be attributed to the difference in the part of hair studied by them.

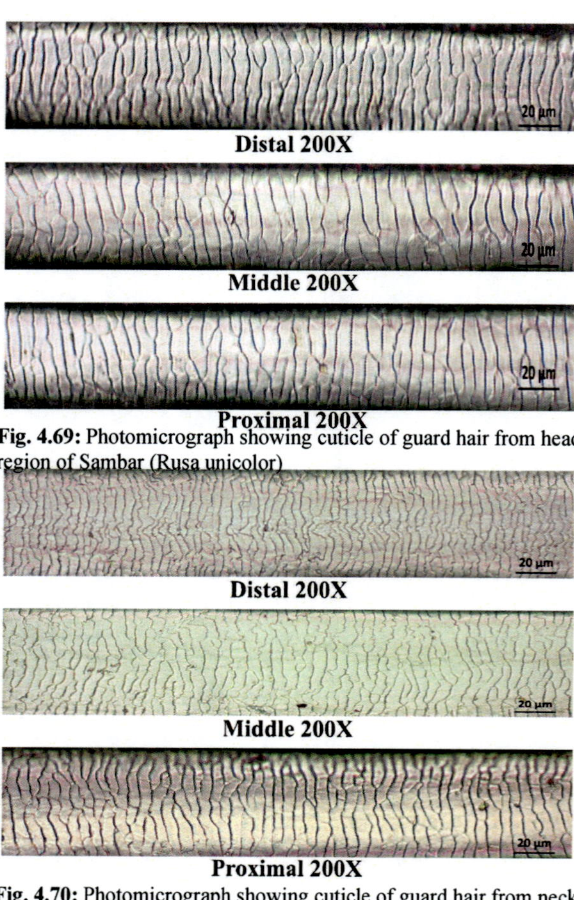

Fig. 4.69: Photomicrograph showing cuticle of guard hair from head region of Sambar (Rusa unicolor)

Fig. 4.70: Photomicrograph showing cuticle of guard hair from neck region of Sambar (Rusa unicolor)

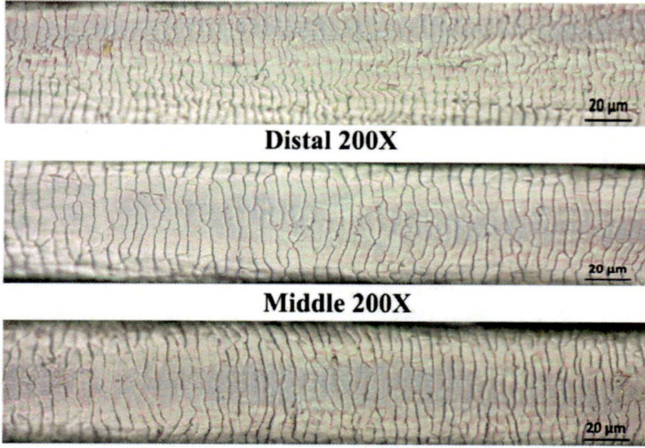

Fig. 4.71: Photomicrograph showing cuticle of guard hair from back region of Sambar (*Rusa unicolor*)

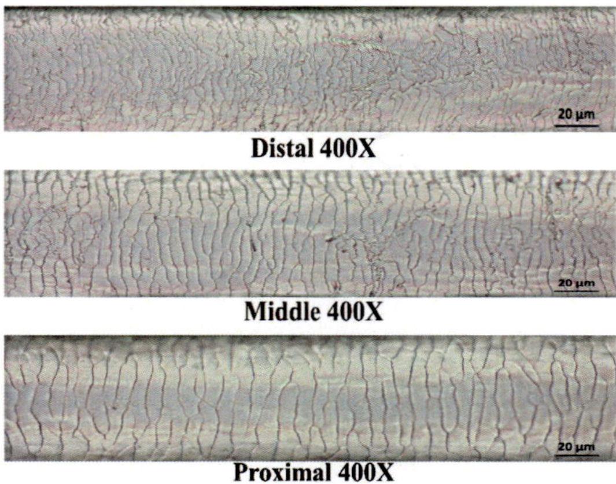

Fig. 4.72: Photomicrograph showing cuticle of guard hair from abdomen region in Sambar (*Rusa unicolor)*

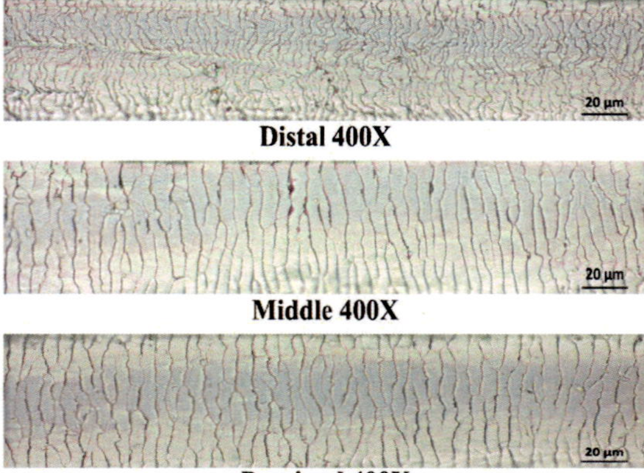

Fig. 4.73: Photomicrograph showing cuticle of guard hair from thigh region of Sambar (*Rusa unicolor*)

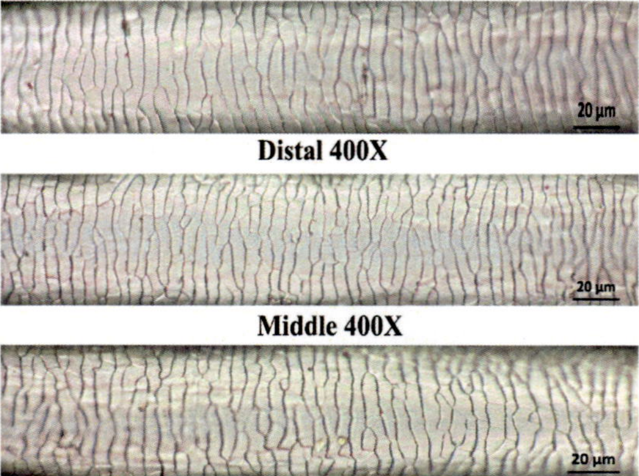

Fig. 4.74: Photomicrograph showing cuticle of guard hair from tail region of Sambar (Rusa unicolor)

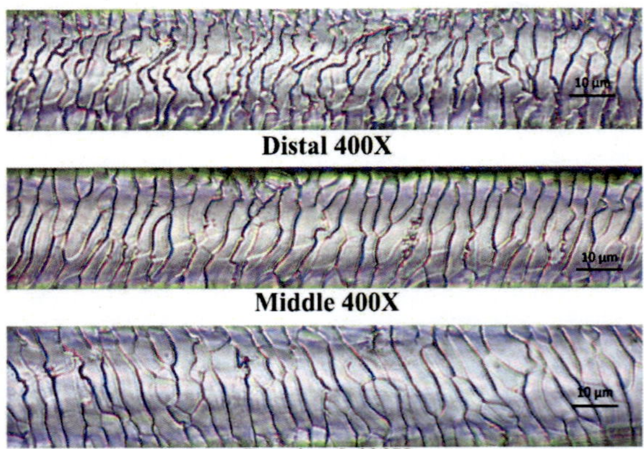

Fig. 4.75: Photomicrograph showing cuticle of guard hair from head region of Nilgai (*Boselaphus tragocalmelus*)

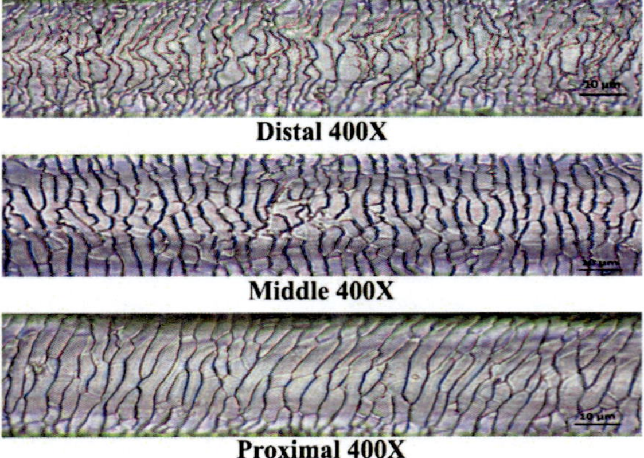

Fig. 4.76: Photomicrograph showing cuticle of guard hair of neck region of Nilgai (*Boselaphus tragocamelus*)

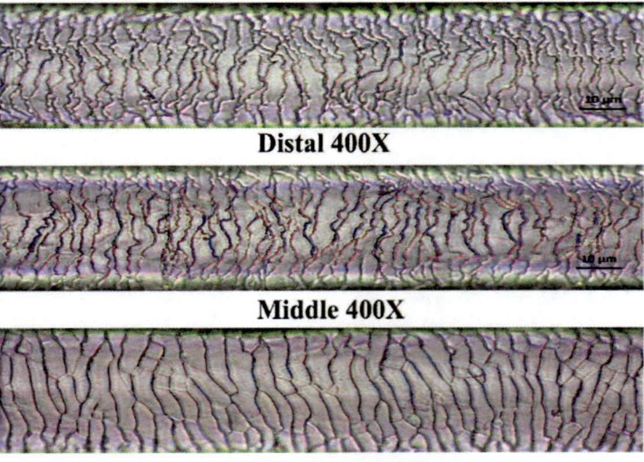

Fig. 4.77: Photomicrograph showing cuticle of guard hair of back region of Nilgai (*Boselaphus tragocamelus*)

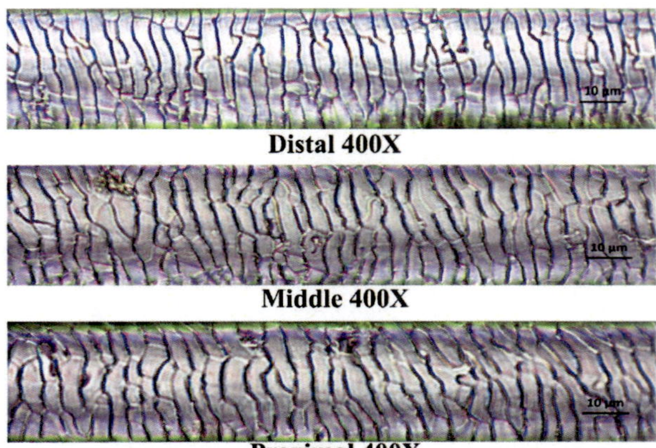

Fig. 4.78: Photomicrograph showing cuticle of guard hair of abdomen region of Nilgai (*Boselaphus tragocamelus)*

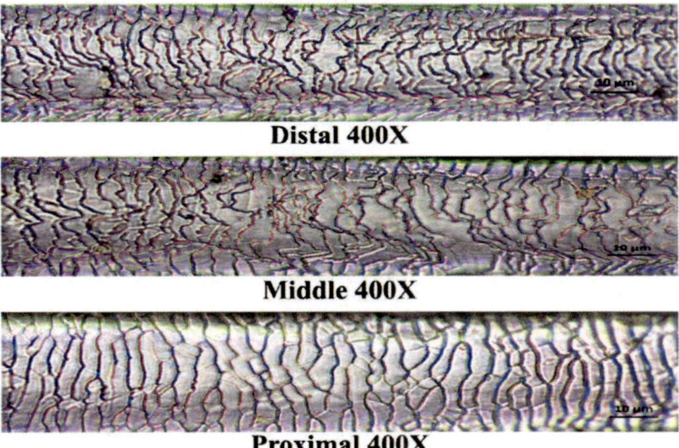

Fig. 4.79: Photomicrograph showing cuticle of guard hair of thigh region of Nilgai (*Boselaphus tragocamelus*)

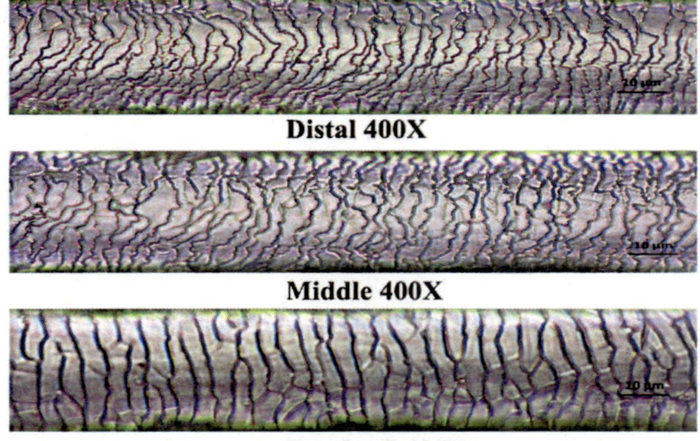

Proximal 400X

Fig. 4.80: Photomicrograph showing cuticle of guard hair of thigh region of Nilgai (*Boselaphus tragocamelus*)

The cuticular scales showed uniformity in position, arrangement and distance between scale margins among all the body regions in Hanuman Langur (*Samnopithecus entellus*). The cuticular scales were arranged in irregular wave pattern with distant scale margins distance among all the body regions and parts of hair (Fig. 4.81. to 4.85) except distal part of hair from tail region, where cuticular scales were arranged in irregular wave pattern with near scale margin distance (Fig. 4.86). The margins of the cuticular scales were slightly crenate at proximal and middle parts of hair, while at distal part of hair, it was found densely crenate.

The present observations are in accordance with Sarkar *et al.* (2011) and Gharu and Trivedi (2013) who reported transversal scale position, irregular wave pattern with crenate scale margin and distant, distance between scale margins in Hanuman Langur. The findings of the Koppikar and Sabnis (1976) were not in accordance with the present findings. They stated that the cuticular scale borders were imbricate at proximal and middle part and smooth at the distal part of hair in Hanuman Langur. Dharaiya and Soni (2012) commented that the dorsal guard hair of Hanuman Langur had regular wave pattern with crenate margin and close distance between scale margins. These findings are also in contradictory with the findings noted during the present study.

The cuticular scale from Tiger (*Panthera tigris*) showed various types of patterns along the longitudinal axis of hair. The cuticular scale patterns observed in Tiger hair during the present study was streaked, regular as well as irregular wave and single chevron patterns. At the proximal part of hair, cuticular scale in six body regions were arranged in regular wave pattern with smooth margin and near scale margins distance. The hair from head region had streaked scale pattern with rippled margin at distal part of hair and regular wave pattern with smooth margin at middle part of hair. Close separation was noted at distal part, while near separation was observed at the middle part of hair (Fig. 4.87). Single chevron cuticular scale pattern with rippled scale margin and close, distance between scale margins was observed at the distal and middle part of hair from neck region of Tiger (Fig. 4.88).

The cuticular scales of distal part of hair from back region of Tiger were found arranged in streaked pattern with rippled scale margin and closed separation between scales. At the middle part of hair two different types of cuticular scale patterns were noted as single chevron and irregular wave and both the pattern had rippled margins with close scale margin distance (Fig. 4.89.). Irregular wave pattern with smooth scale margin and near scale separation was noted at the middle part of hair from the abdomen region of Tiger. Further, it was noted that the distal part of hair from abdomen region had streaked pattern with rippled scale margin and close separation (Fig. 4.90).

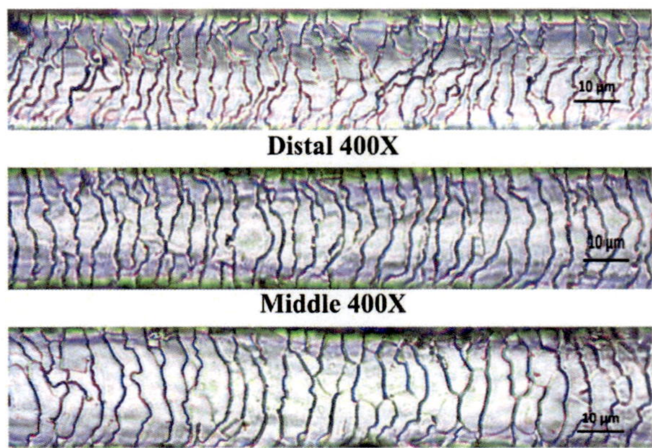

Fig. 4.81: Photomicrograph showing cuticle of guard hair from head region of Hanuman Langur (*Samnopithecus entellus*)

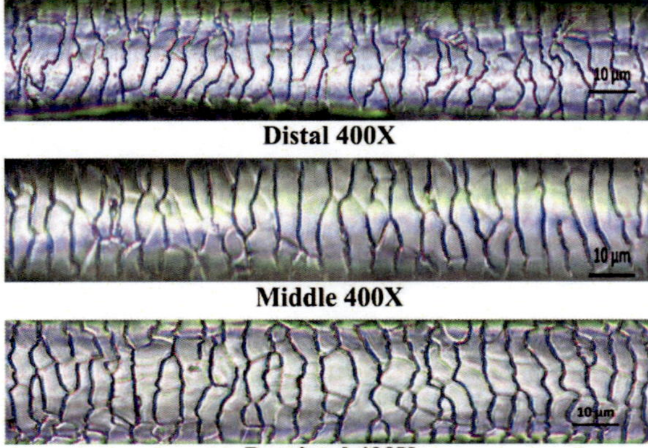

Fig. 4.82: Photomicrograph showing cuticle of guard hair from neck region of Hanuman langur (*Samnopithecus entellus*)

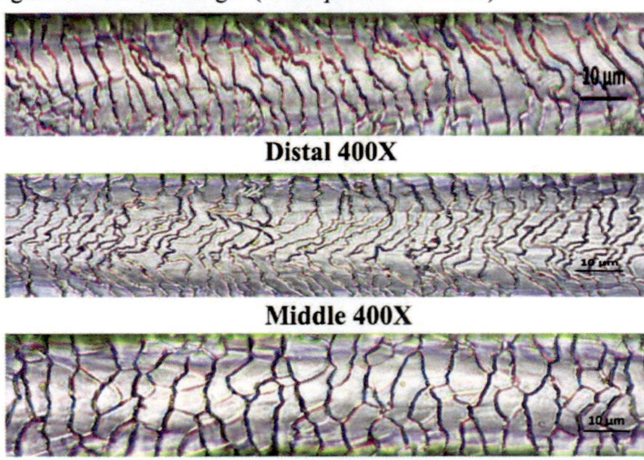

Fig. 4.83: Photomicrograph showing cuticle of guard hair from back region of Hanuman langur (*Samnopithecus entellus*)

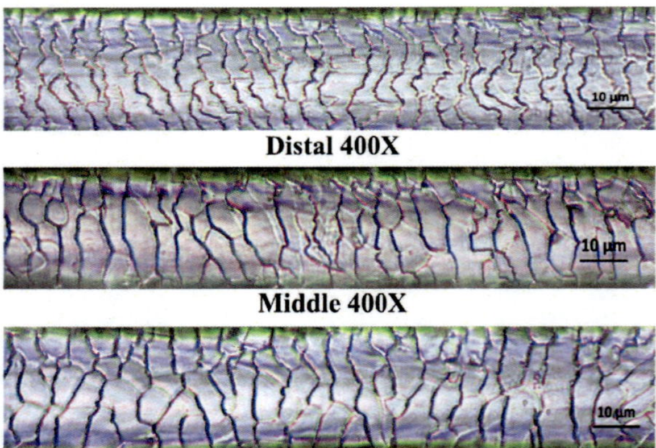

Fig. 4.84: Photomicrograph showing cuticle of guard hair from abdomen region of Hanuman langur (*Samnopithecus entellus*)

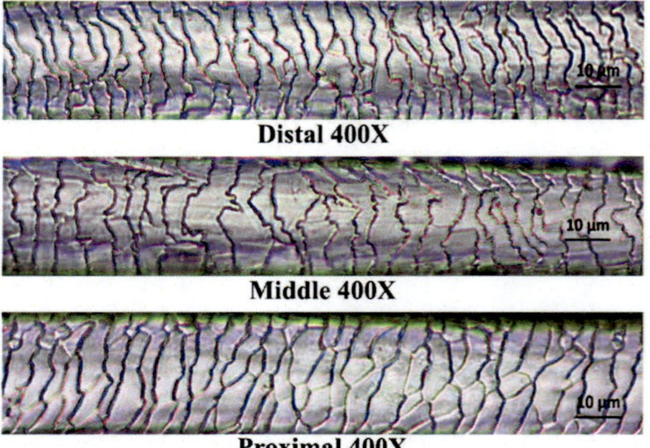

Fig. 4.85: Photomicrograph showing cuticle of guard hair from thigh region of Hanuman langur (*Samnopithecus entellus*)

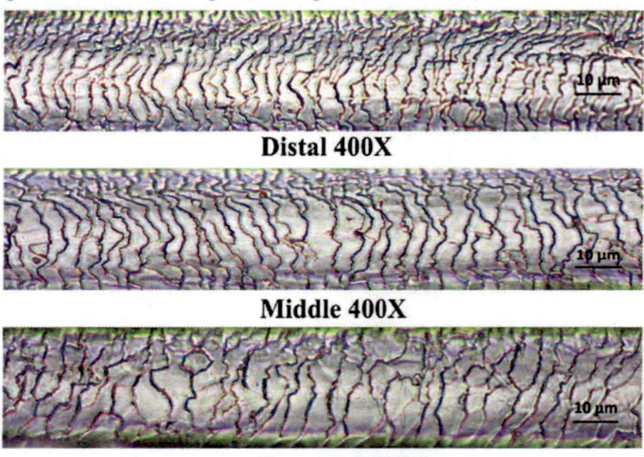

Fig. 4.86: Photomicrograph showing cuticle of guard hair from tail region of Hanuman langur (*Samnopithecus entellus*)

Further, it was noted that the single chevron pattern with rippled margin and close distance between scale margins at the middle part of hair from tail region (Fig. 4.92), while streaked pattern with rippled margin and close separation was found at the distal hair part from thigh region in Tiger (Fig. 4.91).

Kitpipit and Thanakiatkrai (2013) observed five various cuticular scale pattern single chevron, regular wave, irregular wave, streaked and mixture of above patterns. They further stated that the proximal part of hair from all body regions had regular wave pattern with smooth margin and near scale margin distance. While, other four patterns mostly with rippled margin at close separation were noted at middle and distal part of hair. These region wise findings in Tiger hair are in total concurrence with the observations noted during present study. They also mentioned that distant scale margin distance, rarely found in Tiger hair, mainly observed in ungulate animals. The chevron pattern was rarely reported in other species. So, it can be concluded from the present study and findings of the previous workers, that hair could be a very important tool for identification of Tiger from hair characters.

The findings of the present work are also in line with the findings of Soni *et al.* (2004) and Gharu and Trivedi (2015), who reported regular wave pattern with rippled scale margin with near scale margin distance. Koppikar and Sabnis (1976) observed spiny cuticular scale borders at proximal part of hair, while plain border was observed at middle and distal parts of hair from Tiger. These observations are not in agreement with the findings of the present study.

The cuticular scale in Leopard (*Panthera pardus*) was found arranged in regular wave pattern with smooth margin and distant separation at proximal and middle parts of hair (Fig. 4.93 to 4.98.) among all the regions of body. While, at distal hair part, irregular wave pattern with rippled margin and close scale margin distance was noted in all body regions except neck region. The distal part of neck region showed single chevron pattern with rippled margin and close distance between scale margins (Fig. 4.94).

The findings of the present study were not in agreement with the findings of Soni *et al.* (2004), who reported diamond cuticular scale pattern with smooth margins and distant scale separation. Similarly the findings are also not in agreement with the findings of Koppikar and Sabnis (1976) who stated that the crenate border was observed at proximal part middle part while, plain border was noted at distal part of hair of Leopard.

Gharu and Trivedi (2015a) observed regular, chevron, imbricate and petal scale pattern in the hair from Leopard cub. The findings regarding imbricate and petal cuticular pattern were partly in accordance with the findings of present study.

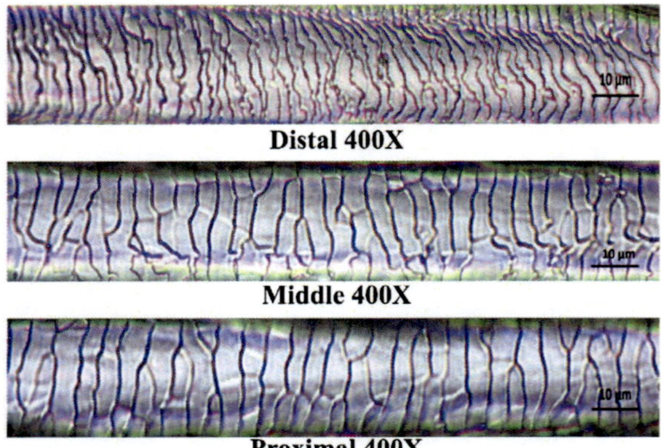

Fig. 4.87: Photomicrograph showing cuticle of guard hair from head region of Tiger (*Panthera tigris*)

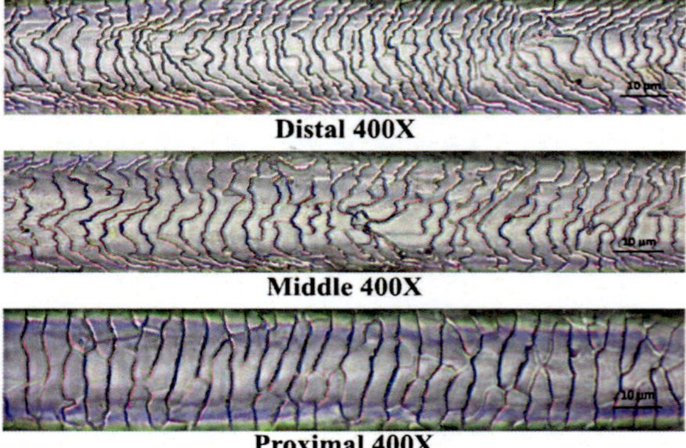

Fig. 4.88: Photomicrograph showing cuticle of guard hair from neck region of Tiger (*Panthera tigris*)

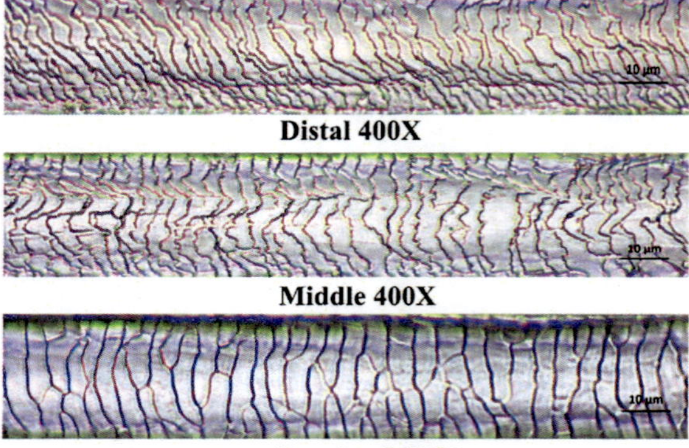

Fig. 4.89: Photomicrograph showing cuticle of guard hair from back region of Tiger (*Panthera tigris*)

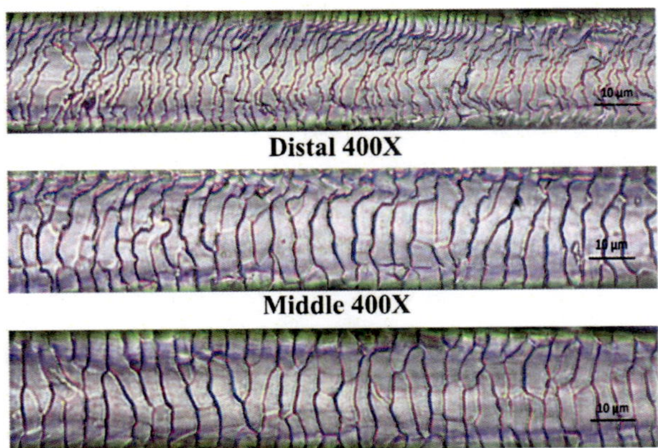

Fig. 4.90: Photomicrograph showing cuticle of guard hair from abdomen region of Tiger (Panthera tigris)

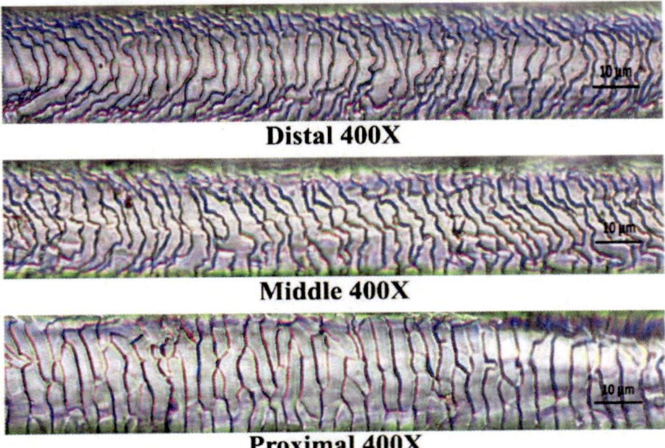

Fig. 4.91: Photomicrograph showing cuticle of guard hair from thigh region of Tiger (*Panthera tigris*)

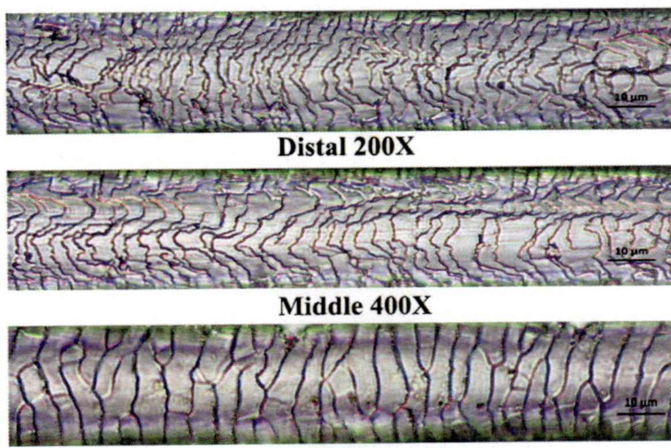

Fig. 4.92: Photomicrograph showing cuticle of guard hair from tail region of Tiger (Panthera tigris)

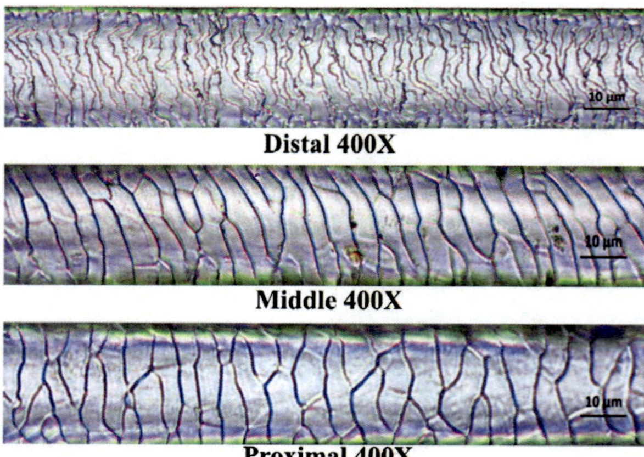

Fig. 4.93: Photomicrograph showing cuticle of guard hair from head region of Leopard (Panthera pardus)

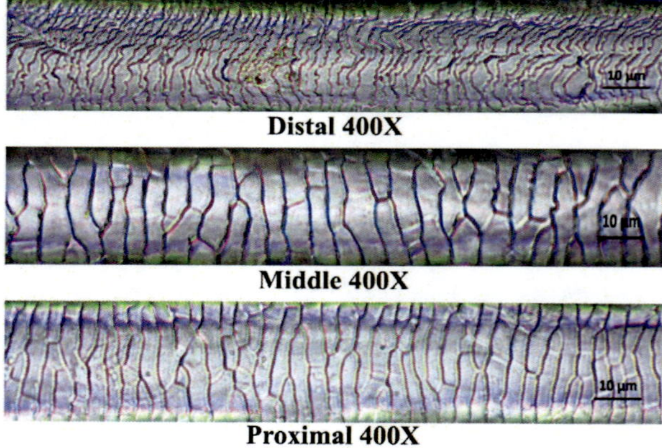

Fig. 4.94: Photomicrograph showing cuticle of guard hair from neck region of Leopard (*Panthera pardus*)

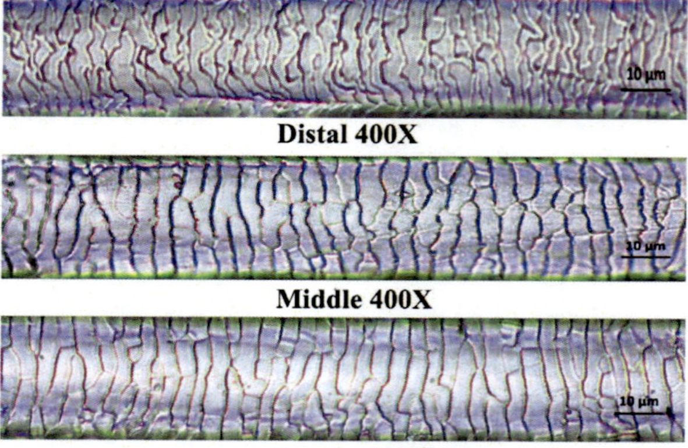

Fig. 4.95: Photomicrograph showing cuticle of guard hair from back region of Leopard (*Panthera pardus*)

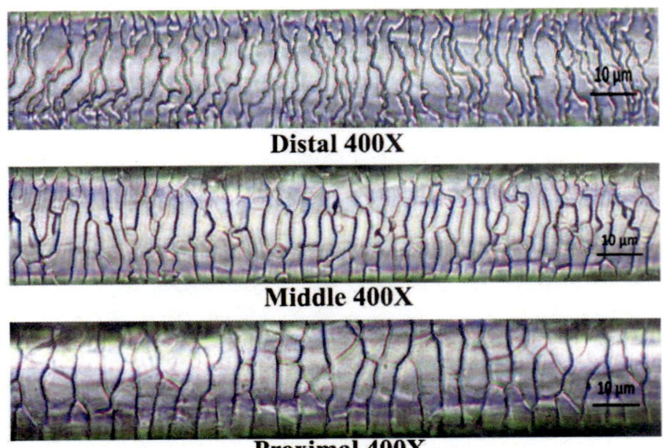

Fig. 4.96: Photomicrograph showing cuticle of guard hair from abdomen region of Leopard (*Panthera pardus*)

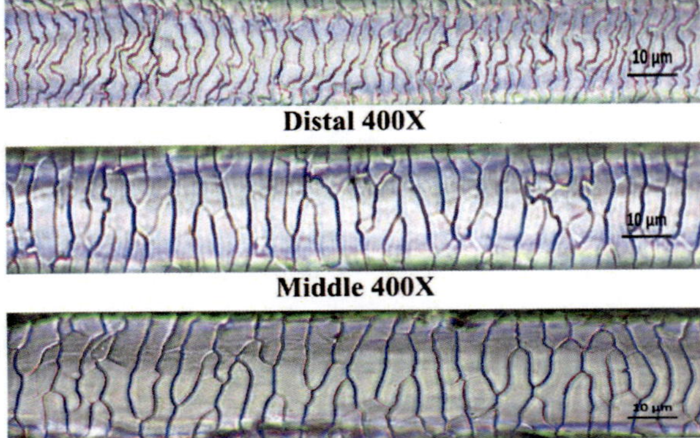

Fig. 4.97: Photomicrograph showing cuticle of guard hair from thigh region of Leopard (*Panthera pardus*)

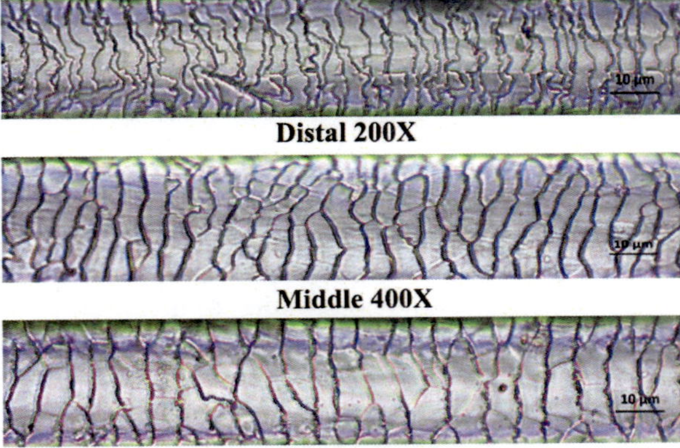

Fig. 4.98: Photomicrograph showing cuticle of guard hair from tail region of Leopard (*Panthera pardus*)

4.3. Characteristics of medulla and cortex of hair

The medulla and cortex characteristics were studied by preparing whole mounts of the hair. It was observed from the whole mount of all hair parts, body regions and all the domestic as well as wild animal species that the medulla was absent at the tip of hair. The medulla tapers from the proximal part of hair towards the tip of hair in all the species.

In Cattle (*Bos indicus*) the fragmental medulla was observed at all the three parts of hair from the head and tail regions (Fig. 4.99 and 4.104) however, hair from neck, back, abdomen and thigh region showed continuous type of medulla. It was also observed that the cells of medulla were arranged in unicellular irregular manner with almost straight medulla border at all three parts of hair from back, abdomen and thigh regions (Fig. 4.101, 4.102 and 4.103). However, the hair from neck region showed fragmental medulla at middle part of hair and continuous medulla at proximal and distal part of hair (Fig. 4.100.). Some amorphous medulla zones were also found in the hair of head and neck regions in cattle.

The cortex from head, neck and tail regions was mummy brown coloured whereas cortex of back, abdomen and thigh regions was raw umber coloured with dark coloured pigment granules. Dense in distribution of large clove brown coloured pigment granules were in uniformly distributed throughout the shaft of hair of tail region (Fig. 4.103) of Cattle.

These findings of the present study are in agreement with those reported by Gharu and Trivedi (2015) stated that the medulla was continuous, simple with straight margin in hair of Cattle at neck, back, abdomen and thigh regions at all parts of hair.

The observations noted during present study were partially in akin with the findings observed by Marinis and Asprea (2016) who noted multicellular, multiseriate appearing, vacuolated and continuous medulla with straight to irregular margins. Patches of amorphous medulla were noted in hair of head and neck regions in Cattle during present study are in accordance with the findings noted by Gharu and Trivedi (2015) who mentioned continuous medulla with some amorphous zones throughout the length of hair shaft.

Sahajpal *et al.* (2009) noted amorphous and wide medulla throughout the length of hair. These findings are not in agreement with the observation of the present study.

Fragmental to scanty medulla was observed at the three parts of hair from head and tail region of Buffalo (*Bubalus bubalis*) (Fig. 4.105 and 4.110). The fine pigment granules were uniformly distributed at middle and distal part of hair, however large sized granules were observed at proximal part. The wool brown coloured cortex was noted at middle and distal parts of hair from head region. However clove brown cortex with dark coloured pigment granules were noted at all the parts of hair from tail region of Buffalo (Fig. 4.110).

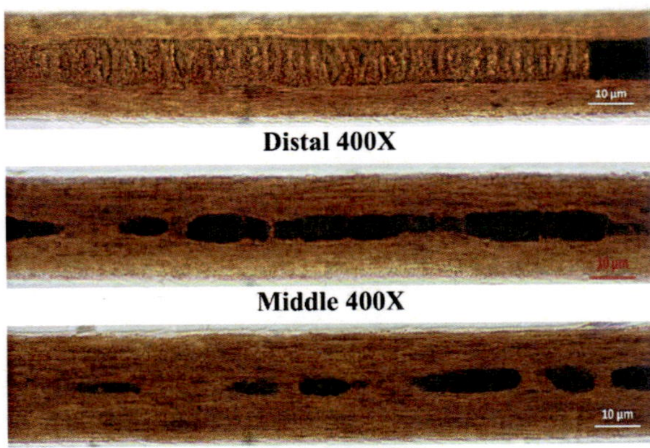

Distal 400X

Middle 400X

Proximal 400X

Fig. 4.99: Photomicrograph showing composition of medulla of guard hair from head region of Cattle (*Bos indicus*)

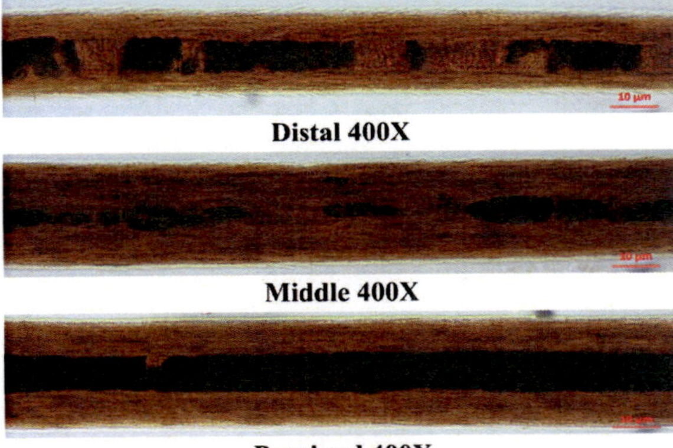

Distal 400X

Middle 400X

Proximal 400X

Fig. 4.100: Photomicrograph showing composition of medulla of guard hair from neck region of Cattle (*Bos taurus*)

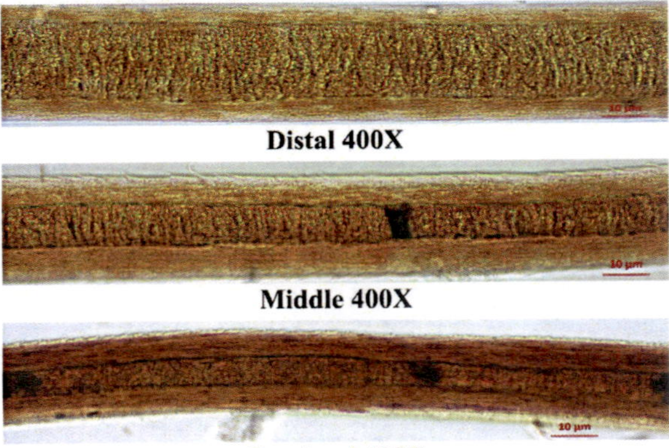

Distal 400X

Middle 400X

Proximal 400X

Fig. 4.101: Photomicrograph showing composition of medulla of guard hair from back region in Cattle *(Bos taurus)*

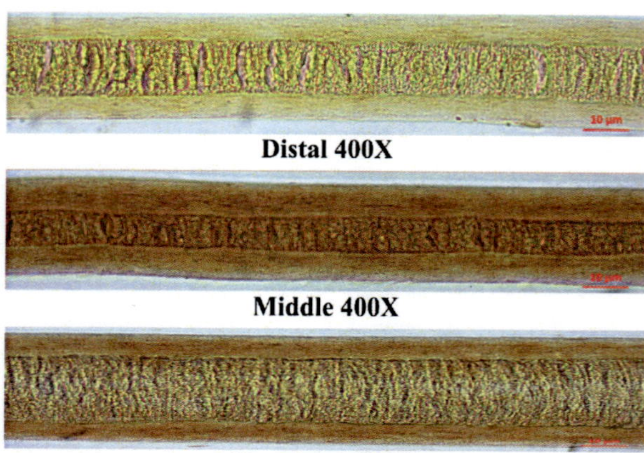

Distal 400X

Middle 400X

Proximal 400X

Fig. 4.102: Photomicrograph showing medulla and cortex of guard hair from abdomen region of Cattle (*Bos taurus*)

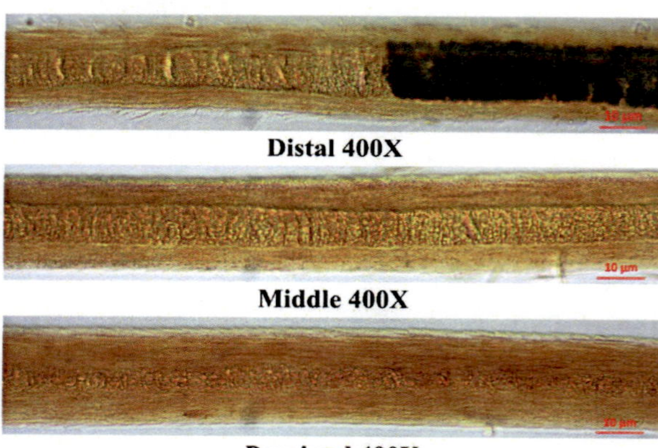

Distal 400X

Middle 400X

Proximal 400X

Fig. 4.103: Photomicrograph showing composition of medulla of guard hair from thigh region of Cattle (*Bos taurus*)

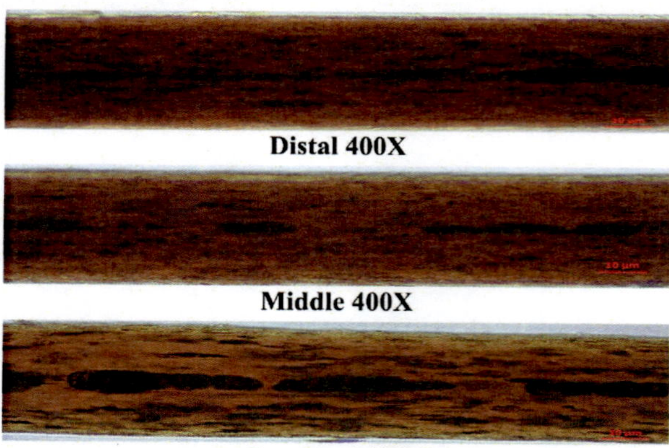

Distal 400X

Middle 400X

Proximal 400X

Fig. 4.104: Photomicrograph showing composition of medulla of guard hair from tail region of Cattle (*Bos taurus*)

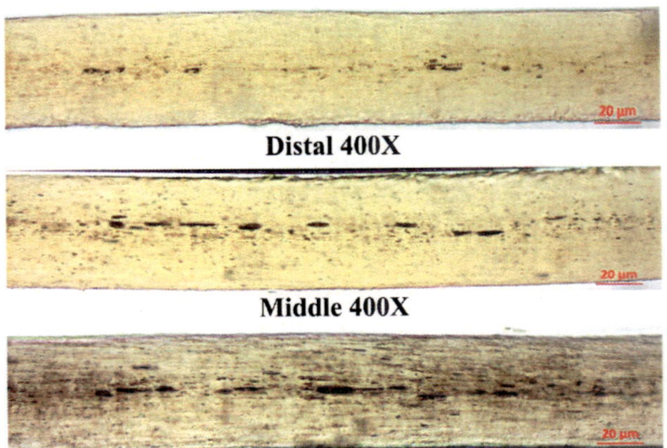

Proximal 400X

Fig. 4.105: Photomicrograph showing composition of medulla of guard hair from head region of Buffalo (*Bubalus bubalis*)

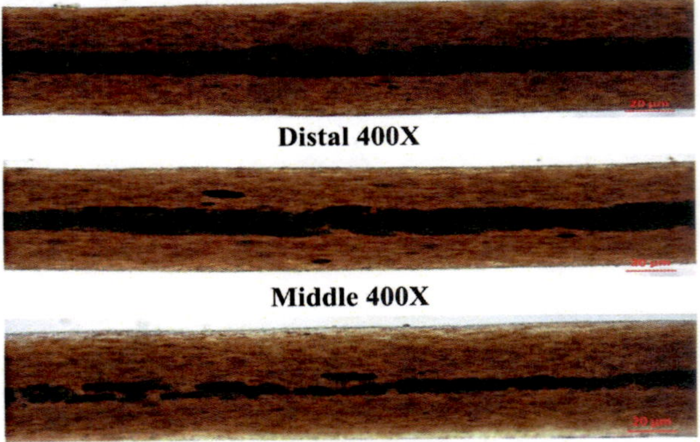

Proximal 400X

Fig. 4.106 Photomicrograph showing composition of medulla of guard hair from neck region of Buffalo (*Bubalus bubalis*)

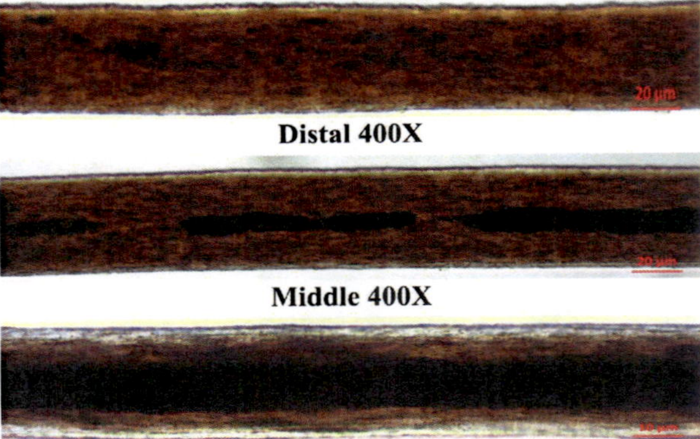

Proximal 400X

Fig. 4.107: Photomicrograph showing composition of medulla of guard hair from back region of Buffalo (*Bubalus bubalis*)

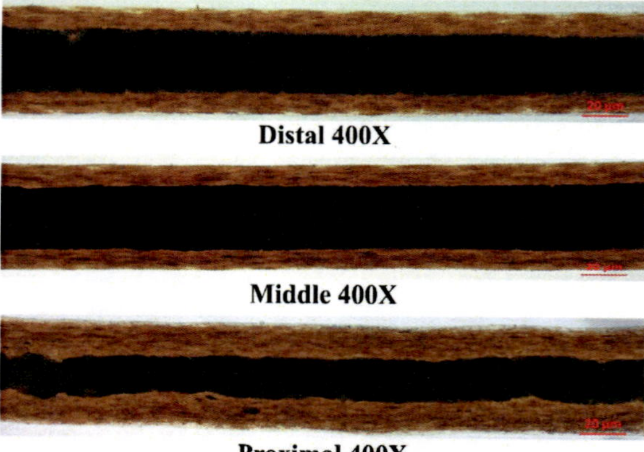

Fig. 4.108: Photomicrograph showing composition of medulla of guard hair from abdomen region of Buffalo (*Bubalus bubalis*)

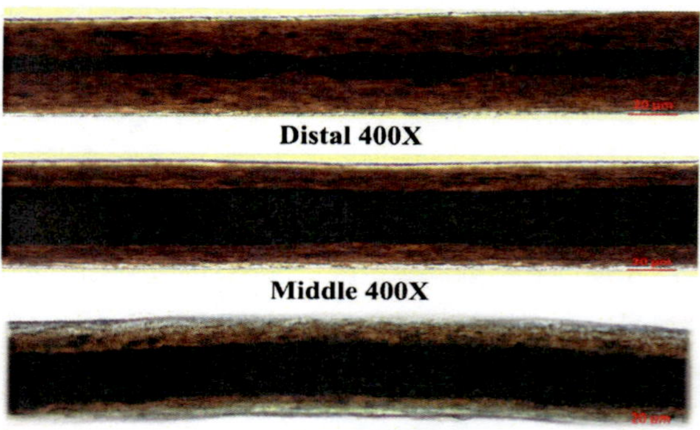

Fig. 4.109: Photomicrograph showing composition of medulla of guard hair from thigh region of Buffalo (*Bubalus bubalis*)

Fig. 4.110: Photomicrograph showing composition of medulla of guard hair from tail region of Buffalo (*Bubalus bubalis*)

The continuous medulla pattern with amorphous structure and irregular medulla border was observed at almost all parts of hair of neck, back, abdomen and thigh regions with the exception of middle and distal parts of hair from back region. At the back region scanty medulla was noted at distal part while, fragmental medulla was found at middle part of hair (Fig. 4.107). The clove brown coloured cortex with uniformly distributed pigment granules were noted at all three parts of hair from neck, back, abdomen and thigh region.

The findings of the present study are in agreement with the observations noted by Mukherjee *et al.* (2016) in Buffalo. They noted continuous amorphous medulla structure with straight to irregular medulla margin in guard hair from mid-dorsal region. Gharu and Trivedi (2015) reported dense, continuous, simple medulla with irregular margin and wide medulla occupying entire hair width. These findings regarding continuous medulla structure with irregular medulla margin found in buffalo are in agreement with the findings noted during the present work.

Sheep (*Ovis aries*) hair from all the three parts of hair and body regions showed continuous medulla with scalloped margin. The width composition of medulla of Sheep hair from all three parts of hair in all the six body regions was multicellular type. The wide lattice / filled medulla structure with scalloped margin was observed at almost all the hair parts of head, neck, back, abdomen and thigh region of Sheep (Fig. 4.111 to 4.115). In the proximal part of hair from head region and all three parts of hair from tail region, narrow medulla lattice structure with scalloped medulla margin was noted (Fig. 4.116).

The cortex showed uniform distribution of pigments throughout the length of hair shaft in all the three parts of hair and six body regions. The burnt umber coloured with dark coloured cortex was observed at head, neck, abdomen and thigh region however, prouts brown coloured cortex was found at all three parts of hair at neck region (Fig. 4.112).

Marinis and Asprea (2005) reported continuous medulla with multicellular composition, multiseriate medulla structure and scalloped margin in Sheep hair. The findings regarding medullary composition and medulla border are in line with the findings noted during the present study however, findings about medullary structure are contradictory with the present observations. Gharu and Trivedi (2015) noted continuous medulla pattern, narrow medulla lattice in Sheep hair in agreement with the observations of the present study. However, their findings regarding irregular margin are not in akin with the findings noted during present work. Similarly the observations of Sahajpal *et al.* (2009) like narrow medulla with vacuoles are not in accordance with the present observations.

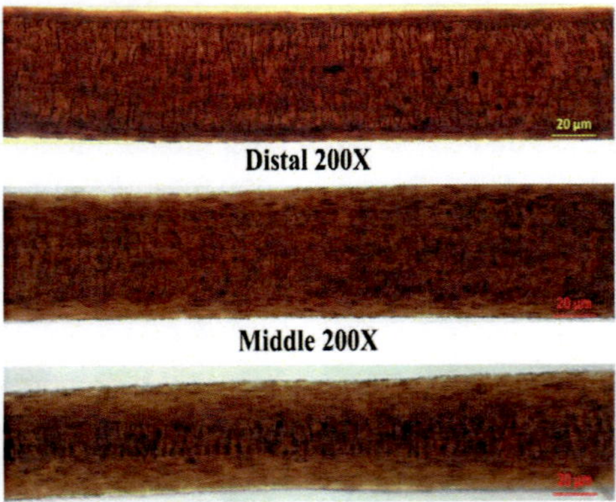

Fig. 4.111: Photomicrograph showing composition of medulla of guard hair from head region of Sheep (*Ovis aries*)

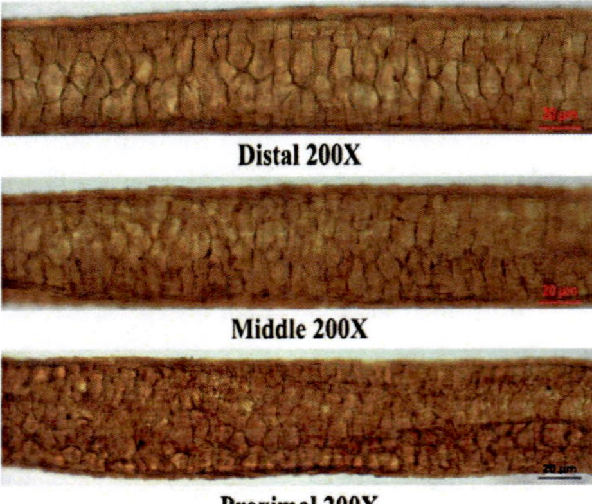

Fig. 4.112: Photomicrograph showing composition of medulla of guard hair from neck region of Sheep (*Ovis aries*)

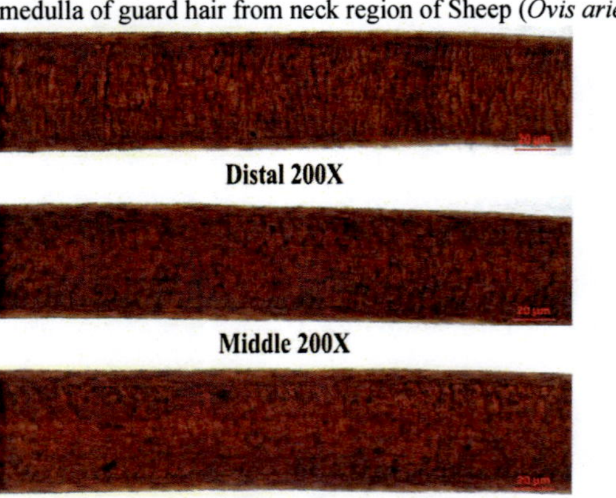

Fig. 4.113: Photomicrograph showing composition of medulla of guard hair from back region of Sheep (Ovis aries)

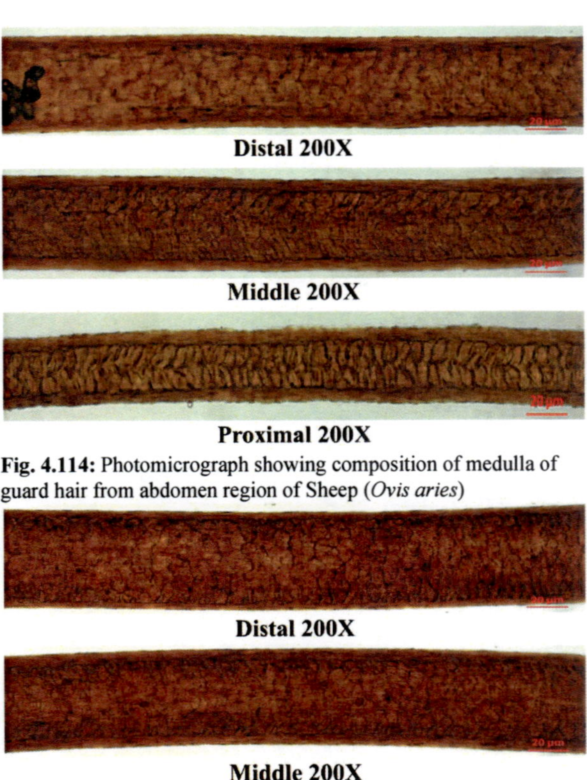

Distal 200X

Middle 200X

Proximal 200X

Fig. 4.114: Photomicrograph showing composition of medulla of guard hair from abdomen region of Sheep (*Ovis aries*)

Distal 200X

Middle 200X

Proximal 200X

Fig. 4.115: Photomicrograph showing composition of medulla of guard hair from thigh region of Sheep (*Ovis aries*)

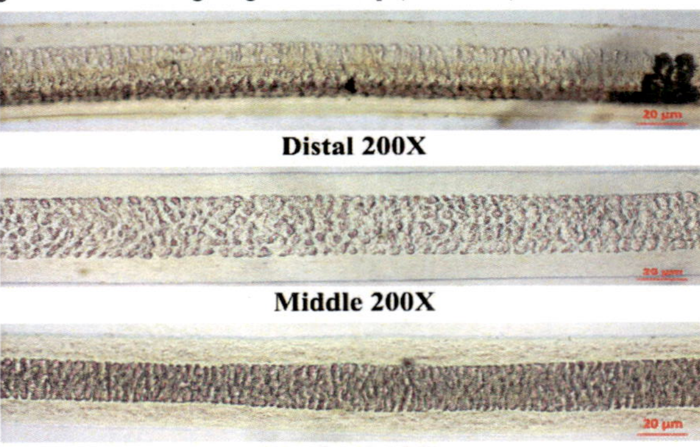

Distal 200X

Middle 200X

Proximal 200X

Fig. 4.116: Photomicrograph showing composition of medulla of guard hair from tail region of Sheep (*Ovis aries*)

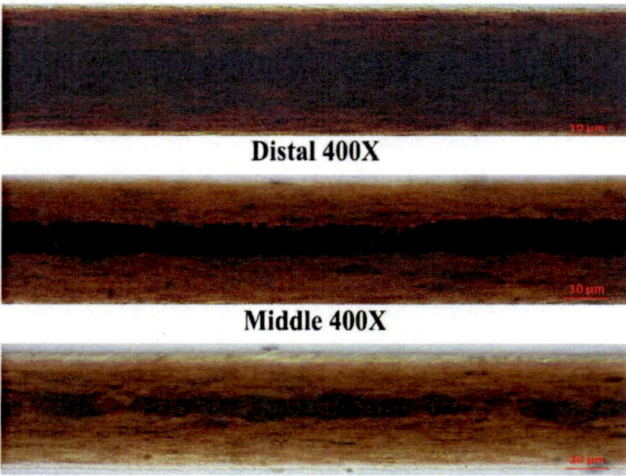

Fig. 4.117: Photomicrograph showing composition of medulla of guard hair from head region of Goat (*Capra hircus*)

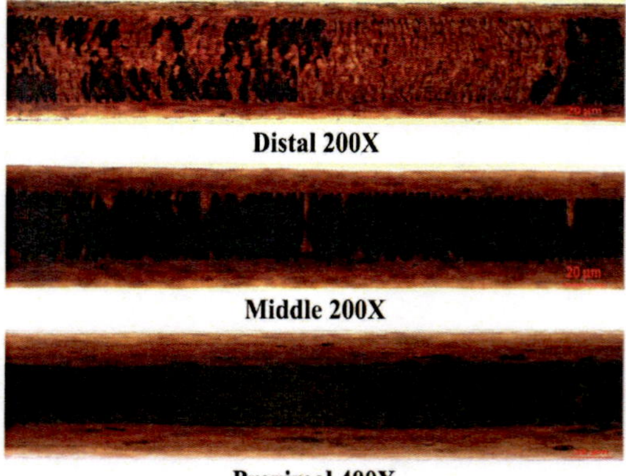

Fig. 4.118: Photomicrograph showing composition of medulla of guard hair from neck region of Goat (*Capra hircus*)

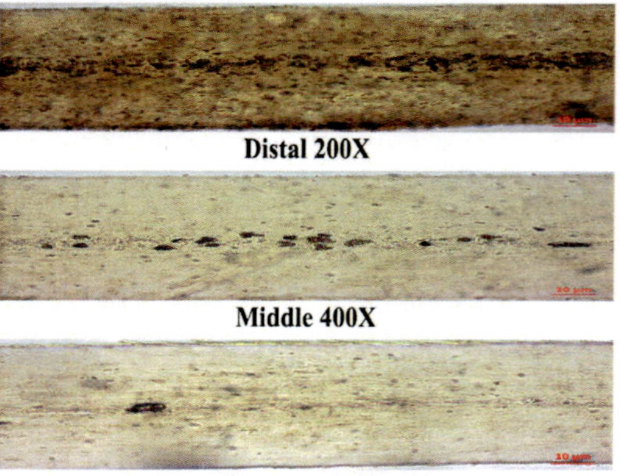

Fig. 4.119: Photomicrograph showing composition of medulla of guard hair from back region of Goat (*Capra hircuss*)

The hair of head, neck, abdomen and thigh regions of Goat (*Capra hircus*) showed continuous medulla, with irregular medulla margin at all the three parts of hair in all six body regions except distal part of hair from tail region where interrupted medulla was observed (Fig. 4.122). The fragmental medulla was observed at all the three parts of hair from back region with hair brown coloured cortex at proximal and middle parts and olive coloured cortex at distal part of hair (Fig. 4.119). The multicellular composition of medulla with multiseriate medulla structure was found in the hair of neck, abdomen and thigh regions (Fig. 4.118, 4.120 and 4.121). The middle and proximal hair parts showed amorphous medulla . Goat hair at all the three parts of hair from head region had continuous, narrow, amorphous medulla with irregular margins (Fig. 4.117.).

The findings of the present work are in agreement with the observation noted by Marinis and Asprea (2005) who reported continuous and fragmental medulla with multicellular composition and multiseriate structure with scalloped to irregular margins. The observations on Goat hair by Koppikar and Sabnis (1976) are in partial agreement with the present observations. Who delineate about fragmental medulla at proximal and distal parts of hair and discoidal medulla at middle part. However, they did not mentioned about body region considered for the study. Similarly the observation reported by Gharu and Trivedi (2015) are in partial agreement with the present findings. They stated that the Goat hair showed simple, continuous, broad medulla with irregular margin.

Mukherjee *et al.* (2016) stated that Goat hair had central core of cells and some hair parts with dark black appearance. These findings are totally contradictory with the present observations.

During the present study, fragmental medulla with clove brown coloured cortex was observed at the proximal part of hair from head, neck and back regions of Horse (*Equus caballus*). However, middle and distal parts of hair from head, neck and back regions showed continuous, multicellular composition, multiseriate structure with irregular medulla borders (Fig. 4.123, 4.124 and 4.125). Narrow, continuous medulla was noted at the distal part of hair however, continuous, multicellular, multiseriate medulla structure with irregular medullary margins was observed at proximal and middle parts of hair from abdomen and thigh region (Fig. 4.126 and 4.127). The hair from tail region of Horse showed very wide medulla which completely occupied the hair width (Fig. 4.128).

The cortex was very thin and mostly invisible due to dark colour of medulla. The cortex of all the three parts of hair from all the body regions are clove brown coloured. These findings are in complete concurrence with the findings of Marinis and Asprea (2006) and Gharu and Trivedi (2015) who reported multicellular composition, multiseriate structure, continuous to fragmented medulla pattern with scalloped to irregular margins.

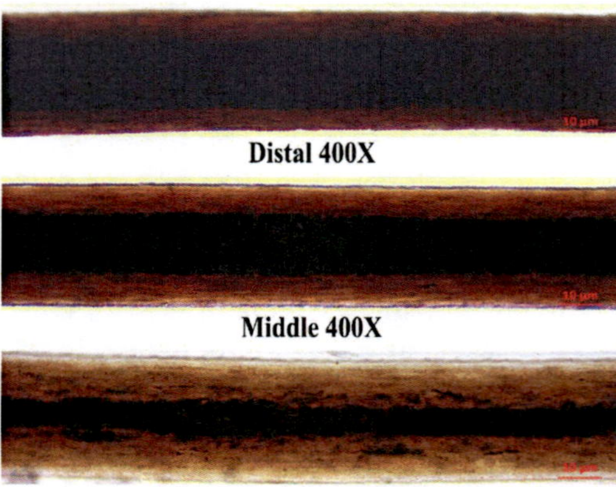

Fig. 4.120: Photomicrograph showing composition of medulla of guard hair from abdomen region of Goat (*Capra hircuss*)

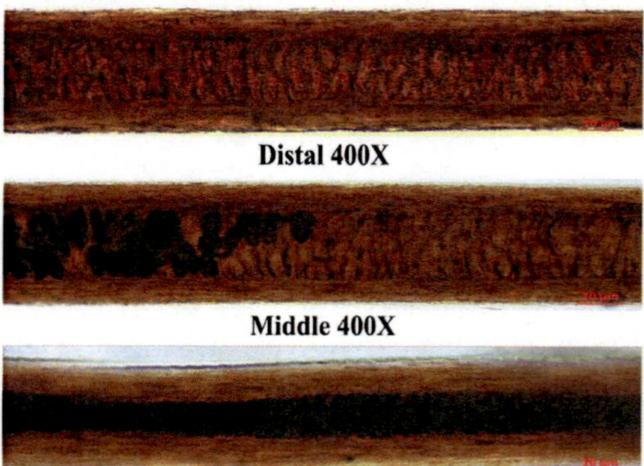

Fig. 4.121: Photomicrograph showing composition of medulla of guard hair from thigh region of Goat (*Capra hircuss*)

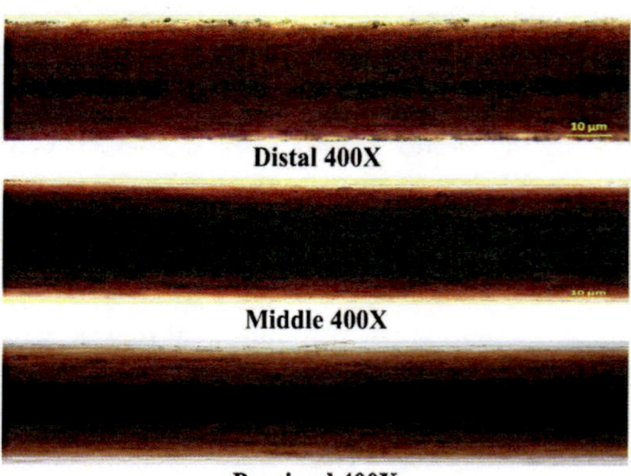

Fig. 4.122 Photomicrograph showing composition of medulla of guard hair from tail region of Goat (Capra hircuss)

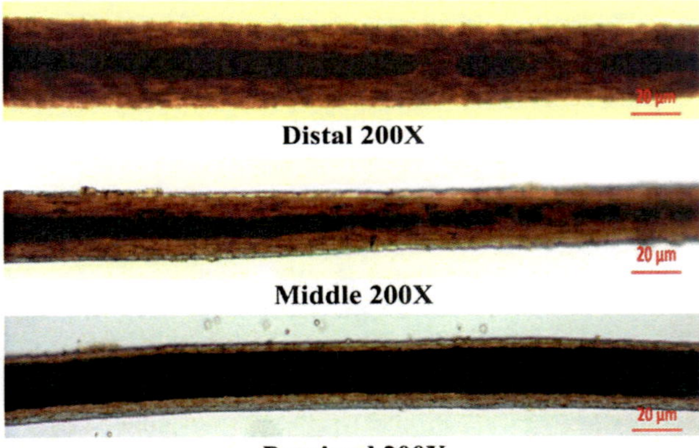

Fig. 4.123: Photomicrograph showing composition of medulla of guard hair from head region of Horse (*Equus caballus*)

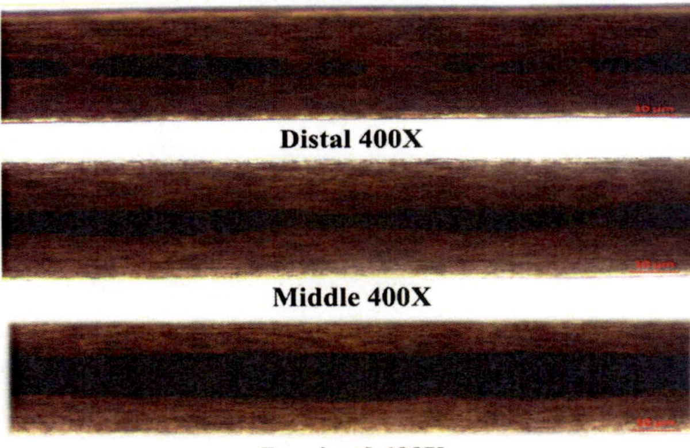

Fig. 4.124: Photomicrograph showing composition of medulla of guard hair from neck region of Horse (*Equus caballus*)

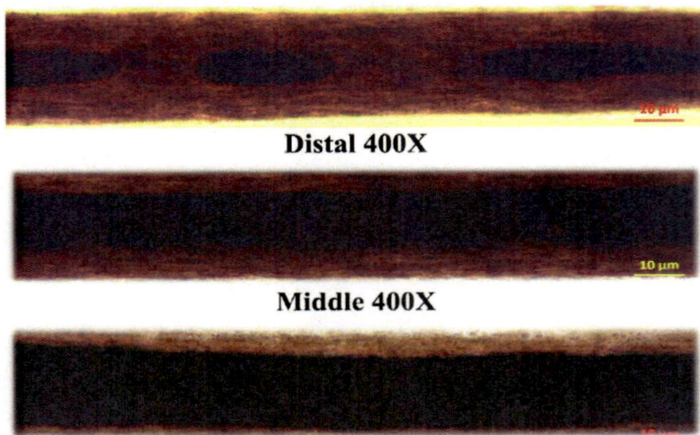

Fig. 4.125: Photomicrograph showing composition of medulla of guard hair from back region of Horse (*Equus caballus*)

The fragmental medulla with amorphous zones and clove brown coloured wide cortex was observed at all the three parts of hair from head, abdomen and thigh region in Domestic Pig (*Sus Scrofa domesticus*) (Fig. 4.129, 4.132 and 4.133). The distal part of hair in neck and back regions had fragmental, thin medulla with some amorphous zones through the hair shaft however, middle and proximal part of hair had continuous, amorphous medulla structure with irregular borders clove brown coloured cortex (Fig. 4.130 and 4.131). During the present study, it was noted that the hair from tail region of Domestic Pig were composed only of two layers the cuticle and cortex. The medulla was absent in the tail region (Fig. 4.134). At this region, wide tawny olive coloured cortex with uniformly distributed pigment granules was noted.

Similar observation were noted in Wild Pig (*Sus scrofa*) by Marinis and Asprea (2006) who reported continuous to fragmental medulla pattern with amorphous structure and irregular margins.

The medulla of Cat (*Felis catus domesticus*) among all the hair parts and body regions was wide and occupied almost one third of hair width. The continuous medulla pattern, unicellular regular composition and uniserial ladder like structure in medulla structure with scalloped margin was seen at proximal, middle and distal parts of hair from head neck and back regions (Fig. 4.135, 4.136 and 4.137). Similar medulla characteristics features were noted at middle and distal hair parts of abdomen region while, proximal part had continuous medulla pattern multicellular composition and multiseriate ladder structure with scalloped medulla border (Fig. 4.138). The hair from thigh and tail region showed continuous, multicellular, multiseriate ladder type of medulla with scalloped medulla margin (Fig. 4.139 and 4.140). The medulla was seen vacuolated at some places of proximal and middle parts of hair from tail region.

These observations are in concurrence with the results noted by Gharu and Trivedi (2015). They reported continuous unicellular ladder type medulla with scalloped margin in Domestic Cat. The observations of Mukherjee *et al.* (2016) regarding relatively wide medulla in Cat are akin with findings during the present research work.

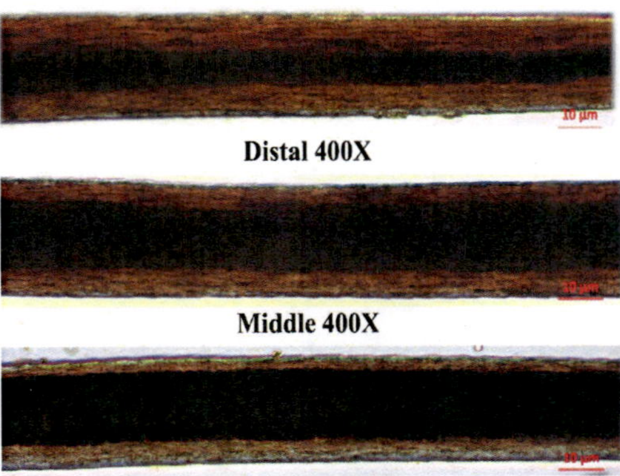

Distal 400X

Middle 400X

Proximal 400X

Fig. 4.126: Photomicrograph showing composition of medulla of guard hair from abdomen region of Horse (*Equus caballus*)

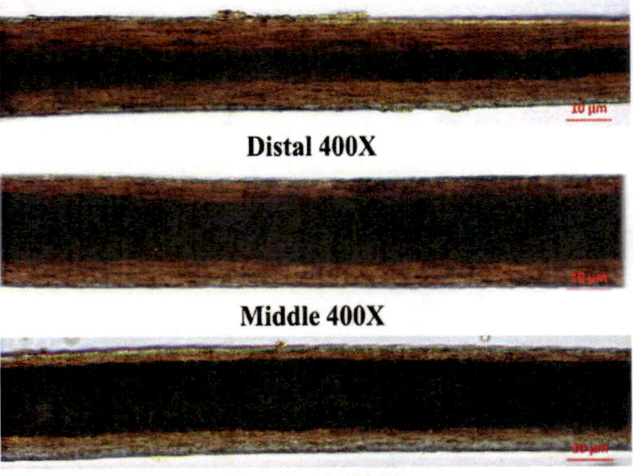

Distal 400X

Middle 400X

Proximal 400X

Fig. 4.127: Photomicrograph showing composition of medulla of guard hair from thigh region of Horse (*Equus caballus*)

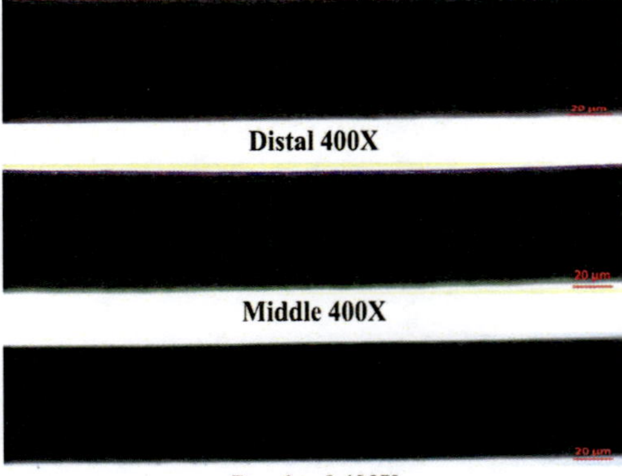

Distal 400X

Middle 400X

Proximal 400X

Fig. 4.128: Photomicrograph showing composition of medulla of guard hair from tail region of Horse (*Equus caballus*)

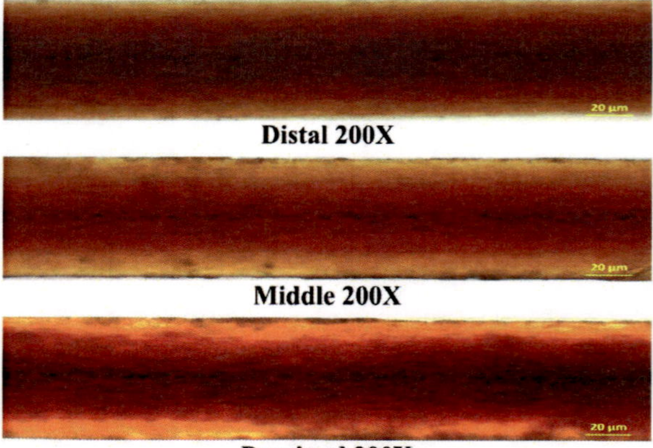

Fig. 4.129: Photomicrograph showing composition of medulla of guard hair from head region of Domestic pig (*Sus scrofa domestica*)

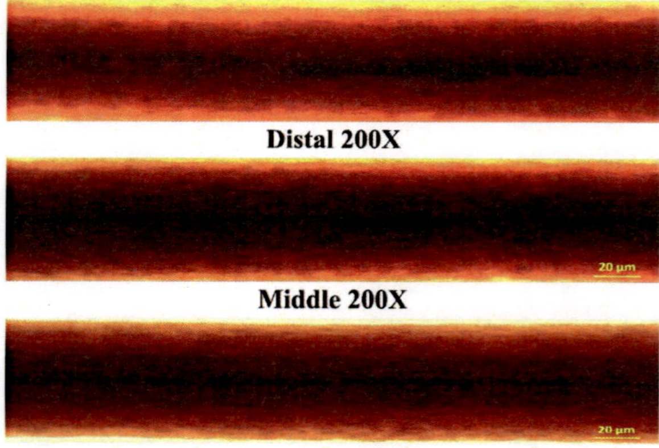

Fig. 4.130: Photomicrograph showing cuticle of guard hair from neck region of Domestic pig (*Sus scrofa domestica*)

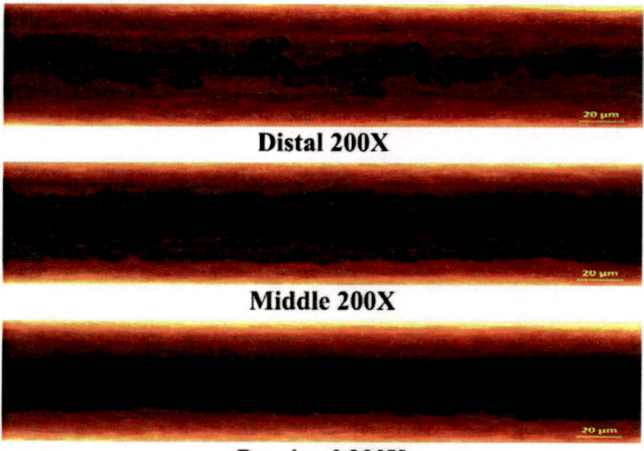

Fig. 4.131: Photomicrograph showing composition of medulla of guard hair from back region of Domestic pig (*Sus scrofa domestica*)

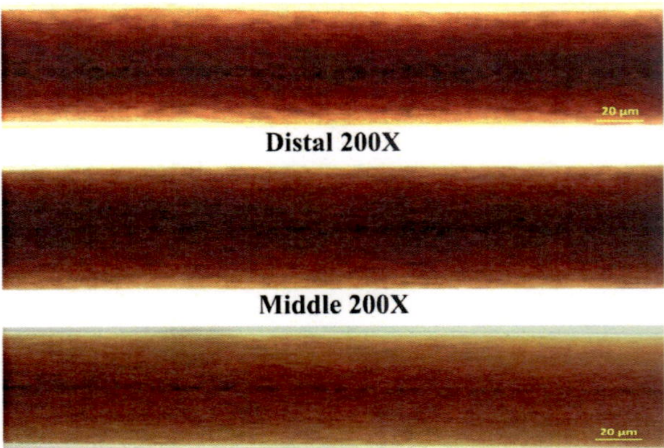

Fig. 4.132: Photomicrograph showing cuticle of guard hair from abdomen region of Domestic pig (*Sus scrofa domestica*)

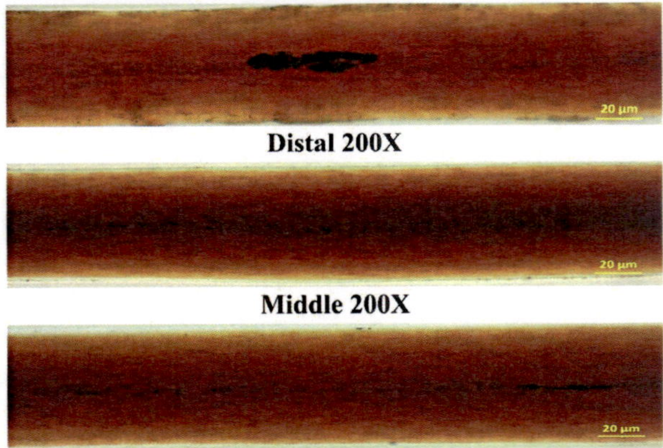

Fig. 4.133: Photomicrograph showing composition of medulla of guard hair from thigh region of Domestic pig (*Sus scrofa domestica*)

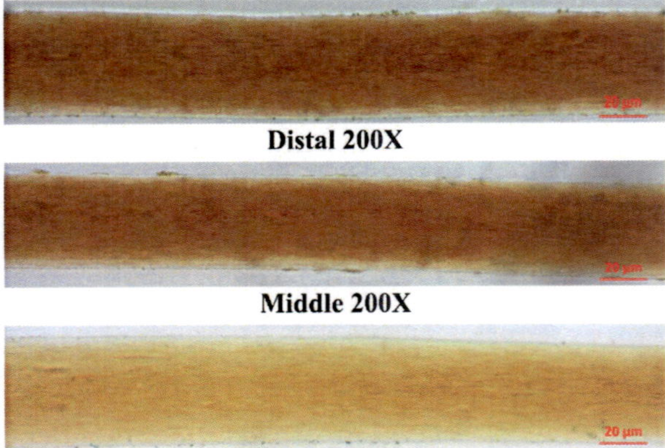

Fig. 4.134: Photomicrograph showing composition of medulla of guard hair from tail region of Domestic pig (*Sus scrofa domestica*)

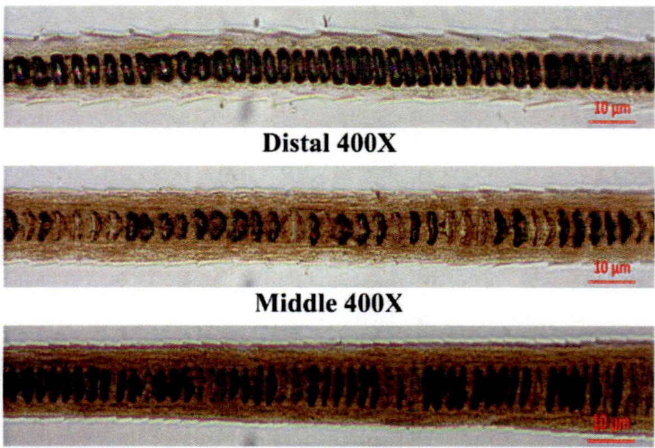

Fig. 4.135: Photomicrograph showing composition of medulla of guard hair from head region of Cat (*Felis catus*)

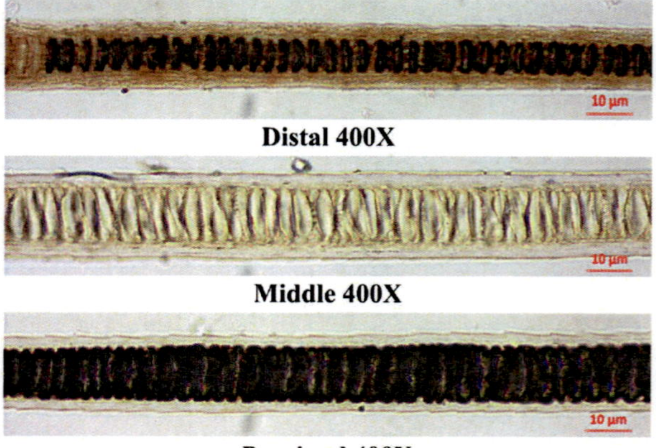

Fig. 4.136: Photomicrograph showing composition of medulla of guard hair from neck region of Cat (*Felis catus*)

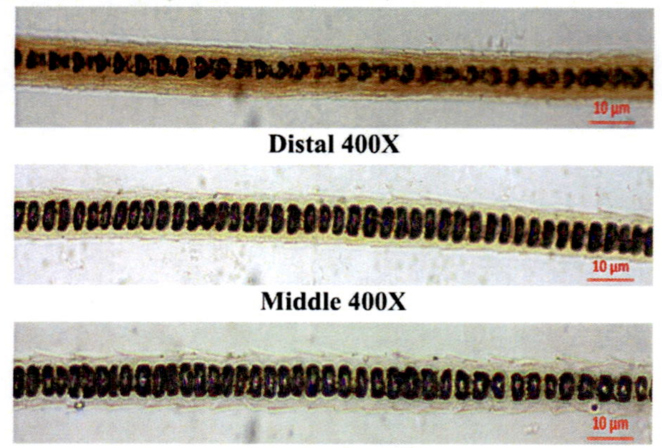

Fig. 4.137: Photomicrograph showing composition of medulla of guard hair from back region in Cat (*Felis catus*)

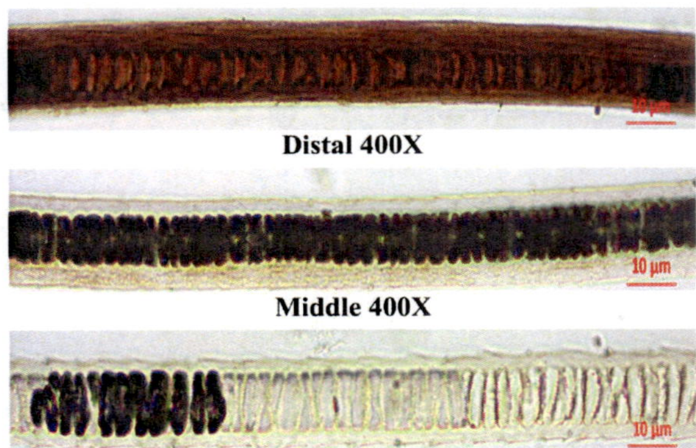

Proximal 400X

Fig. 4.138: Photomicrograph showing medulla and cortex of guard hair from abdomen region of Cat (*Felis catus*)

Proximal 400X

Fig. 4.139: Photomicrograph showing composition of medulla of guard hair from thigh region of Cat (*Felis catus*)

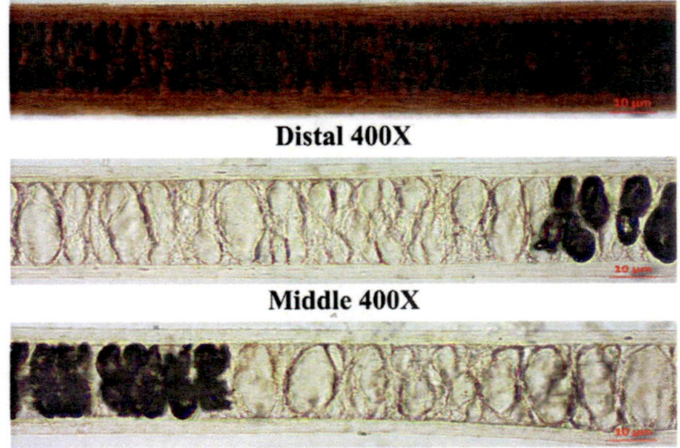

Proximal 400X

Fig. 4.140: Photomicrograph showing composition of medulla of guard hair from tail region of Cat (*Felis catus*)

The thickness of medulla was found varied amongst the body regions and parts of hair in Dog (*Canis lupus familiaris*). However, uniform medulla pattern was noted at all the six regions considered during present work. It was noted that the medulla of proximal part of hair from all the body regions was thinnest while that of distal part of hair was thickest among the three parts of hair. The medulla showed continuous pattern, multicellular composition and vacuolated structure with scalloped to irregular margin at all the parts of hair and body regions (Fig. 4.141 to 4.146).

Gharu and Trivedi (2015) noted continuous, broad vacuolated medulla with occasional very broad medulla in dorsal guard hair of Dog. These findings are in total agreement with the findings noted during the present study. Mukherjee *et al.* (2016) also noted vacuolated type medulla in Dog which was in full concurrence with present observations.

The corticular colour at proximal and middle parts of hair was noted as French gray however, distal part showed cinnamon coloured cortex with uniformly distributed thin fiber like pigments.

The hair of spotted Deer (*Axis axis*) from all the body regions showed unique structure at tip and root of hair. At the root of hair medulla showed wine glass like structure, which was a characteristic feature of hair of animal from cervidae family (Fig. 4.183). The hair tip was observed as broken or bifurcated. The continuous medulla with scalloped medulla margin was noted at all the three parts of hair and body regions of Spotted Deer. The proximal, middle and distal parts of hair from head region of Spotted Deer exhibited multicellular composition of medulla and narrow medulla lattice structure (Fig. 4.147). Further it was noted that the proximal and middle parts of hair from neck, back, abdomen, thigh and tail regions had multicellular, cloisonné or wide medulla lattice structure of medulla however, the distal part of hair showed narrow medulla lattice (Fig. 4.148 to 4.152) with the exception of distal part of neck region, where it showed amorphous medulla structure (Fig. 4.148).

The cortex at all the regions except head region was very thin almost one third of the hair width. The colour of cortex at all the three parts of hair from head, neck, back, middle and distal parts of hair from abdomen regions was cinnamon coloured. However, all the three parts of hair from thigh and tail region and proximal part of hair from abdomen region had lilac gray colour.

The findings reported by Joshi *et al.* (2012) are in accordance with the present observations. They reported wide and narrow medulla lattice structure at thigh, back and neck region in Spotted Deer. In contrast with the present findings Koppikar and Sabnis (1976) reported continuous medulla pattern at proximal and middle part of hair while fragmental at distal part of hair in Spotted Deer.

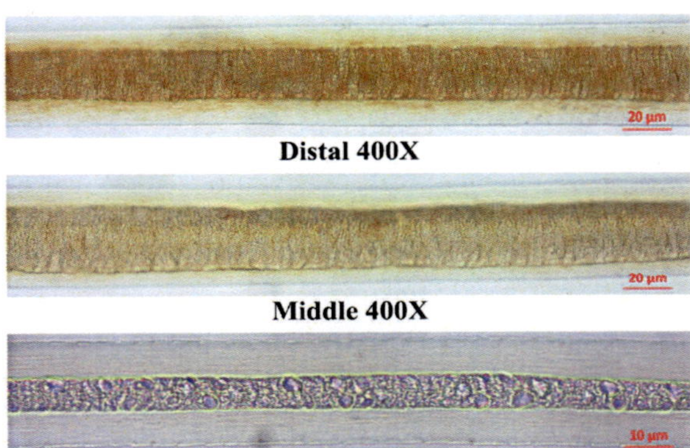

Distal 400X

Middle 400X

Proximal 400X

Fig. 4.141: Photomicrograph showing composition of medulla of guard hair from head region of Dog (*Canis familiaris*)

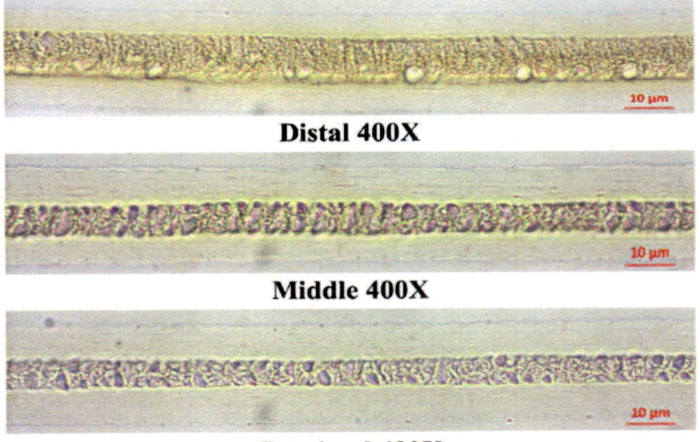

Distal 400X

Middle 400X

Proximal 400X

Fig. 4.142: Photomicrograph showing composition of medulla of guard hair from neck region of Dog (*Canis familiaris*)

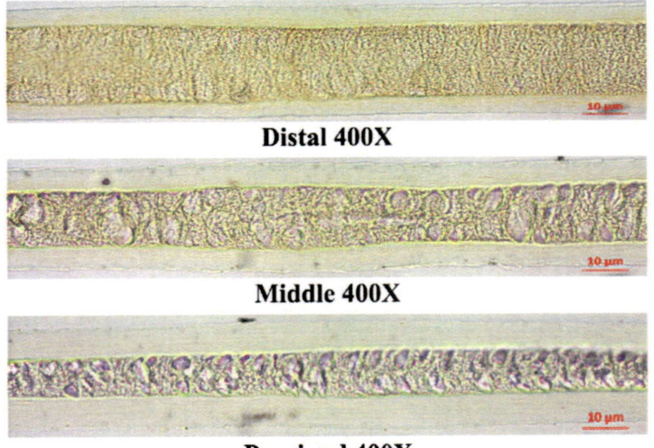

Distal 400X

Middle 400X

Proximal 400X

Fig. 4.143: Photomicrograph showing composition of medulla of guard hair from back region of Dog (*Canis familiaris*)

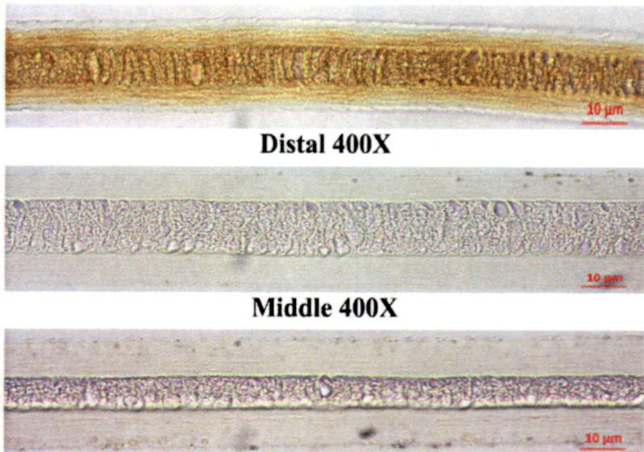

Fig. 4.144: Photomicrograph showing composition of medulla of guard hair from abdomen region of Dog (*Canis familiaris*)

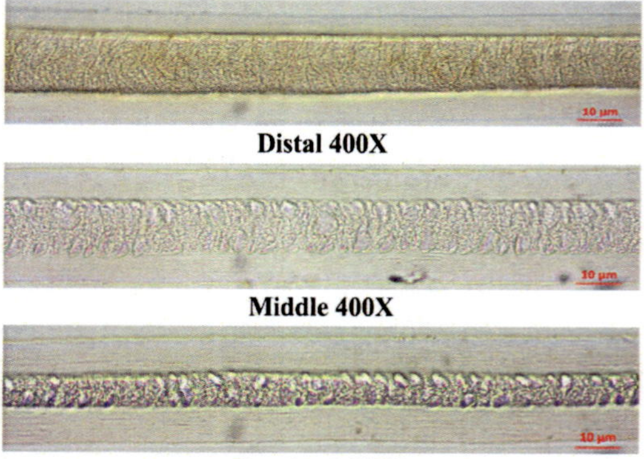

Fig. 4.145: Photomicrograph showing composition of medulla of guard hair from thigh region of Dog (*Canis familiaris*)

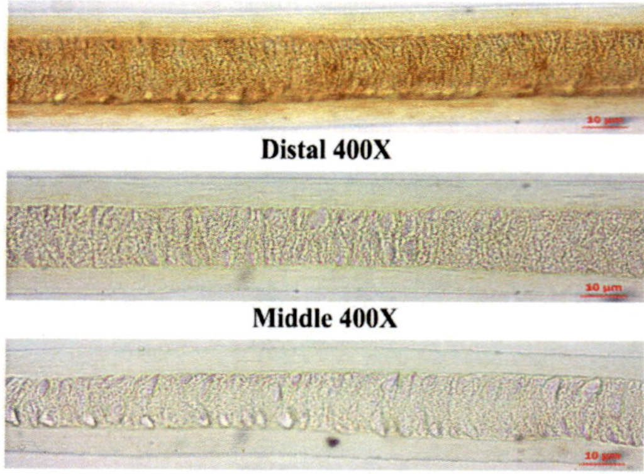

Fig. 4.146: Photomicrograph showing composition of medulla of guard hair from tail region of Dog (*Canis fmiliaris*)

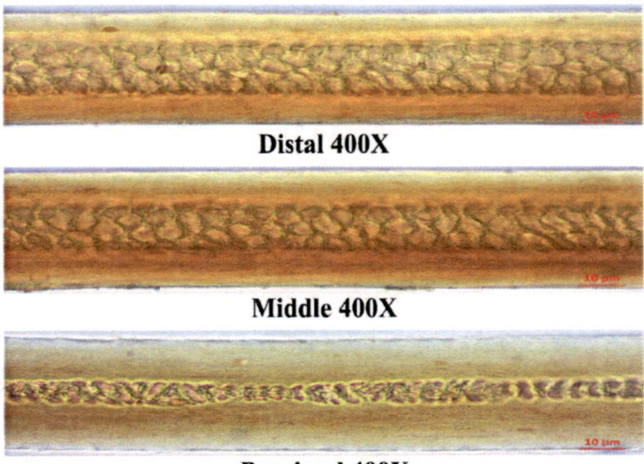

Fig. 4.147: Photomicrograph showing composition of medulla of guard hair from head region of Spotted deer (*Axis Axis*)

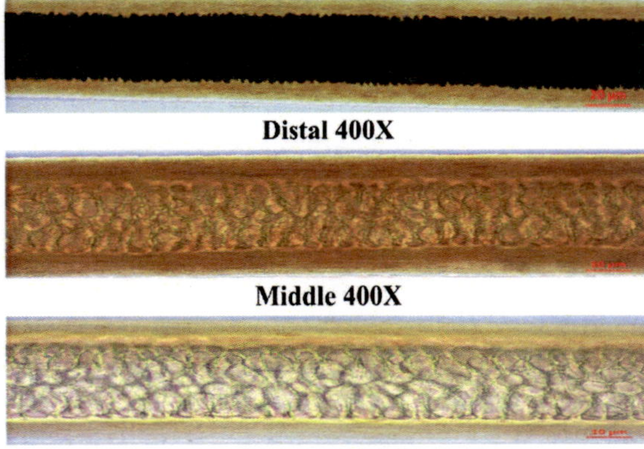

Fig. 4.148: Photomicrograph showing composition of medulla of guard hair from neck region of Spotted deer (*Axis axis*)

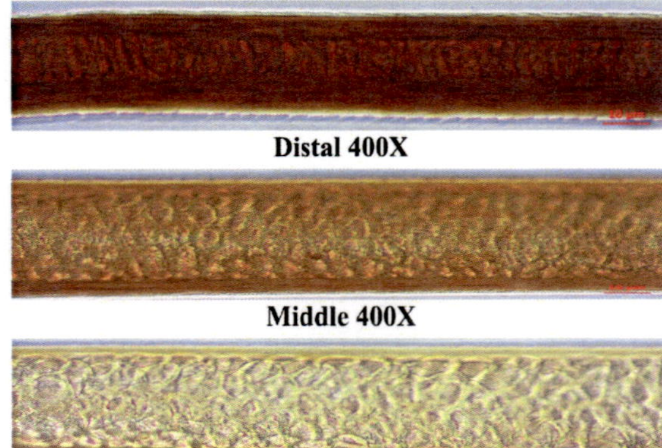

Fig. 4.149: Photomicrograph showing composition of medulla of guard hair from back region of Spotted deer (*Axis axis*)

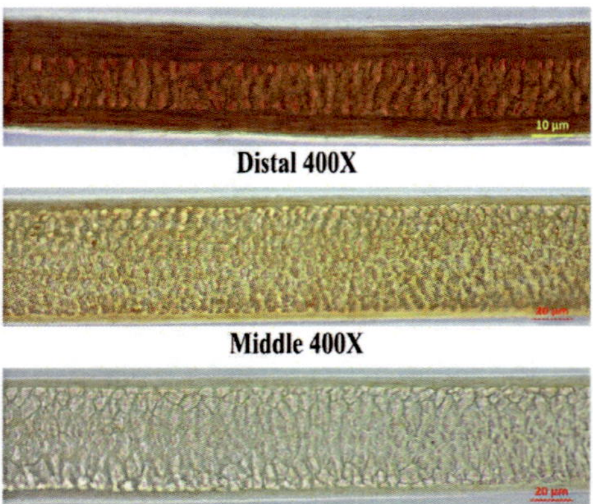

Fig. 4.150: Photomicrograph showing composition of medulla of guard hair from abdomen region of Spotted deer (*Axis axis*)

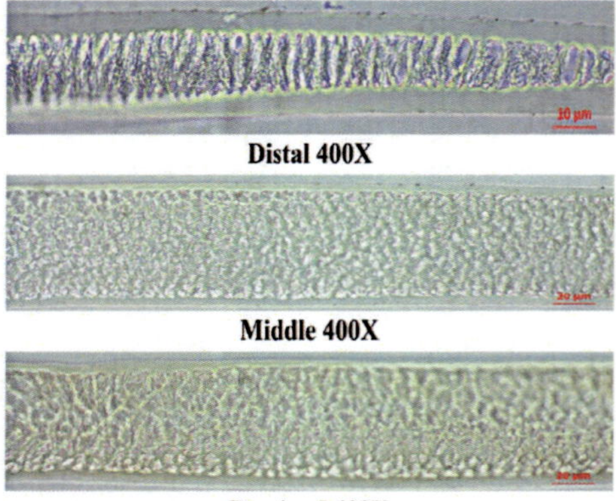

Fig. 4.151: Photomicrograph showing composition of medulla of guard hair from thigh region of Spotted deer (*Axis axis*)

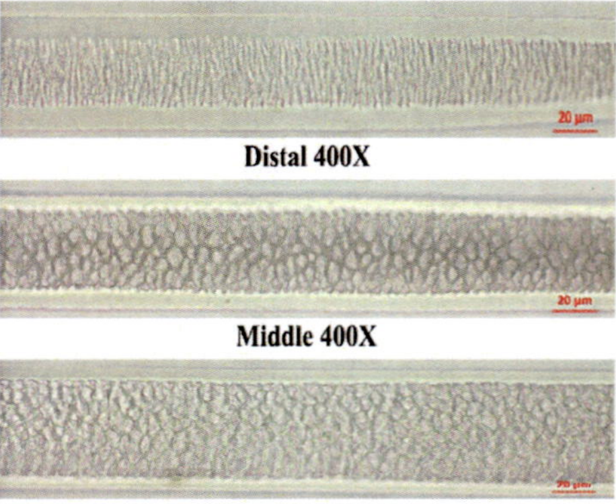

Fig. 4.152: Photomicrograph showing composition of medulla of guard hair from tail region of Spotted deer (*Axis axis*)

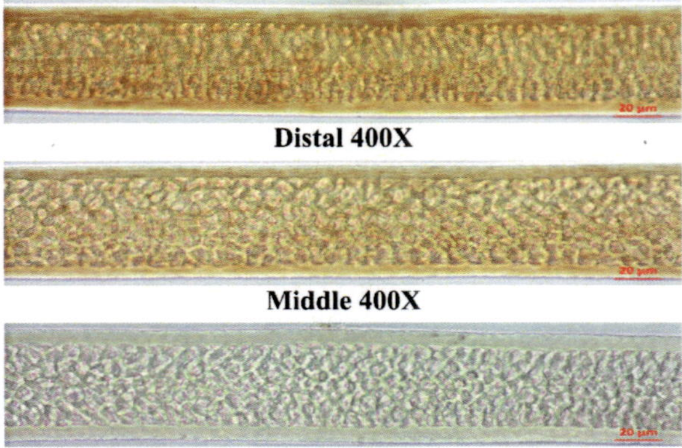

Fig. 4.153: Photomicrograph showing composition of medulla of guard hair from head region of Sambar (*Rusa unicolor*)

Fig. 4.154: Photomicrograph showing composition of medulla of guard hair from neck region of Sambar (*Rusa unicolor*)

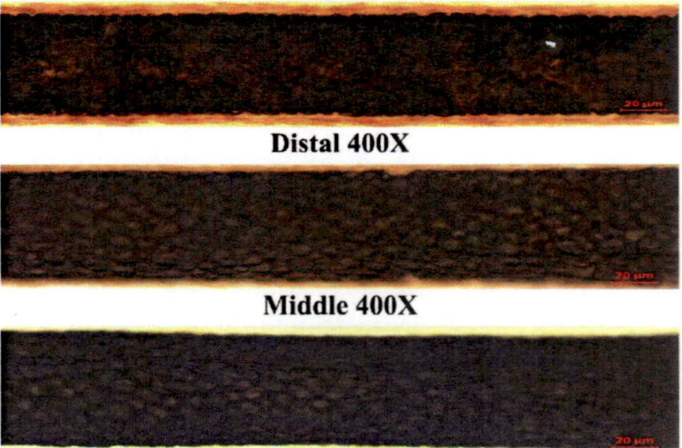

Fig. 4.155: Photomicrograph showing composition of medulla of guard hair from back region of Sambar (*Rusa unicolor*)

The hair of all the six body regions in Sambar (*Rusa unicolor*) showed unique broad / wide form of medulla at the base / root of hair (Fig. 4.183). At all the three parts of hair and body regions, continuous, multicellular composition, wide medulla lattice structure with scalloped margin was noted in Sambar (Fig. 4.153 to 4.158). The medulla cells were polygonal in shape and are filled with slate black coloured pigments. It was observed that the medulla occupied almost entire hair width. The colour of medulla was found slate black particularly at the proximal and middle parts of hair from neck, back, abdomen, thigh and tail region. However middle and distal part of hair from head region, distal part of other five regions was cinnamon coloured while, proximal part of head region was olive gray coloured.

The cortex was hardly demarcated from the medulla. The findings of Koppikar and Sabnis (1976) are in partial agreement with the observations of present study. They noted reticulate polygonal medulla at proximal and middle parts of hair of Sambar. Joshi *et al.* (2012) reported wide medulla lattice pattern which is in accordance with the findings of the present study.

The hair from all the body regions of Nilgai (*Bosellaphus tragocamellus*) showed wine glass structure of medulla at the base of hair (Fig. 4.183). Continuous type of medulla was observed at all three parts of hair and body regions. The multicellular composition, narrow medulla lattice structure with straight medulla border was observed at proximal, middle and distal parts of hair from head, abdomen, thigh and tail regions. In addition these characters vacuolated medulla structure was noted at some places in all three parts of hair from thigh and tail regions (Fig. 4.163 and 4.164). However hair of neck and back regions showed continuous, amorphous medulla at all the three hair parts (Fig. 4.160 and 4.161). The proximal part of hair from back region showed some zones of narrow medulla lattice type.

The cortex of all three parts of hair from head region was russet coloured with uniformly distributed dark fine pigments. Further, it was noted that the cortex of all three parts from neck region, middle and distal parts of back region and distal part of thigh region was clove brown coloured while, that of proximal part of hair from back region was olive gray coloured cortex. Olive gray coloured cortex with cinnamon coloured fine pigments were found at all three parts of hair from abdomen and tail regions and proximal and middle parts of hair from thigh region.

De and Chakraborty (2012) reported multiserial ladder type medulla in hair of Nilgai. These observations are not in agreement with the present findings. The observations noted by Koppikar and Sabnis (1976) are in partial accordance with the observations of present study. They stated that the medulla was continuous at proximal and middle parts of hair, while medulla was not visible at distal part. The observations by Bhat *et al.* (2014) are contradictory with the findings of Kamalakannan (2017d) who noted unicellular regular pattern and simple medulla with straight margin. The findings regarding medulla margin in Nilgai was in accordance with observations of the present work.

Fig. 4.156: Photomicrograph showing composition of medulla of guard hair from abdomen region in Sambar (*Rusa unicolor*)

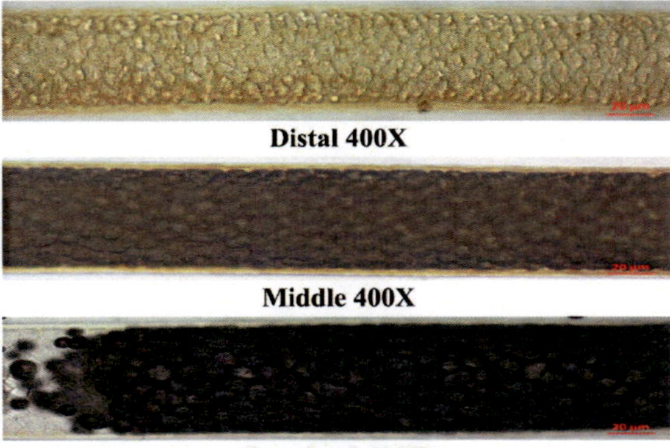

Fig. 4.157: Photomicrograph showing composition of medulla of guard hair from thigh region of Sambar (*Rusa unicolor*)

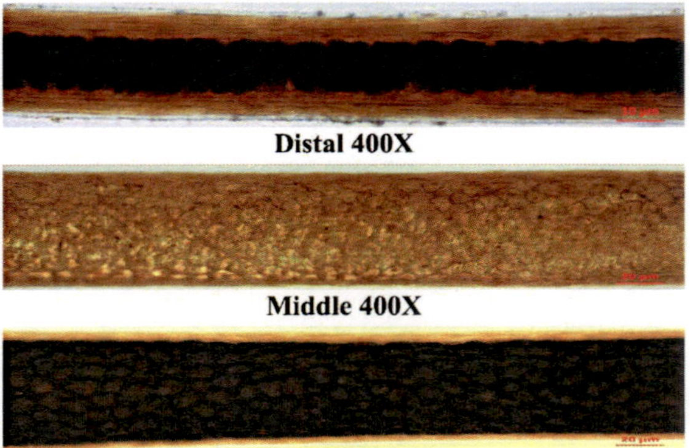

Fig. 4.158: Photomicrograph showing composition of medulla of guard hair from tail region of Sambar (*Rusa unicolor*)

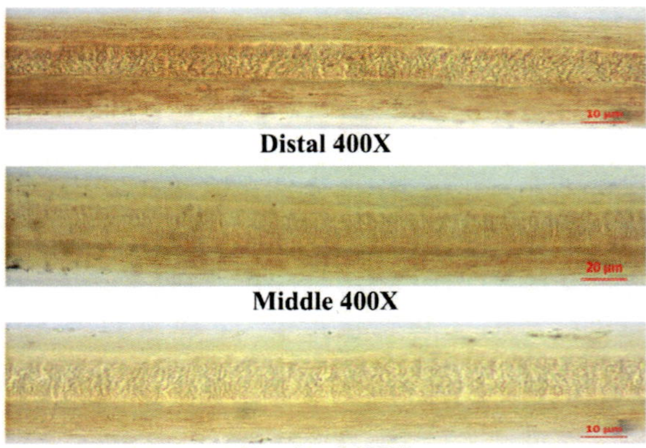

Fig. 4.159: Photomicrograph showing composition of medulla of guard hair from head region of Nilgai (*Boselaphus tragocalmelus*)

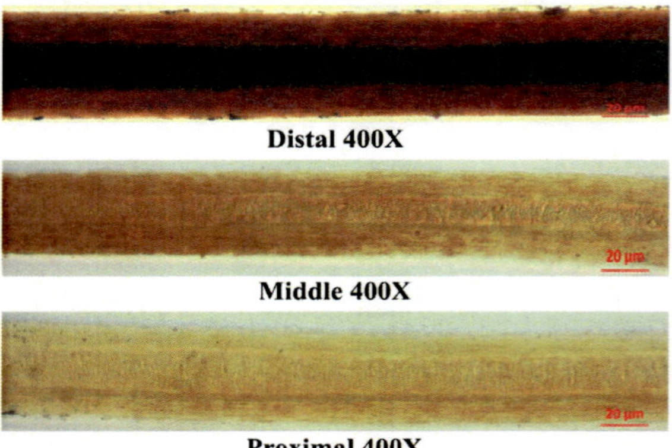

Fig. 4.160: Photomicrograph showing composition of medulla of guard hair of neck region of Nilgai (*Boselaphus tragocamelus*)

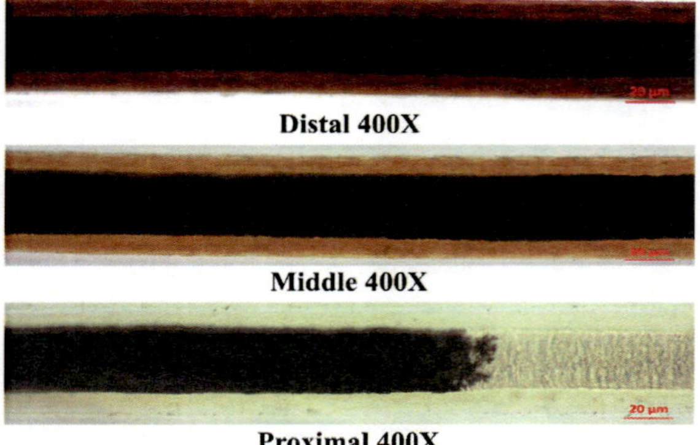

Fig. 4.161: Photomicrograph showing composition of medulla of guard hair of back region of Nilgai (*Boselaphus tragocamelus*)

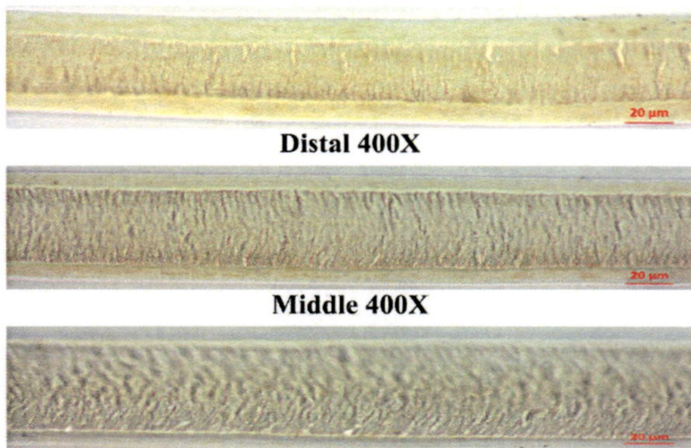

Fig. 4.162: Photomicrograph showing composition of medulla of guard hair of abdomen region of Nilgai (*Boselaphus tragocamelus*)

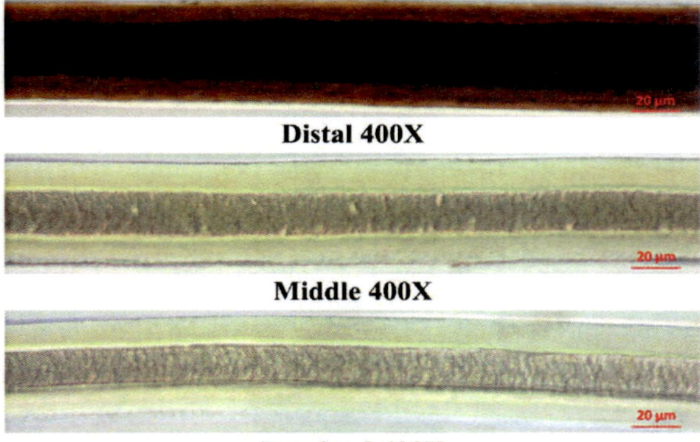

Fig. 4.163: Photomicrograph showing composition of medulla of guard hair from thigh region of Nilgai (*Boselaphus tragocamelus*)

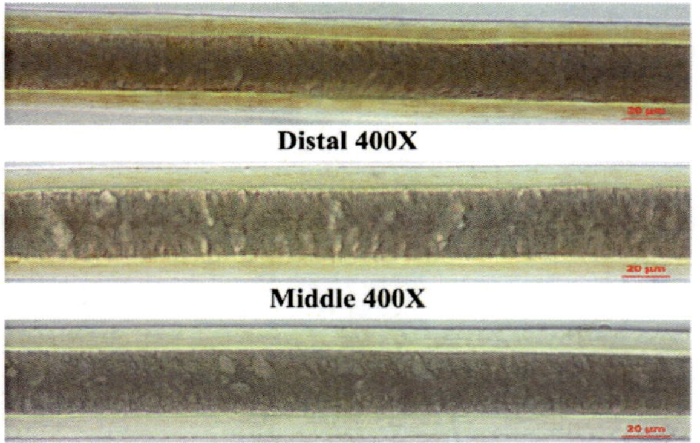

Fig. 4.164: Photomicrograph showing composition of medulla of guard hair from tail region of Nilgai (*Boselaphus tragocamelus*)

The hair of Hanuman Langur (*Samnopithecus entellus*) showed broccoli brown colouration throughout the length of hair shaft with uniformly distributed pigments which were little darker and thread like. A very unique structure feature i e. cortical fusi were noted in the cortex. The cortical fusi were observed at all the parts of hair amongst all the body regions (Fig. 4.165).

The fragmental medulla pattern with uniserial ladder medulla with scalloped margin was found at middle part of hair from head region, however, it was not visible at the proximal and distal parts of hair from head region in Hanuman Langur (Fig. 4.165). The medulla was not evident throughout the length of hair from neck region (Fig. 4.166). In the proximal part of hair from back region of Hanuman Langur, medulla was not found however, fragmental, uniserial ladder type medulla with scalloped margin was evident at middle and distal parts of hair (Fig. 4.167). At the middle and distal parts of hair, medulla was of continuous medulla pattern and scalloped margin, where as it was absent at the proximal part of hair from abdomen region (Fig. 4.168). The fragmental medulla pattern and uniserial ladder type medulla structure with scalloped margin was observed throughout the length of hair in thigh and tail regions of Hanuman Langur (Fig. 4.169 and 4.170).

Gharu and Trivedi (2013) reported fragmental medulla along with light gray colour pigments in the hair of Hanuman Langur. These observations are in partial agreement with the findings of present work. Koppikar and Sabnis (1976) stated that there was uniform distribution of threadlike pigmentation throughout the length of hair with no evidence of medulla. These findings are partially in concurrence with the findings reported during present study. In contrast with the present observations Sarkar *et al.* (2011) reported presence of simple medulla in Hanuman Langur.

Continuous pattern of medulla was noticed throughout the shaft of hair from all the six body regions in Tiger (*Panthera tigris*). Multicellular composition, multiseriate medulla structure with irregular margin was observed in head, neck and back (Fig. 4.171, 4.172 and 4.173) regions, proximal and distal parts of hair from abdomen region showed unicellular composition and uniserial ladder type of medulla with scalloped margin (Fig. 4.174). However, the middle part of hair from thigh region had globular medulla structure (Fig. 4.175).

The findings of the present research work are in agreement with the findings noted by Gharu and Trivedi (2015) in Tiger cub. Who mentioned uniserial ladder type medulla in the hair of Tiger cub. These findings of the present study are in partial concurrence with the observations of Koppikar and Sabnis (1976). They stated that continuous medulla was present throughout the length of hair, except distal par of hair, where it was fragmental.

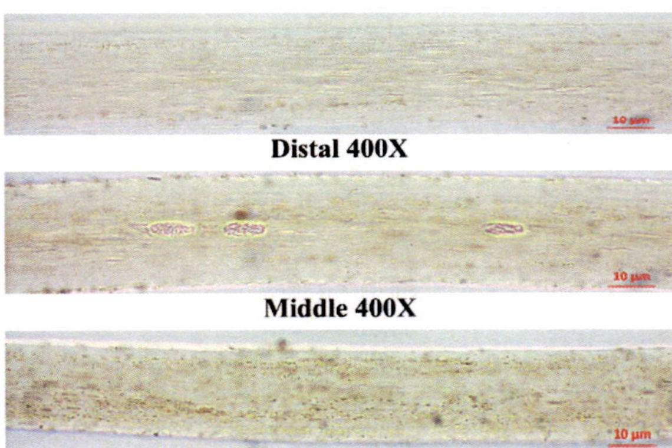

Fig. 4.165: Photomicrograph showing composition of medulla of guard hair from head region of Hanuman langur (*Samnopithecus entellus*)

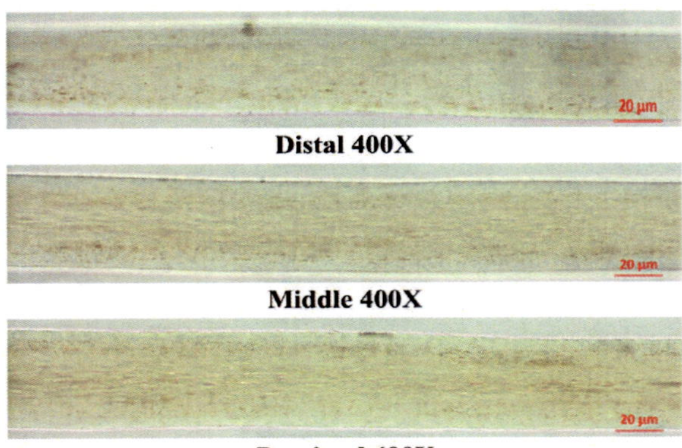

Fig. 4.166: Photomicrograph showing composition of medulla of guard hair from neck region of Hanuman langur (*Samnopithecus entellus*)

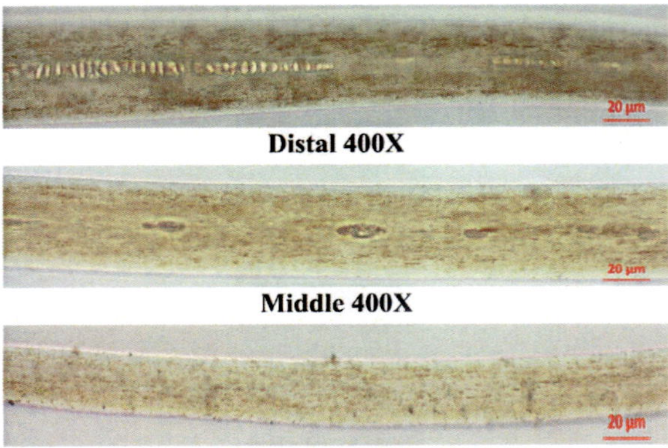

Fig. 4.167: Photomicrograph showing composition of meulla of guard hair from back region of Hanuman langur (*Samnopithecus entellus*)

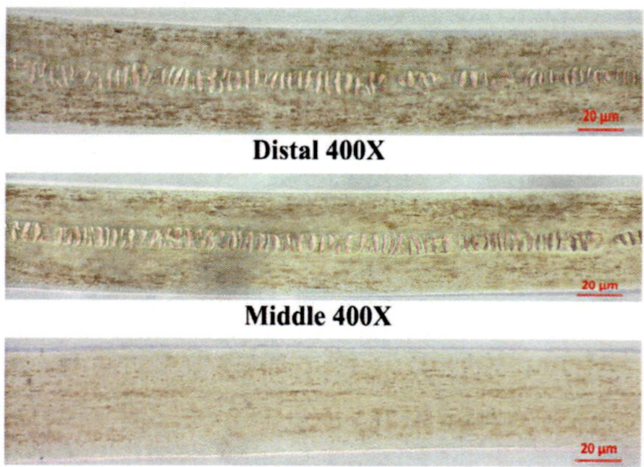

Distal 400X

Middle 400X

Proximal 400X

Fig. 4.168: Photomicrograph showing composition of medulla of guard hair from abdomen region of Hanuman langur (*Samnopithecus entellus*)

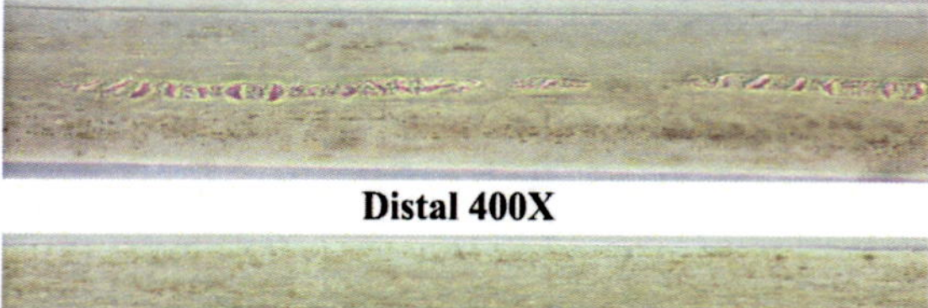

Distal 400X

Middle 400X

Proximal 400X

Fig. 4.169: Photomicrograph showing composition of medulla of guard hair from thigh region of Hanuman langur (*Samnopithecus entellus*)

Distal 400X

Middle 400X

Proximal 400X

Fig. 4.170: Photomicrograph showing composition of medulla of guard hair from tail region of Hanuman langur (*Samnopithecus entellus*)

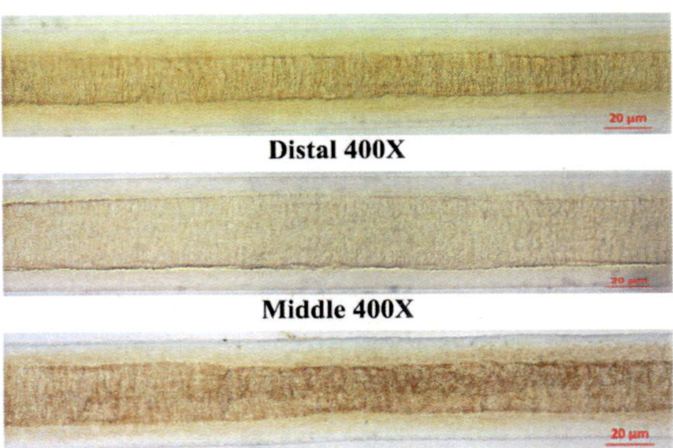

Fig. 4.171: Photomicrograph showing composition of medulla of guard hair from head region of Tiger (*Panthera tigris*)

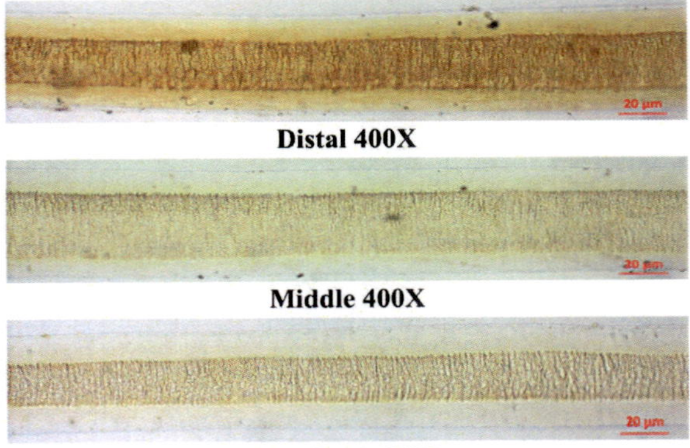

Fig. 4.172: Photomicrograph showing composition of medulla of guard hair from neck region of Tiger (*Panthera tigris*)

Fig. 4.173: Photomicrograph showing composition of medulla of guard hair from back region of Tiger (*Panthera tigris*)

Kitpipit and Thanakiatkrai (2013) observed three types of medulla in Tiger hair such as simple, uniserial ladder and mixed. These observations are in akin with the present findings regarding uniserial ladder.

Leopard (*Panthera pardus*) hair from all the body regions and parts of the hair showed similar medulla structure with same medulla width. Continuous, multicellular composition, multiserial ladder type of medulla with irregular margins was observed in the hair of leopard (Fig. 4.177 to 4.182).

These observations are in partial agreement with the observations noted by Koppikar and Sabnis (1976). They noted continuous medulla at proximal and middle fragmental type of medulla at distal part. The observations of Gharu and Trivedi (2015) are in contradicted with the present findings. Who noted amorphous and fragmental medulla.

4.4 Characteristics of cross section of hair

The cross section of hair from head, neck, back and abdomen regions in Cattle (*Bos indicus*) was oval shape with medium sized medulla, while hair from thigh and tail regions were circular with medium sized medulla (Fig. 4.184). These findings are in agreement with the findings noted by Dharaiya and Soni (2012) in Domestic Cow. Sahajpal *et al.* (2009) also reported oval shaped cross section of hair along with oval medulla in the hair of Cattle.

Uniform distribution of pigments was also observed in the cortex of hair from all the six body regions of Cattle. The medulla margin was smooth.

The cross sections of hair from head, neck, back, abdomen and thigh region in Buffalo (*Bubalis bubalus*) was oval shaped with quite large oval medulla (Fig. 4.185). The medullary margin in cross section of hair was irregular in all the regions. The tail region showed circular cross section with uniformly distributed large sized pigment granules. The hair of the tail region did not show presence of medulla in its cross section.

The cross section of hair from the head region of Sheep (*Ovis aries*) showed oblong shape with oblong medulla located in the center of the hair (4.186). The dumb-bell shaped cross section with large sized medulla was observed in hair from neck, abdomen, thigh and tail regions (Fig. 4.186). The cross section of hair from back region of Sheep showed oval shape with large sized medulla. In agreement with the findings of the present study, Sahajpal *et al.* (2009) reported oval shaped cross section and oval shaped medulla in the hair of Sheep.

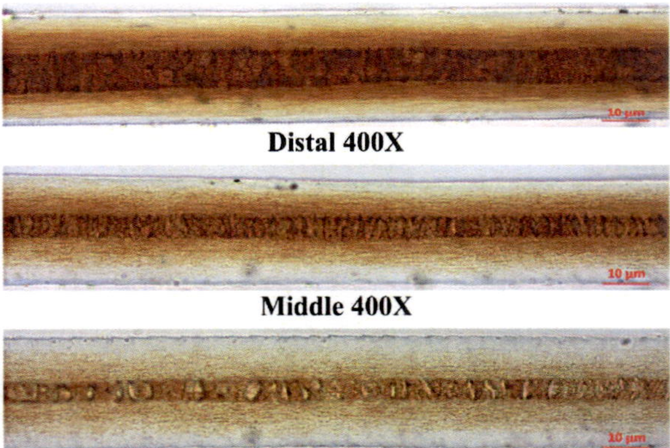

Fig. 4.174: Photomicrograph showing composition of medulla of guard hair from abdomen region of Tiger (*Panthera tigris*)

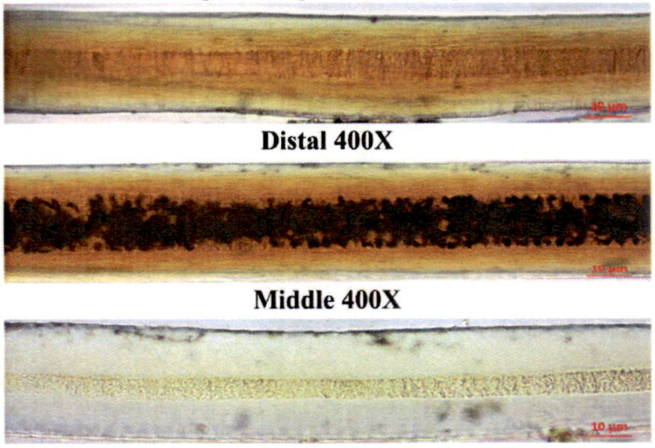

Fig. 4.175: Photomicrograph showing composition of medulla of guard hair from thigh region of Tiger (*Panthera tigris*)

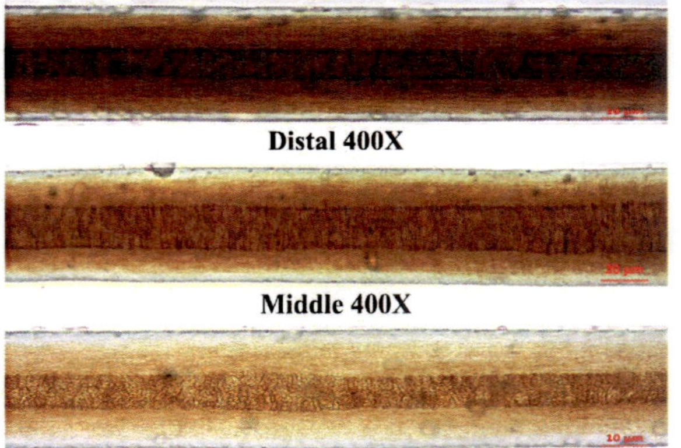

Fig. 4.176: Photomicrograph showing composition of medulla of guard hair from tail region of Tiger (*Panthera tigris*)

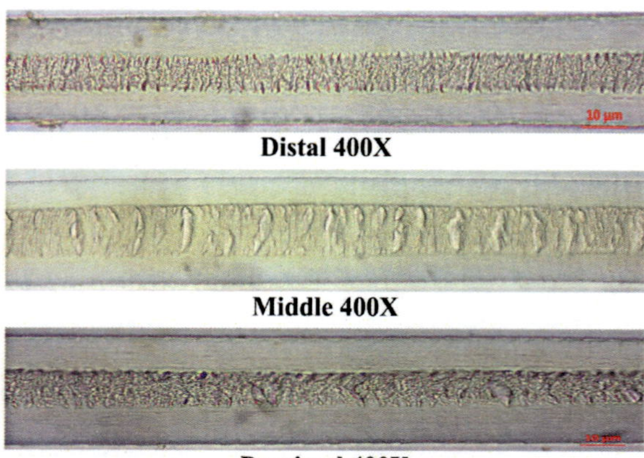

Fig. 4.177: Photomicrograph showing composition of medulla of guard hair from head region of Leopard (*Panthera pardus*)

Fig. 4.178: Photomicrograph showing composition of medulla of guard hair from neck region of Leopard (*Panthera pardus*)

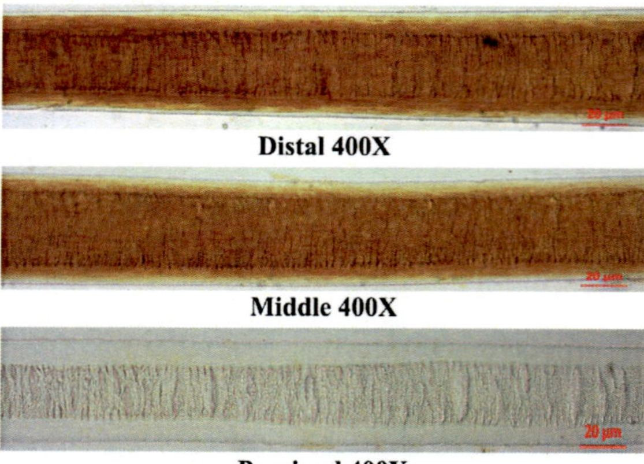

Fig. 4.179: Photomicrograph showing composition of medulla of guard hair from back region of Leopard (*Panthera pardus*)

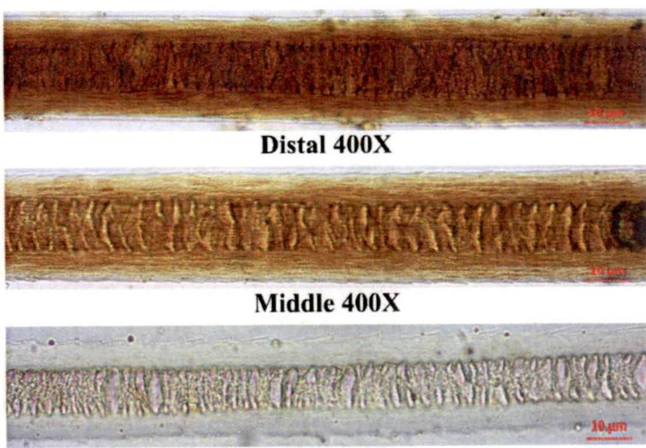

Fig. 4.180: Photomicrograph showing composition of medulla of guard hair from abdomen region of Leopard (*Panthera pardus*)

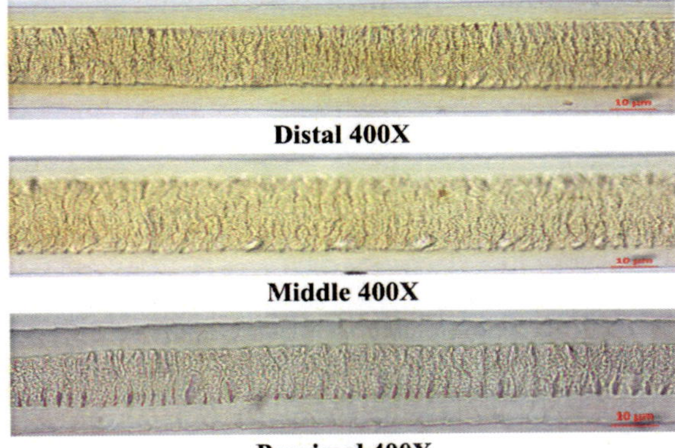

Fig. 4.181: Photomicrograph showing composition of medulla of guard hair from thigh region of Leopard (*Panthera pardus*)

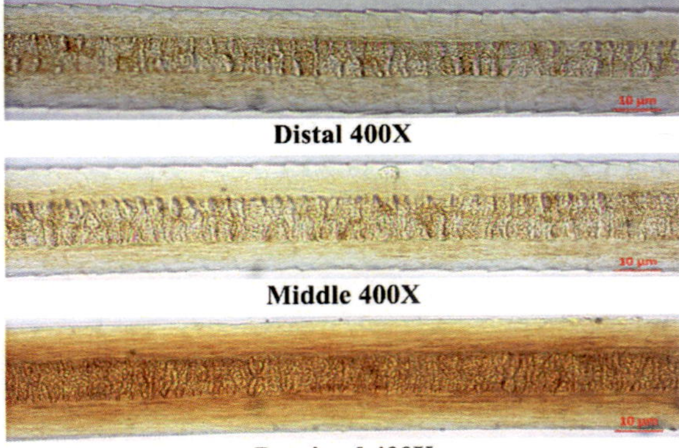

Fig. 4.182: Photomicrograph showing composition of medulla of guard hair from tail region of Leopard (*Panthera pardus*)

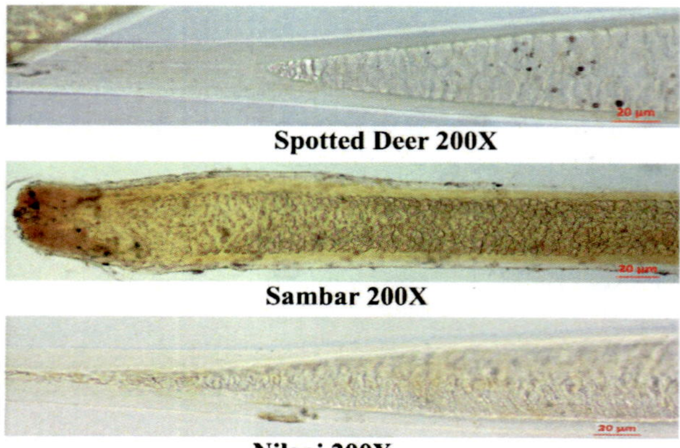

Fig. 4.183: Photomicrograph showing structure of medulla at the base of hair

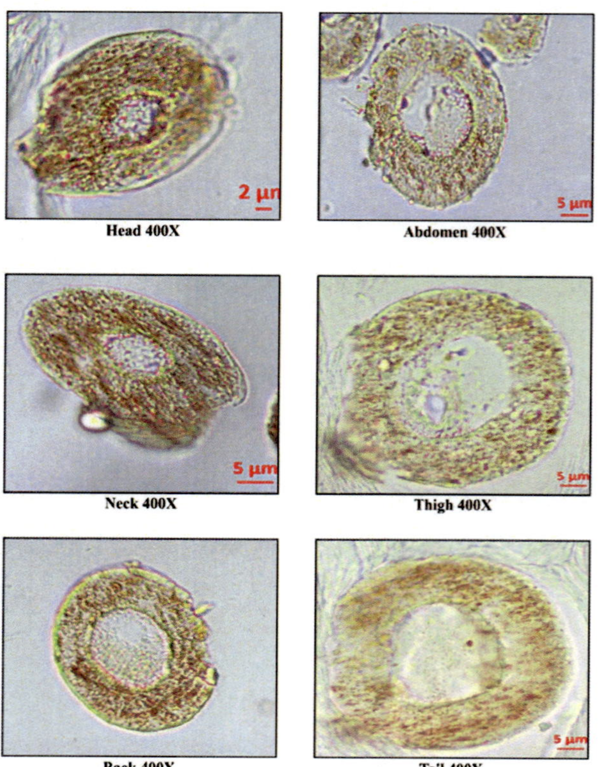

Fig. 4.184: Microphotograph showing cross sections of hair from various body regions of Cattle (*Bos indicus*)

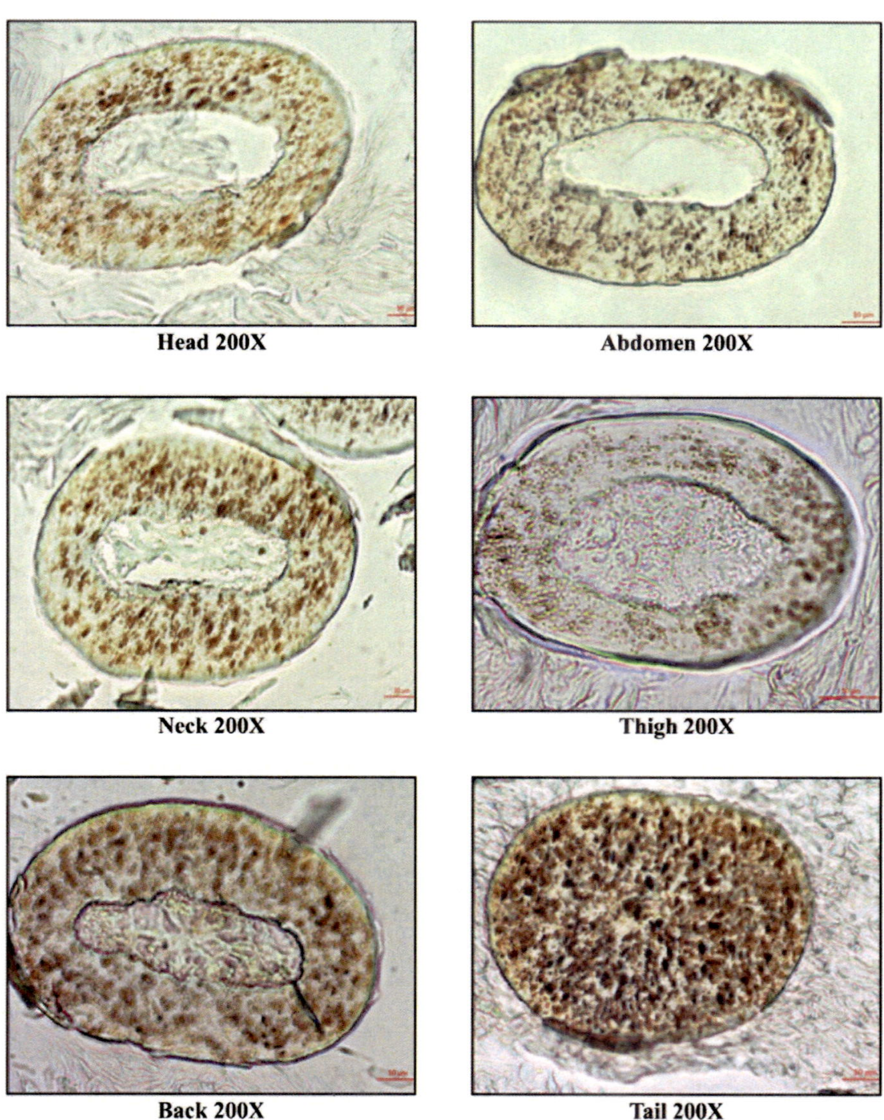

Fig. 4.185: Microphotograph showing cross sections of hair from various body regions of Buffalo (*Bubalus bubalis*)

The cross section of hair in Goat (*Capra hircus*) showed oval shape with oval, medium sized medulla in head, neck and abdomen region (Fig. 4.187). Whereas, cross section of hair of back and thigh region had circular shape with circular and medium sized medulla. The medulla margin of cross section from above five body regions was irregular. The cross section of hair from tail region was oval shaped with very small medulla. The large sized pigment granules were found uniformly distributed in the cortex. These findings noted during the present study were in partial agreement with the findings reported by Sahajpal *et al.* (2009) in Goat. They reported oval to oblong shaped cross section of hair with oval to oblong medulla. The cortex of all the regions except tail region showed fine uniformly distributed pigments.

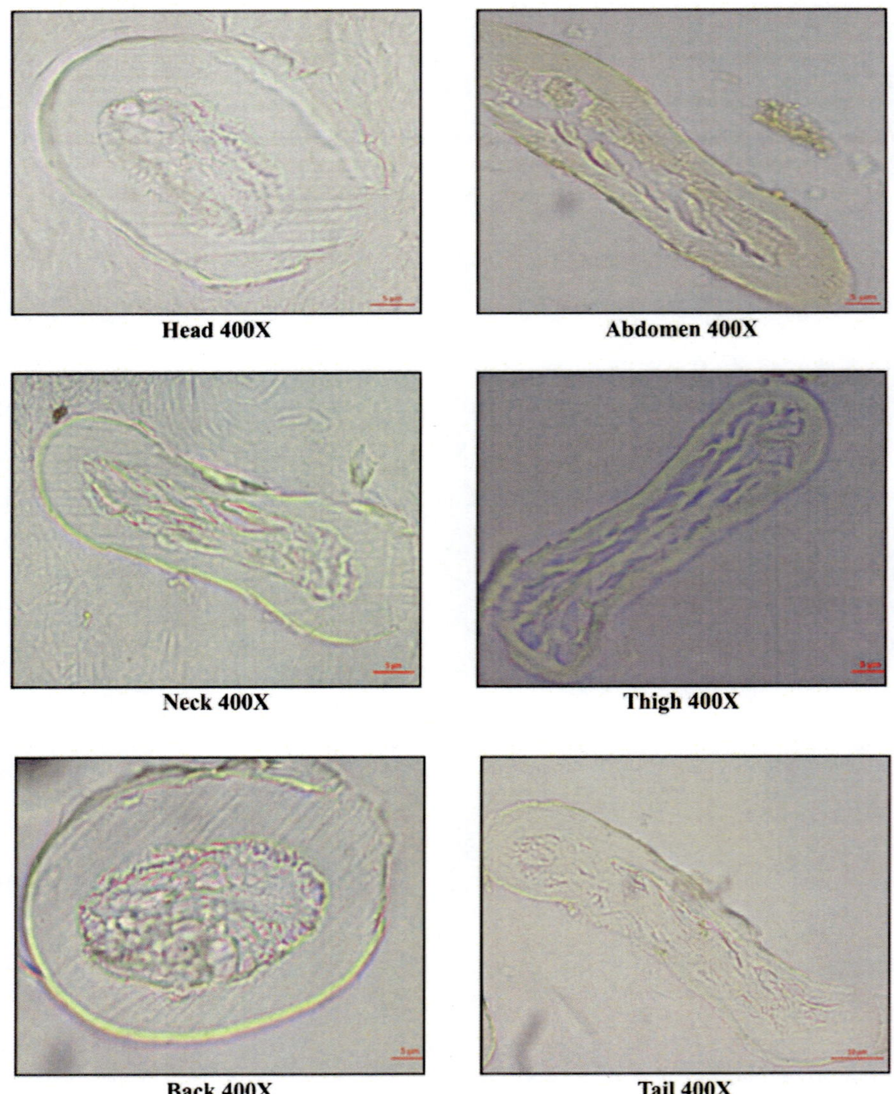

Fig. 4.186: Microphotograph showing cross sections of hair from various body regions of Sheep (*Ovis aries)*

In Horse (*Equus caballus*) the cross section of hair from head region showed oblong shape with medium circular medulla. The medulla margin was irregular in the cross section of head region. The cross section of hair of neck and thigh regions were oval in shape with medium sized oval medulla (Fig. 4.188). However, hair from back, abdomen and tail region had circular cross section with medium sized medulla. The pigments were found distributed uniformly in the cortex of all the regions of body.

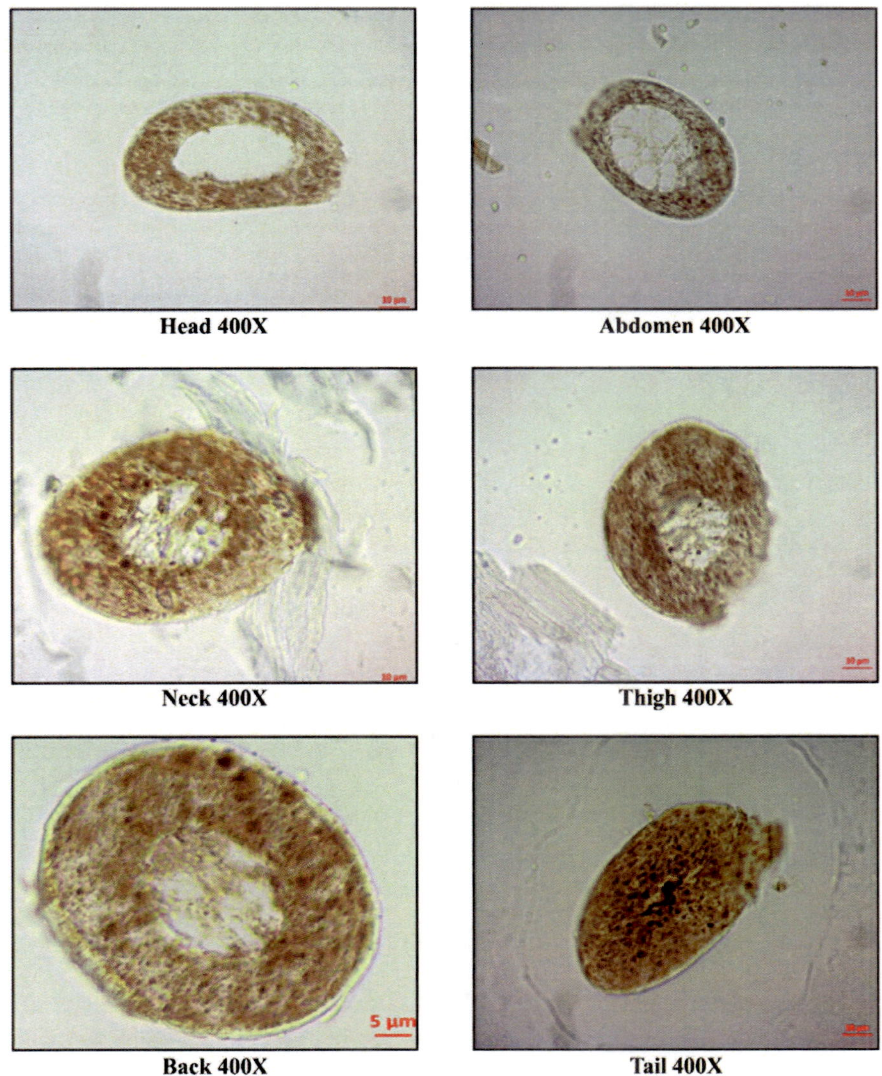

Fig. 4.187: Microphotograph showing cross sections of hair from various body regions of Goat (*Capra hircus*)

The shape of cross section of hair from head, neck, back, abdomen and tail regions of Domestic Pig (*Sus scrofa domesticus*) was circular. The medulla was very small dot like found centrally located in all the above regions except tail region. The medulla was absent in the tail region (Fig. 4.189). The variable sizes of pigments were found aggregated towards the medulla margin. The cross section of hair from thigh region was found oblong shaped with small dot like centrally located medulla (Fig. 4.189).

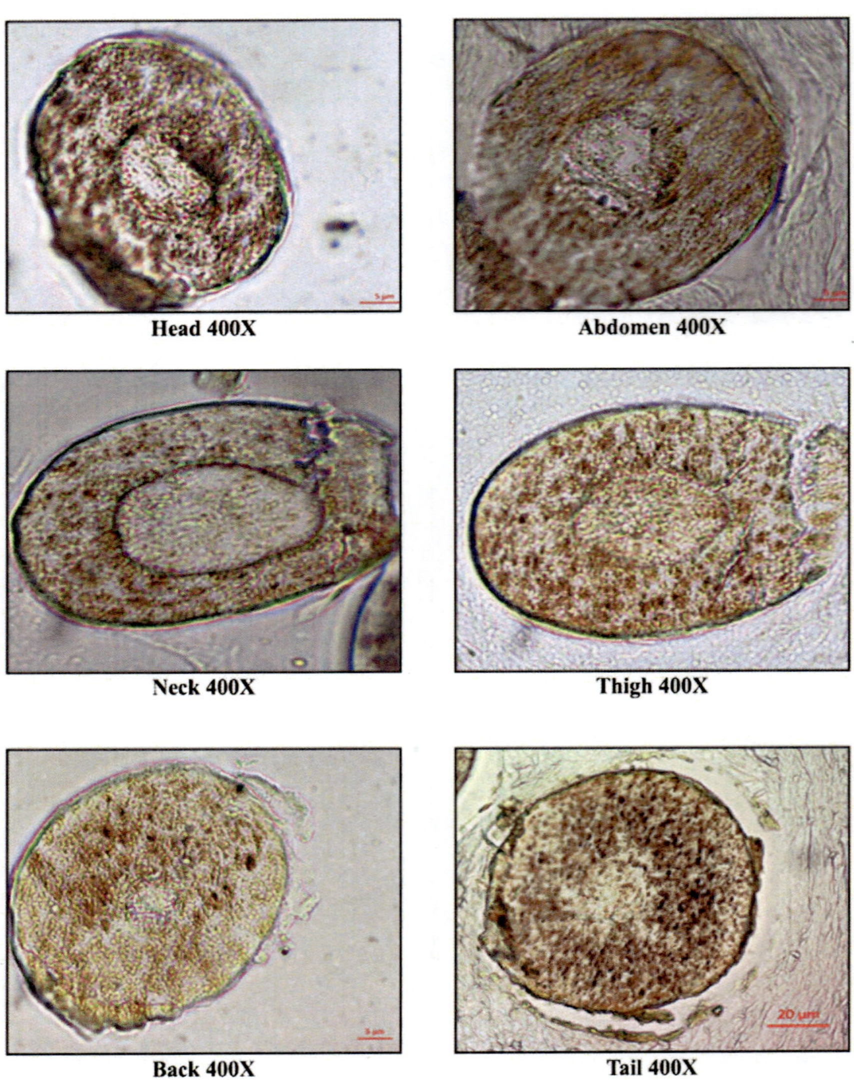

Fig. 4.188: Microphotograph showing cross sections of hair from various body regions of Horse (*Equus caballus*)

The cross section of hair from head, neck, thigh and tail regions of Cat (*Felis catus familiaris*) was oval shaped. The medulla from head, neck and tail regions was oval, medium sized with smooth medulla margin. Whereas, medu.la from the thigh region was also oval shaped but was large sized (Fig. 4.190). the cross section of hair from back and abdomen regions was circular and had medium sized medulla. The brown coloured pigment granules were uniformly distributed in the cortex of hair from head, neck, back and abdomen regions. Whereas, these pigment granules were found concentrated more towards the medulla margin in thigh and tail regions.

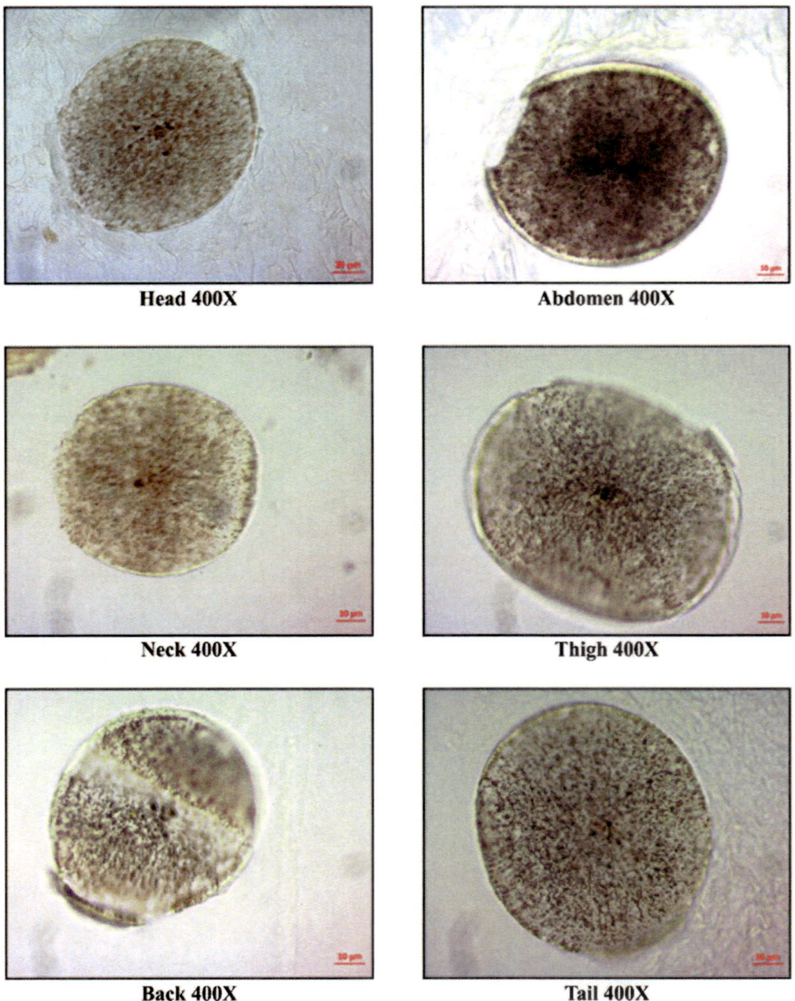

Fig. 4.189: Microphotograph showing cross sections of hair from various body regions of Domestic Pig (*Sus scrofa domesticus*)

In Dog (*Canis lupus domesticus*) the cross section of hair from head, neck, back, abdomen and tail regions was circular shaped. Its medulla was medium sized circular shape. The medulla of hair showed irregular margin in all the abovenbody regions. However, the cross sections of hair from thigh region were oval shaped with medium sized medulla (Fig. 4.191).

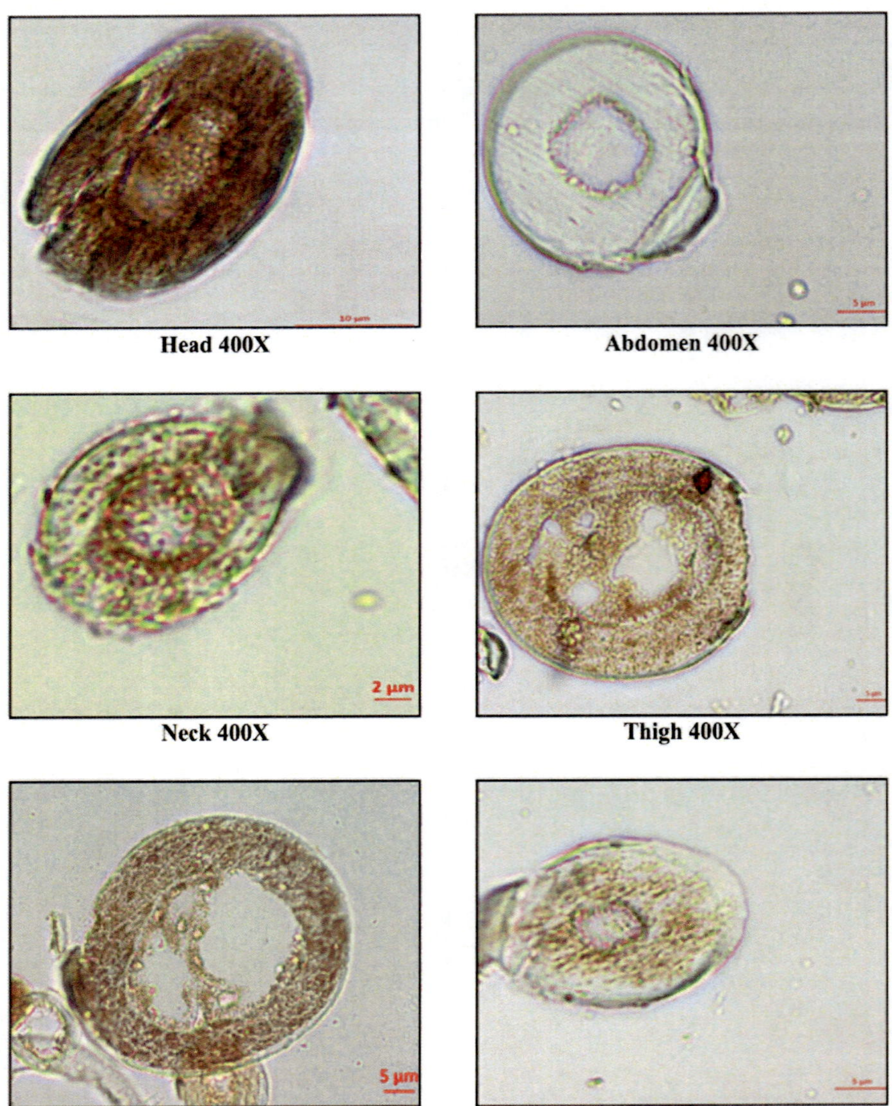

Fig. 4.190: Microphotograph showing cross sections of hair from various body regions of Cat (*Felis catus familiaris*)

The circular cross section eith large medulla was observed in the hair of head, neck, back and thigh regions of Spotted Deer (*Axis axis*) (Fig. 4.192). The cross section of hair from the abdomen and tail regions had oval shape and large medulla. The irregular medulla margin was noted in all the body regions. The medulla from all six body regions showed strand like cells arranged in a particular manner forming net like structure. The stranded arrangement of medullary cells was seen only in Spotted Deer and Sambar.

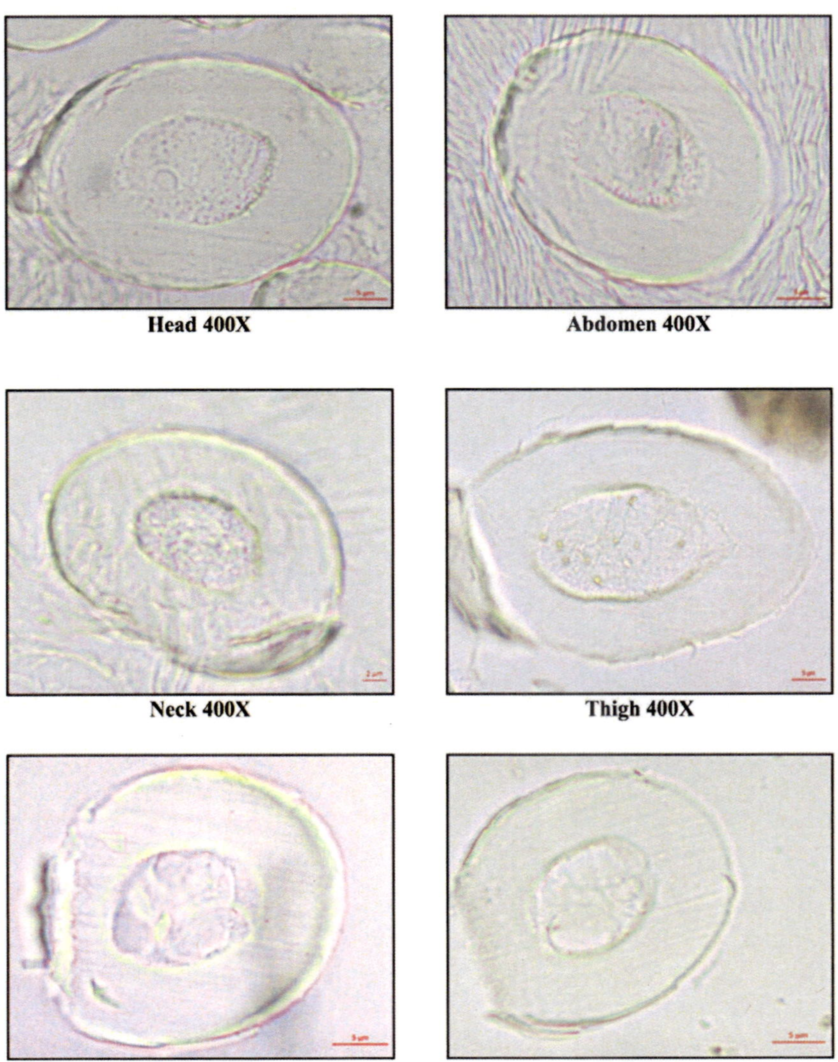

Fig. 4.191: Microphotograph showing cross sections of hair from various body regions of Dog (*Canis lupus domesticus*)

The cross section of Sambar (*Rusa unicolor*) hair was typically cigar shaped with a large sized medulla, which occupied almost two third of hair diameter. The cigar shaped cross section s were noted in the hair from all the six body regions. The medulla cells were present in the form of strands, which were arranged in a web pattern (Fig. 4.193). the fine pigment granules were found uniformly distributed throughout the cortex. In contrast with the findings of the present stidy Dharaiya and Soni (2012) mentioned presence of large oval shaped hair in its cross section in Sambar.

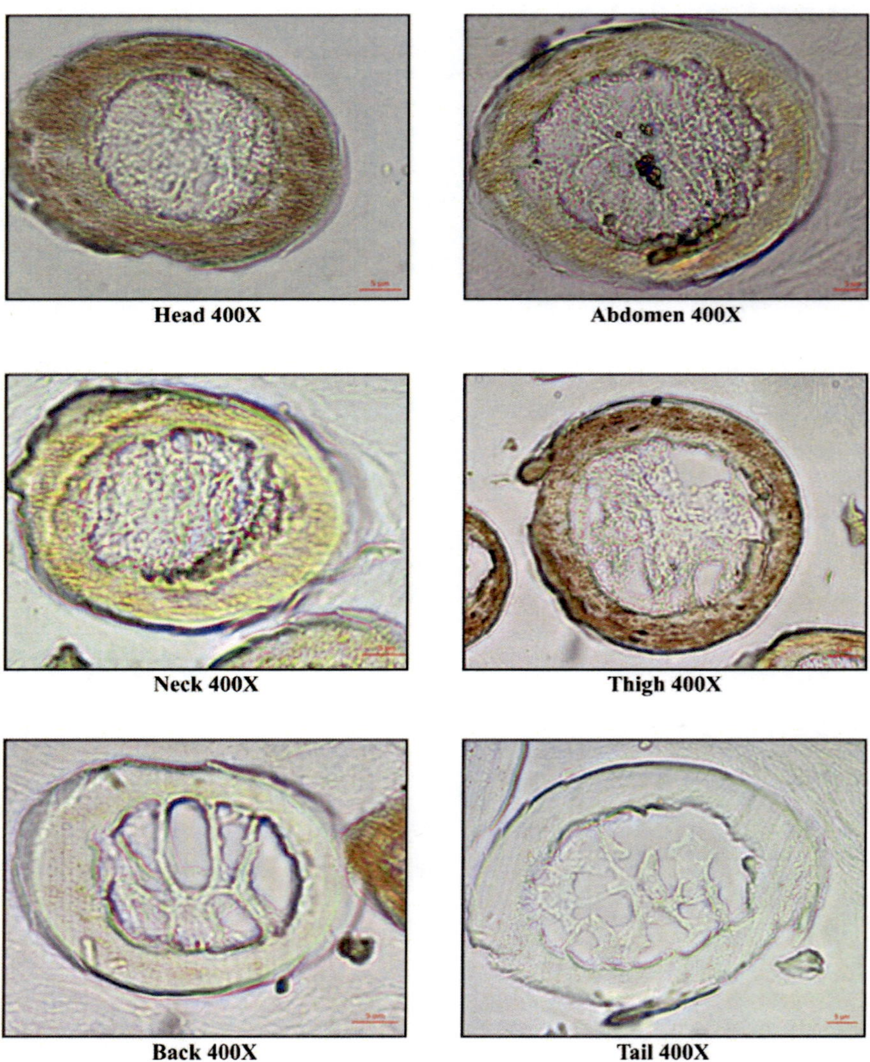

Fig. 4.192: Microphotograph showing cross sections of hair from various body regions of Spotted Deer (*Axis axis*)

Hair of Nilgai (*Boselaphus tragocamelus*) from head, neck, back and abdomen region was circular shape in cross section with circular, medium sized, centrally located medulla (Fig. 4.194). While, it was oval shaped in thigh and tail regions with medium sized, oval medulla. The medulla showed irregular margins in all the six body regions. The pigment granules, which were uniformly distributed in cortex were fine in head, neck, back, abdomen and thigh regions, while large granules were observed in the cortex of hair from tail region. The findings reported by De and Chakraborty (2012) about circular shape of cross section of hair from Nilgai are in complete agreement with the observations noted during present work.

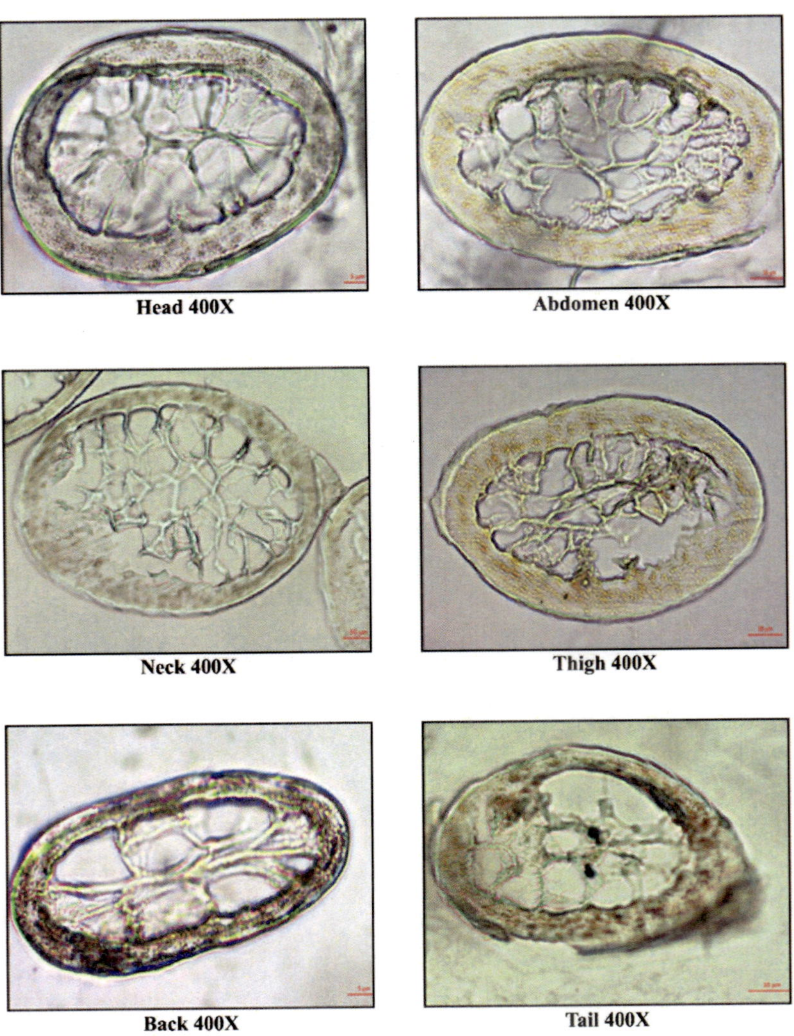

Fig. 4.193: Microphotograph showing cross sections of hair from various body regions of Sambar (*Rusa unicolor*)

The circular shaped cross sections of hair were observed in all the six body regions of Tiger (*Panthera tigris*). The medulla in the head, neck, back, abdomen and thigh region were medium sized and were circular shaped. While it was small, circular shaped and centrally placed in tail region (Fig. 4.195). fine pigment granules were found uniformly distributed in the cortex of head, neck and tail regions, while in back, abdomen and thigh region it was concentrated more towards the periphery of medulla.

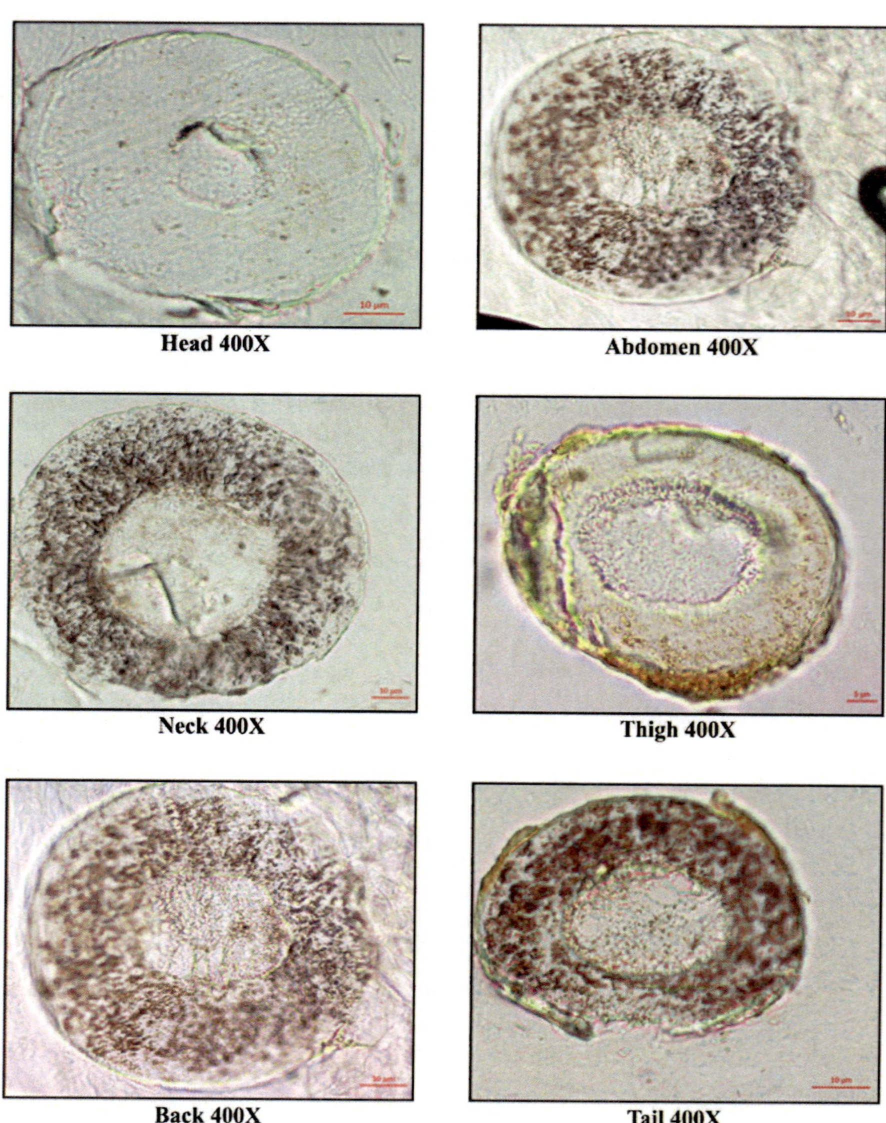

Fig. 4.194: Microphotograph showing cross sections of hair from various body regions of Nilgai (*Boselaphus tagocamellus*)

The hair from head and back regions in Leopard (*Panthera pardus*) had oval shaped cross section with medium sized oval medulla. The shape of cross section of hair from neck, abdomen, thigh and tail regions was circular in cross section with centrally located medium sized circular shaped medulla (Fig. 4.196). the medulla margin was observed irregular in all the six body regions. Dharaiya and Soni (2012) reported large oval shaped hair in cross section in Leopard. These observations are partial agreement with the findings noted during the present work.

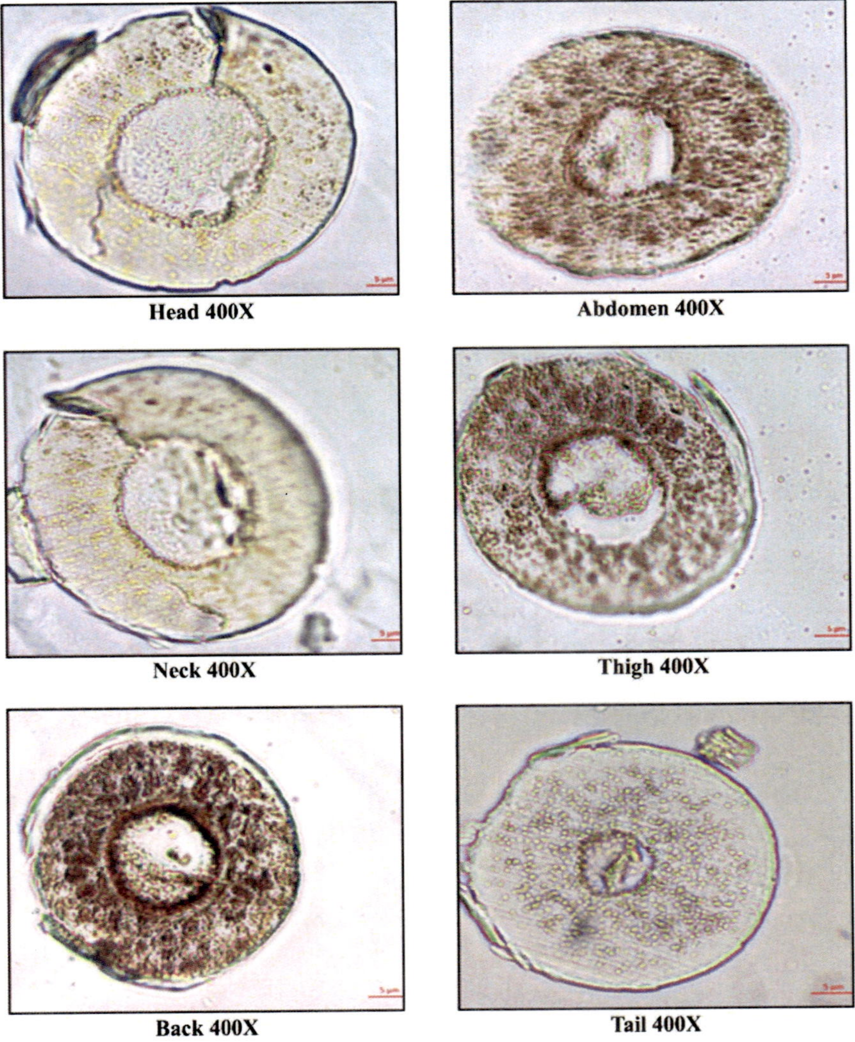

Fig. 4.195: Microphotograph showing cross sections of hair from various body regions of Tiger (*Panthera tigris*)

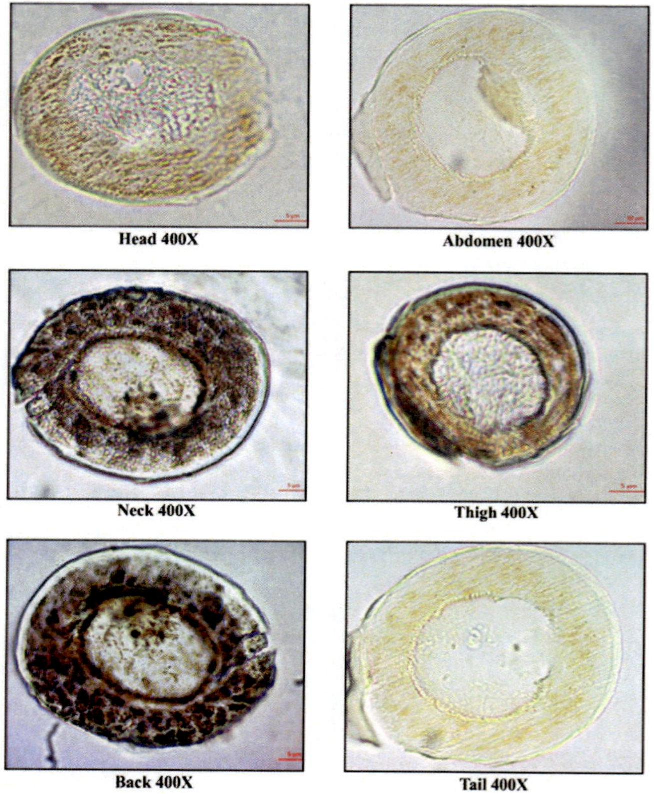

Fig. 4.196. Microphotograph showing cross sections of hair from various body regions of Leopard (*Panthera pardus*)

4.5. Micrometrical observations of hair

4.5.1. Cuticular scale height

The cuticular scales are the transparent keratinized cell layer arranged in overlapping manner one above the other. The cuticular scale height at three parts (proximal, middle and distal) was measured at 400X magnification and was measured from free margin of one scale to the free margin of successive scale.

The cuticular scale height from the proximal part of hair was significantly lowest in Nilgai (3.71 ± 0.08 μm), and in Pig (3.72 ± 0.16 μm), followed by Dog (4.42 ± 0.14 μm), while significantly higher scale height was noted in Hanuman Langur Table 4.6 (6.61 ± 0.15 μm) as compared with all the species considered for the present study. Amongst all species with respect to various body regions, the cuticular scale height at proximal part of hair was lowest in the hair from the back region of Pig (2.80 ± 0.26 μm), while highest scale height was noted in hair from neck region of Hanuman Langur (6.91 ± 0.35 μm).

The Means and SE for scale height of hair (μ) at middle part in various region of body in different species are presented in Table 4.7 During the present study, it was observed that the scale height at middle part of hair was significantly lowest in Nilgai (3.48 ± 0.09 μm) followed by Buffalo (3.39 ± 0.08 μm) and Tiger (3.94 ± 0.13 μm), while significantly highest scale height was noted in Sheep (7.61 ± 0.12 μm) followed by Dog (6.77 ± 0.76 μm) as compared to all other species considered for the present work. Amongst all the species with respect to various regions of body the cuticular scale height was noted lowest in Nilgai (3.48 ± 0.09 μm) at tail region, while highest scale height was observed in tail region of Dog (16.49± 0.81 μm).

Significantly lowest cuticular scale height was noted at the distal part of the hair in Tiger (2.49 ± 0.07 μm), while highest mean cuticular scale height was found in Sheep (10.72 ± 0.08 μm) as compared to all other species under present research work. With respect to the cuticular scale height at the distal part of hair, amongst all the mammalian species considered for the study the minimum scale height was found in Buffalo (2.12 ± 0.06 μm) at back region, while maximum scale height was noted in Sheep (10.72 ± 0.08 μm) at tail region Table 4.8.

No specific trends were noted regarding cuticular scale height at all the three parts of hair, i e. proximal middle and distal at various body regions in different species under study (Table 4.6, 4.7 and 4.8).

Table 4.6. Means and SE for scale height of hair (μ) at proximal part of hair from various regions of body in different species

Species→ Region ↓	Cattle	Buf-falo	Sheep	Goat	Horse	Domes-tic Pig	Cat	Dog	Spot-ted Deer	Sam-bar	Nil-gai	Hanu-man Langur	Tiger	Leop-ard	Pooled Mean± SE
Head	4.32± 0.29^{a}	4.14± 0.30^{c}	4.65 ± 0.08	4.69 ± 0.36^{ab}	4.49± 0.21^{a}	3.57± 0.46^{b}	4.34± 0.31^{a}	4.69± 0.60^{c}	4.31± 0.15^{ab}	5.31± 0.03^{b}	3.49± 0.16^{a}	6.44± 0.17^{b}	4.81 ± 0.08	4.98± 0.22^{a}	**4.59** ± 0.19
Neck	4.55 ± 0.35^{ab}	4.34± 0.11^{ab}	4.69 ± 0.16	4.89 ± 0.40^{b}	4.93 ± 0.20^{c}	3.75 ± 0.20^{bc}	5.00± 0.24^{bc}	4.11± 0.23^{a}	4.16± 0.17^{a}	5.19± 0.06^{b}	4.14 ± 0.36^{c}	6.91± 0.35^{d}	4.63 ± 0.13	5.22± 0.43^{b}	**4.75** ± 0.20
Back	4.24± 0.32^{a}	4.59^{b}± 0.11^{b}	4.69 ± 0.13	4.57 ± 0.41^{ab}	4.86 ± 0.11^{c}	2.80± 0.26^{a}	5.18 ± 0.17^{c}	4.41± 0.32^{b}	4.45± 0.21^{bc}	4.65± 0.02^{a}	3.64± 0.14 ^{ab}	6.60± 0.51^{bc} . 0.35	4.76 ± 0.18	5.31± 0.65^{b}	**4.63** ± 0.22
Abdomen	4.70 ± 0.39^{bc}	4.42± 0.21^{ab}	4.59 ± 0.15	4.57 ± 0.26^{ab}	4.79 ± 0.16^{bc}	3.83± 0.46^{c}	4.90± 0.28^{b}	4.54± 0.16^{bc}	4.73± 0.12^{d}	6.77± 0.06^{d}	3.73± 0.08^{b}	6.73± 0.35^{cd}	4.76 ± 0.16	4.91± 0.21^{a}	**4.86** ± 0.23
Thigh	4.49^{ab} ± 0.35^{ab}	4.57± 0.07^{b}	4.67 ± 0.06	4.69 ± 0.47^{ab}	4.64 ± 0.36^{ab}	3.75± 0.43^{bc}	5.41± 0.26^{d}	4.36± 0.33^{b}	4.94± 0.15^{e}	5.58± 0.15^{c}	3.69± 0.10 ^{ab}	6.80± 0.37^{cd}	4.66 ± 0.23	4.80± 0.27^{a}	**4.79** ± 0.20
Tail	4.91 ± 0.53^{c}	4.91± 0.17^{c}	4.68 ± 0.14	4.52 ± 0.27^{a}	4.49± 0.22^{a}	4.59± 0.19^{d}	5.19 ± 0.20^{c}	4.40± 0.31^{b}	4.64± 0.30^{cd}	4.54± 0.09^{a}	3.56± 0.24 ^{ab}	6.17± 0.43^{a}	4.66 ± 0.13	5.22± 0.26^{b}	**4.75** ± 0.15
Pooled Mean±SE	**4.54**± 0.15^{BD}	**4.49** ± 0.08 ^{BC}	**4.66** ± 0.05^{CD}	**4.65** ± 0.14 ^{CD}	**4.70** ± 0.09^{CD}	**3.72**± 0.16^{A}	**5.01** ± 0.11^{E}	**4.42**± 0.14^{B}	**4.54** ± 0.09 ^{BD}	**5.34** ± 0.18^{F}	**3.71**± 0.08^{A}	**6.61** ± 0.15^{G}	**4.71** ± 0.06^{D}	**5.07** ± 0.15^{E}	

Means bearing same superscript in a column do not differ significantly
CD for Species is 0.33 and CD for Region is 0.21

Table 4.6. Means and SE for scale height of hair (μ) at proximal part of hair from various regions of body in different species

Species→ Region ↓	Cattle	Buffalo	Sheep	Goat	Horse	Domestic Pig	Cat	Dog	Spotted Deer	Sambar	Nilgai	Hanuman Langur	Tiger	Leopard	Pooled Mean± SE
Head	4.32± 0.29^{a}	4.14± 0.30^{c}	4.65 ± 0.08	4.69 ± 0.36^{ab}	4.49± 0.21^{a}	3.57± 0.46^{b}	4.34± 0.31^{a}	4.69± 0.60^{c}	4.31± 0.15^{ab}	5.31± 0.03^{b}	3.49± 0.16^{a}	6.44± 0.17^{b}	4.81 ± 0.08	4.98± 0.22^{a}	**4.59** ± 0.19
Neck	4.55 ± 0.35^{ab}	4.34± 0.11^{ab}	4.69 ± 0.16	4.89 ± 0.40^{b}	4.93 ± 0.20^{c}	3.75 ± 0.20^{bc}	5.00± 0.24^{bc}	4.11± 0.23^{a}	4.16± 0.17^{a}	5.19± 0.06^{b}	4.14 ± 0.36^{c}	6.91± 0.35^{d}	4.63 ± 0.13	5.22± 0.43^{b}	**4.75** ± 0.20
Back	4.24± 0.32^{a}	4.59^{b}± 0.11^{b}	4.69 ± 0.13	4.57 ± 0.41^{ab}	4.86 ± 0.11^{c}	2.80± 0.26^{a}	5.18 ± 0.17^{c}	4.41± 0.32^{b}	4.45± 0.21^{bc}	4.65± 0.02^{a}	3.64± 0.14^{ab}	6.60± 0.51^{bc} . 0.35	4.76 ± 0.18	5.31± 0.65^{b}	**4.63** ± 0.22
Abdomen	4.70 ± 0.39^{bc}	4.42± 0.21^{ab}	4.59 ± 0.15	4.57 ± 0.26^{ab}	4.79 ± 0.16^{bc}	3.83± 0.46^{c}	4.90± 0.28^{b}	4.54± 0.16^{bc}	4.73± 0.12^{d}	6.77± 0.06^{d}	3.73± 0.08^{b}	6.73± 0.35^{cd}	4.76 ± 0.16	4.91± 0.21^{a}	**4.86** ± 0.23
Thigh	4.49^{ab} ± 0.35^{ab}	4.57± 0.07^{b}	4.67 ± 0.06	4.69 ± 0.47^{ab}	4.64 ± 0.36^{ab}	3.75± 0.43^{bc}	5.41± 0.26^{d}	4.36± 0.33^{b}	4.94± 0.15^{e}	5.58± 0.15^{c}	3.69± 0.10^{ab}	6.80± 0.37^{cd}	4.66 ± 0.23	4.80± 0.27^{a}	**4.79** ± 0.20
Tail	4.91 ± 0.53^{c}	4.91± 0.17^{c}	4.68 ± 0.14	4.52 ± 0.27^{a}	4.49± 0.22^{a}	4.59± 0.19^{d}	5.19 ± 0.20^{c}	4.40± 0.31^{b}	4.64± 0.30^{cd}	4.54± 0.09^{a}	3.56± 0.24^{ab}	6.17± 0.43^{a}	4.66 ± 0.13	5.22± 0.26^{b}	**4.75** ± 0.15
Pooled Mean±SE	**4.54**± 0.15^{BD}	**4.49** ± 0.08^{BC}	**4.66** ± 0.05^{CD}	**4.65** ± 0.14^{CD}	**4.70** ± 0.09^{CD}	**3.72**± 0.16^{A}	**5.01** ± 0.11^{E}	**4.42**± 0.14^{B}	**4.54** ± 0.09^{BD}	**5.34** ± 0.18^{F}	**3.71**± 0.08^{A}	**6.61** ± 0.15^{G}	**4.71** ± 0.06^{D}	**5.07** ± 0.15^{E}	

Means bearing same superscript in a column do not differ significantly

CD for Species is 0.33 and CD for Region is 0.21

Table 4.7: Means and SE for scale height of hair (μ) at middle part of hair from various regions of body in different species.

Species→ Region ↓	Cattle	Buffalo	Sheep	Goat	Horse	Domestic Pig	Cat	Dog	Spotted Deer	Sambar	Nilgai	Hanuman Langur	Tiger	Leopard	Pooled Mean± SE
Head	4.09 ± 0.23	3.98 ± 0.23^{ab}	7.61 ± 0.27^{b}	3.88 ± 0.15^{a}	4.41 ± 0.17^{a}	3.72 ± 0.52^{a}	5.07 ± 0.33^{d}	4.39 ± 0.47^{a}	4.77 ± 0.35^{b}	6.12 ± 0.18^{d}	3.40 ± 0.22^{b}	6.65 ± 0.38^{d}	4.16 ± 0.48^{b}	4.41 ± 0.17^{bc}	**4.76 ±** 0.31
Neck	4.14 ± 0.42	3.99 ± 0.22^{ab}	7.13 ± 0.38^{a}	4.92 ± 0.66^{c}	5.10 ± 0.16^{b}	4.09 ± 0.91^{b}	4.51 ± 0.10^{bc}	5.52 ± 0.64^{c}	4.39 ± 0.19^{a}	5.03 ± 0.09^{a}	3.78 ± 0.16^{c}	6.48 ± 0.60^{bc}	3.85 ± 0.25^{a}	4.69 ± 0.28^{c}	**4.83 ±** 0.25
Back	4.35 ± 0.26	3.74 ± 0.07^{a}	7.56 ± 0.29^{b}	4.91 ± 0.40^{c}	4.88 ± 0.14^{b}	3.93 ± 0.70^{a}	4.31 ± 0.18^{bc}	4.87 ± 0.43^{b}	4.96 ± 0.18^{b}	5.34 ± 0.13^{bc}	3.76 ± 0.26^{c}	6.27 ± 0.29^{bc}	3.94 ± 0.17^{ab}	4.09 ± 0.16^{a}	**4.78 ±** 0.27
Abdomen	4.20 ± 0.28	3.92 ± 0.21^{ab}	7.72 ± 0.29^{b}	4.68 ± 0.27^{bc}	4.56 ± 0.20^{a}	4.11 ± 0.66^{b}	3.83 ± 0.19^{a}	4.77 ± 0.35^{b}	4.93 ± 0.13^{b}	5.19 ± 0.07^{ab}	3.53 ± 0.12^{bc}	6.10 ± 0.29^{b}	4.21 ± 0.24^{b}	4.11 ± 0.20^{a}	**4.70 ±** 0.28
Thigh	4.11 ± 0.22	3.79 ± 0.12^{a}	7.78 ± 0.13^{b}	4.59 ± 0.25^{b}	4.51 ± 0.17^{a}	3.82 ± 0.72^{a}	4.76 ± 0.31^{c}	4.60 ± 0.43^{ab}	4.92 ± 0.16^{b}	5.53 ± 0.16^{c}	3.35 ± 0.14^{b}	6.10 ± 0.60^{b}	3.77 ± 0.39^{a}	4.12 ± 0.11^{ab}	**4.70 ±** 0.30
Tail	4.38 ± 0.25	4.17 ± 0.29^{b}	7.84 ± 0.34^{b}	4.06 ± 0.31^{a}	4.93 ± 0.30^{b}	6.41 ± 3.18^{b}	4.79 ± 0.48^{cd}	16.49 ± 0.81^{d}	4.87 ± 0.24^{b}	5.35 ± 0.06^{bc}	3.05 ± 0.29^{a}	5.46 ± 0.60^{a}	3.69 ± 0.36^{a}	4.23 ± 0.18^{ab}	**6.09 ±** 0.86
Pooled Mean± SE	**4.21 ±** 0.11^{BC}	**3.93 ±** 0.08^{AB}	**7.61 ±** 0.12^{I}	**4.51^{E} ±** 0.16^{CD}	**4.73 ±** 0.09^{DE}	**4.35 ±** 0.37^{BDC}	**4.55 ±** 0.13^{CDE}	**6.77 ±** 0.76^{H}	**4.81 ±** 0.09^{E}	**5.42 ±** 0.10^{F}	**3.48 ±** 0.09^{A}	**6.18 ±** 0.19^{G}	**3.94 ±** 0.13^{B}	**4.28 ±** 0.08^{BCD}	

Means bearing same superscript in a column do not differ significantly

CD for Species is 0.45 and CD for Region is 0.29

Table 4.8. Means and SE for scale height of hair (μ) at distal part of hair in various regions of body in different species.

Species→ Region ↓	Cattle	Buffalo	Sheep	Goat	Horse	Domestic Pig	Cat	Dog	Spotted Deer	Sambar	Nilgai	Hanuman Langur	Tiger	Leopard	Pooled Mean± SE
Head	3.39± 0.19^{b}	3.50± 0.32^{d}	10.74± 0.32^{ab}	3.75± 0.33^{a}	4.62 ± 0.29^{c}	3.17± 0.26^{a}	3.01± 0.33^{b}	3.41± 0.33^{a}	4.20± 0.32^{a}	5.67± 0.21^{d}	2.99± 0.23^{a}	6.25± 0.85^{e}	2.76± 0.26^{c}	2.96± 0.27^{a}	**4.32** ± 0.55
Neck	3.23± 0.21^{ab}	2.83 ± 0.21^{b}	10.50 ± 0.25^{a}	4.31± 0.56^{b}	4.60 ± 0.40^{c}	3.43± 0.42^{b}	3.13± 0.38^{bc}	3.97± 0.25^{c}	4.12± 0.28^{a}	4.22± 0.19^{ab}	2.86± 0.10^{cd}	5.99 ± 0.68^{d}	2.60± 0.20^{bc}	2.93 ± 0.19^{a}	**4.19** ± 0.52
Back	3.21± 0.22^{ab}	2.12± 0.06^{a}	10.77± 0.08^{b}	3.90± 0.55^{ab}	3.70± 0.24^{a}	2.94± 0.46^{a}	3.33± 0.76^{c}	3.50± 0.32^{ab}	4.19± 0.29^{a}	4.37± 0.07^{b}	2.46± 0.26^{ab}	5.47 ± 0.55^{c}	2.43± 0.12^{ab}	3.08± 0.21^{ab}	**3.96** ± 0.55
Abdomen	3.24± 0.22^{ab}	2.86 ± 0.36	10.73± 0.14^{ab}	4.37± 0.33^{b}	4.02± 0.16^{b}	3.16± 0.48^{a}	3.01± 0.11^{b}	3.71± 0.29^{bc}	4.26± 0.12^{a}	4.11± 0.07^{a}	2.74 ± 0.22^{c}	5.27± 0.62^{c}	2.46± 0.13^{b}	2.93± 0.09^{a}	**4.06** ± 0.53
Thigh	3.14± 0.08^{a}	2.38± 0.07^{a}	10.78± 0.22^{b}	4.14± 0.38^{b}	4.40± 0.23^{c}	3.11± 0.53^{a}	2.83± 0.15^{ab}	3.99± 0.24^{c}	4.59± 0.21^{b}	4.06± 0.05^{a}	2.70 ± 0.12^{bc}	4.85± 0.33^{b}	2.48 ± 0.20^{b}	3.30± 0.42^{b}	**4.05** ± 0.54
Tail	4.10± 0.24^{c}	3.22± 0.49^{c}	10.82± 0.16^{b}	4.13 ± 0.17^{b}	3.51± 0.36^{a}	3.54± 0.34^{b}	2.68± 0.12^{a}	3.93± 0.19^{c}	4.26± 0.24^{a}	4.81± 0.10^{c}	2.29 ± 0.18^{a}	4.42± 0.25^{a}	2.19± 0.06^{a}	3.31± 0.24^{b}	**4.09** ± 0.54
Pooled Mean± SE	**3.38**± 0.09^{EF}	**2.82** ± 0.14^{ABC}	**10.72** ± 0.08^{J}	**4.10**± 0.16^{G}	**4.14**± 0.13^{G}	**3.22**± 0.16^{DE}	**3.00** ± 0.15^{BCD}	**3.75**± 0.11^{F}	**4.27**± 0.10^{GH}	**4.54**^{H} ±0.15	**2.67**± 0.08^{AB}	**5.37** ± 0.24^{I}	**2.49**± 0.07^{A}	**3.09**± 0.10^{CDE}	

Means bearing same superscript for particular effect in a column do not differ significantly

CD for Species is 0.37 and CD for Region is 0.24

Table 4.9. Means and SE for scale height at different parts of hair from various regions of body in different species.

Species→ Region ↓	Hair Parts	Cattle	Buffalo	Sheep	Goat	Horse	Domes-tic Pig	Cat	Dog	Spotted Deer	Sam-bar	Nilgai	Hanu-man Langur	Tiger	Leop-ard	Pooled Mean± SE
Head	**Proximal**	4.32^{b} ± 0.29	4.14^{c} ± 0.30	4.65^{a} ±0.08	4.69^{b} ±0.36	4.49 ±0.21	3.57^{b} ±0.46	4.34^{b} ±0.31	4.69^{c} ±0.60	4.31^{a} ±0.15	5.31^{a} ±0.03	3.49^{b} ± 0.16	6.44^{ab} ±0.17	4.81^{c} ±0.08	4.98^{c} ±0.22	**4.59 ± 0.19**
	Middle	4.09^{b} ± 0.23	3.98^{b} ±0.23	7.61^{b} ±0.27	3.88^{a} ±0.15	4.41 ±0.17	3.72^{b} ±0.52	5.07^{c} ±0.33	4.39^{b} ±0.47	4.77^{b} ±0.35	6.12^{c} ±0.18	3.40^{b} ± 0.22	6.65^{b} ±0.38	4.16^{b} ±0.48	4.41^{b} ±0.17	**4.76 ± 0.31**
	Distal	3.39^{a} ± 0.19	3.50^{a} ±0.32	10.74^{c} ± 0.32	3.75^{a} ±0.33	4.62 ±0.29	3.17^{a} ±0.26	3.01^{a} ±0.33	3.41^{a} ±0.33	4.20^{a} ± 0.32	5.67^{b} ±0.21	2.99^{a} ± 0.23	6.25^{a} ±0.85	2.76^{a} ±0.26	2.96^{a} ±0.27	**4.32 ± 0.55**
	Mean ±SE	**3.93^{BCD} ± 0.12**	**3.87^{BC} ± 0.12**	**7.66^{H} ±0.44**	**4.11^{CDE} ± 0.13**	**4.51^{E} ± 0.09**	**3.49^{AB} ± 0.17**	**4.14^{CDE} ± 0.19**	**4.17^{CDE} ± 0.21**	**4.43^{DE} ± 0.12**	**5.70^{F} ±0.11**	**3.29^{A} ± 0.09**	**6.44^{G} ±0.21**	**3.91^{BCD} ± 0.19**	**4.12^{CDE} ± 0.17**	
Neck	Proximal	4.55^{c} ±0.35	4.34^{c} ±0.11	4.69^{a} ±0.16	4.89^{b} ±0.40	4.93^{b} ±0.20	3.75^{b} ±0.20	5.00^{c} ± 0.24	4.11^{a} ±0.23	4.16 ±0.17	5.19^{c} ±0.06	4.14^{c} ±0.36	6.91^{c} ±0.35	4.63^{c} ±0.13	5.22^{c} ±0.43	**4.75 ± 0.20**
	Middle	4.14^{b} ±0.42	3.99^{b} ±0.22	7.13^{b} ±0.38	4.92^{b} ±0.66	5.10^{c} ±0.16	4.09^{c} ±0.91	4.51^{b} ± 0.10	5.52^{b} ±0.64	4.39 ±0.19	5.03^{b} ±0.09	3.78^{b} ±0.16	6.48^{b} ±0.60	3.85^{b} ±0.25	4.69^{b} ±0.28	**4.83 ± 0.25**
	Distal	3.23^{a} ±0.21	2.83^{a} ±0.21	10.50^{c} ±0.25	4.31^{a} ±0.56	4.60^{a} ±0.40	3.43^{a} ±0.42	3.13^{a} ±0.38	3.97^{a} ±0.25	4.12 ±0.28	4.22^{a} ±0.19	2.86^{a} ±0.10	5.99^{a} ±0.68	2.60^{a} ±0.20	2.93^{a} ±0.19	**4.19 ± 0.52**
	Mean ±SE	**3.97^{ABC} ± 0.16**	**3.72^{AB} ±0.13**	**7.44^{F} ±0.42**	**4.71^{CD} ± 0.22**	**4.88^{D} ±0.11**	**3.75^{AB} ± 0.23**	**4.22^{BC} ±0.17**	**4.53^{CD} ±0.20**	**4.22^{BC} ±0.09**	**4.81^{CD} ± 0.12**	**3.59^{A} ±0.13**	**6.46^{E} ±0.22**	**3.69^{AB} ± 0.16**	**4.28^{BC} ±0.21**	

Species→ Region ↓	Hair Parts	Cattle	Buffalo	Sheep	Goat	Horse	Domestic Pig	Cat	Dog	Spotted Deer	Sambar	Nilgai	Hanuman Langur	Tiger	Leopard	Pooled Mean± SE
Back	Proximal	4.24^{b} ±0.32	4.59^{c} ±0.11	4.69^{c} ±0.13	4.57^{b} ±0.41	4.86^{b} ±0.11	2.80^{a} ±0.26	5.18^{c} ±0.17	4.41^{b} ±0.32	4.45^{a} ±0.21	4.65^{b} ±0.02	3.64^{b} ±0.14	6.60^{c} ±0.51	4.76^{c} ±0.18	5.31^{c} ±0.65	**4.63 ± 0.22**
	Middle	4.35^{b} ±0.26	3.74^{b} ±0.07	7.56^{b} ±0.29	4.91^{c} ±0.40	4.88^{b} ±0.14	3.93^{b} ±0.70	4.31^{b} ±0.18	4.87^{c} ±0.43	4.96^{b} ±0.18	5.34^{c} ±0.13	3.76^{b} ±0.26	6.27^{b} ±0.29	3.94^{b} ± 0.17	4.09^{b} ±0.16	**4.78 ± 0.27**
	Distal	3.21^{a} ±0.22	2.12^{a} ±0.06	10.77^{a} ±0.08	3.90^{a} ±0.55	3.70^{a} ±0.24	2.94^{a} ±0.46	3.33^{a} ±0.76	3.50^{a} ±0.32	4.19^{a} ±0.29	4.37^{a} ±0.07	2.46^{a} ±0.26	5.47^{a} ±0.55	2.43^{a} ±0.12	3.08^{a} ±0.21	**3.96 ± 0.55**
	Mean ±SE	**3.94^{BCD} ± 0.14**	**3.49^{AB} ±0.18**	**7.67^{H} ±0.43**	**4.46^{DEF} ±0.19**	**4.48^{DEF} ±0.12**	**3.23^{A} ±0.21**	**4.28^{DEF} ±0.22**	**4.26^{DEF} ± 0.17**	**4.53^{EF} ±0.10**	**4.79^{F} ± 0.11**	**3.28^{A} ±0.13**	**6.11^{G} ±0.20**	**3.71^{ABC} ±0.18**	**4.16^{CDE} ±0.22**	
Abdomen	**Proximal**	4.70 ± 0.39^{c}	4.42 ± 0.21^{c}	4.59 ± 0.15^{a}	4.57 ± 0.26^{ab}	4.79 ± 0.16^{c}	3.83 ± 0.46^{b}	4.90 ± 0.28^{c}	4.54 ± 0.16^{b}	4.73 ± 0.12^{b}	6.77 ± 0.06^{c}	3.73 ± 0.08^{b}	6.73 ± 0.35^{c}	4.76 ± 0.16^{c}	4.91 ± 0.21^{c}	**4.86 ±** 0.23
	Middle	4.20 ± 0.28^{b}	3.92 ± 0.21^{b}	7.72 ± 0.29^{b}	4.68 ± 0.27^{b}	4.56 ± 0.20^{b}	4.11 ± 0.66^{c}	3.83 ± 0.19^{b}	4.77 ± 0.35^{c}	4.93 ± 0.13^{b}	5.19 ± 0.07^{b}	3.53 ± 0.12^{b}	6.10 ± 0.29^{b}	4.21 ± 0.24^{b}	4.11 ± 0.20^{b}	**4.70 ±** 0.28
	Distal	3.24 ± 0.22^{a}	2.86 ± 0.36^{a}	10.73± 0.14^{c}	4.37 ± 0.33^{a}	4.02 ± 0.16^{a}	3.16 ± 0.48^{a}	3.01 ± 0.11^{a}	3.71 ± 0.29^{a}	4.26 ± 0.12^{a}	4.11 ± 0.07^{a}	2.74 ± 0.22^{a}	5.27 ± 0.62^{a}	2.46 ± 0.13^{a}	2.93 ± 0.09^{a}	**4.06 ±** 0.53
	Mean ±SE	**4.05 ± 0.16 BCD**	**3.73 ± 0.15^{AB}**	**7.68 ± 0.44^{H}**	**4.54 ± 0.11^{E}**	**4.46 ± 0.09^{DE}**	**3.70 ± 0.22^{AB}**	**3.91 ± 0.15^{BC}**	**4.34 ± 0.13 CDE**	**4.64 ± 0.07^{E}**	**5.36 ± 0.28^{F}**	**3.34± 0.09^{A}**	**6.03 ± 0.20^{G}**	**3.81± 0.18^{B}**	**3.98 ± 0.16^{BC}**	

Species→ Region ↓	Hair Parts	Cattle	Buffalo	Sheep	Goat	Horse	Domestic Pig	Cat	Dog	Spotted Deer	Sambar	Nilgai	Hanuman Langur	Tiger	Leopard	Pooled Mean± SE
Thigh	4.49 ± 0.35^{c}	4.57 ± 0.07^{c}	4.67 ± 0.06^{a}	4.69 ± 0.47^{b}	4.64 ± 0.36^{b}	3.75 ± 0.43^{b}	5.41 ± 0.26^{c}	4.36 ± 0.33^{b}	4.94 ± 0.15^{b}	5.58 ± 0.15^{b}	3.69 ± 0.10^{c}	6.80 ± 0.37^{c}	4.66 ± 0.23^{c}	4.80 ± 0.27^{c}	**4.79 ± 0.20**	
	Middle	4.11 ± 0.22^{b}	3.79 ± 0.12^{b}	7.78 ± 0.13^{b}	4.59 ± 0.25^{b}	4.51± 0.17^{ab}	3.82 ± 0.72^{b}	4.76 ± 0.31^{b}	4.60 ± 0.43^{c}	4.92 ± 0.16^{b}	5.53 ± 0.16^{b}	3.35 ± 0.14^{b}	6.10 ± 0.60^{b}	3.77 ± 0.39^{b}	4.12 ± 0.11^{b}	**4.70 ± 0.30**
	Distal	3.14 ± 0.08^{a}	2.38 ± 0.07^{a}	10.78± 0.22^{c}	4.14 ± 0.38^{a}	4.40 ± 0.23^{a}	3.11 ± 0.53^{a}	2.83 ± 0.15^{a}	3.99 ± 0.24^{a}	4.59 ± 0.21^{a}	4.06 ± 0.05^{a}	2.70 ± 0.12^{a}	4.85 ± 0.33^{a}	2.48 ± 0.20^{a}	3.30 ± 0.42^{a}	**4.05 ± 0.54**
	Mean ±SE	**3.91 ± 0.14^{BC}**	**3.58 ± 0.16^{AB}**	**7.74 ± 0.43^{H}**	**4.47 ± 0.15^{DE}**	**4.52 ± 0.10^{DE}**	**3.56 ± 0.23^{AB}**	**4.33 ± 0.21^{CDE}**	**4.32 ± 0.14^{CDE}**	**4.82 ± 0.07^{EF}**	**5.05 ± 0.19^{F}**	**3.25 ± 0.08^{A}**	**5.92 ± 0.22^{G}**	**3.64 ± 0.19^{AB}**	**4.08 ± 0.15^{BCD}**	
Tail	**Proximal**	4.91 ± 0.53^{b}	4.91 ± 0.17^{c}	4.68 ± 0.14^{a}	4.52 ± 0.27^{b}	4.49^{b} ±0.22	4.59^{b} ±0.19	5.19^{c} ±0.20	4.40^{b} ±0.31	4.64^{b} ±0.30	4.54^{a} ±0.09	3.56^{c} ±0.24	6.17^{c} ±0.43	4.66^{c} ±0.13	5.22^{c} ±0.26	**4.75 ±0.15**
	Middle	4.38 ± 0.25^{a}	4.17 ± 0.29^{b}	7.84 ± 0.34^{b}	4.06 ± 0.31^{a}	4.93^{c} ±0.30	6.40^{c} ±1.43	4.79^{b} ±0.48	16.49^{c} ±0.81	4.87^{b} ±0.24	5.35^{b} ±0.06	3.05^{b} ±0.29	5.46^{b} ±0.60	3.69^{b} ±0.36	4.23^{b} ±0.18	**6.09 ±0.96**
	Distal	4.10 ± 0.24^{a}	3.22 ± 0.49^{a}	10.82± 0.16^{c}	4.13 ± 0.17^{a}	3.51^{a} ±0.36	3.54^{a} ±0.34	2.68^{a} ±0.12	3.93^{a} ±0.19	4.26^{a} ±0.24	4.81^{a} ±0.10	2.29^{a} ±0.18	4.42^{a} ±0.25	2.19^{a} ±0.06	3.31^{a} ±0.24	**4.09 ±0.54**
	Mean ±SE	**4.46 ± 0.15^{CDE}**	**4.10 ± 0.18^{BC}**	**7.78 ± 0.44^{G}**	**4.23± 0.11^{CD}**	**4.31^{CDE} ±0.15**	**4.84^{DEF} ±0.39**	**4.22^{CD} ±0.22**	**8.27^{H} ±1.02**	**4.59^{CDE} ±0.11**	**4.90^{EF} ±0.09**	**2.96^{A} ±0.13**	**5.35^{F} ±0.21**	**3.51^{AB} ±0.19**	**4.25^{CD} ±0.16**	

Means bearing same superscript for different region (for hair part) in a column do not differ significantly/Means bearing same superscript for different species (hair part of different region) in a row do not differ significantly/Head Region-CD for Species is 0.54 and CD for hair part is 0.25/Neck Region-CD for Species is 0.59 and CD for hair part is 0.27/Back Region-CD for Species is 0.54 and CD for hair part is 0.25/Abdomen Region-CD for Species is 0.45 and CD for hair part is 0.21/ Thigh Region-CD for Species is 0.50 and CD for hair part is 0.23/Tail Region-CD for Species is 0.63 and CD for hair part is 0.29

In comparison with respect to scale height at distal part of hair from head region, the lowest cuticular scale height was seen in all the species except Horse, Sheep and Sambar (Table 4.9). In case of Sheep and Sambar, lowest cuticular scale height was found at proximal part of hair, while in Horse, it was noted in middle part of hair. Highest cuticular scale height was noted at the proximal part of hair in Buffalo, Goat, Cattle, Pig, Dog, Nilgai, Tiger and Leopard, at middle part of hair in Pig, Cat, Spotted Deer, Hanuman Langur and Sambar and at distal part of hair in Horse and Sheep (Table 4.9).

From the present study, it was observed that, the cuticular scale height was found lowest in the distal of hair from neck region in all the species except Sheep (Table 4.9). Highest cuticular scale height was noted at the proximal part of hair from neck region in Buffalo, Cattle, Cat, Nilgai, Tiger, Leopard, Hanuman Langur and Sambar, while it was highest at middle part of hair in Goat, Horse, Pig, Dog and Spotted Deer and at the distal part of hair in Sheep.

The lowest cuticular scale height was noted at the distal part of hair from the back region of all the species under present study, except Pig and Sheep (Table 4.9).

Table 4.10. Means and SE for scale width of hair (μ) at proximal part of hair from various regions of body in different species.

Species→ Region ↓	Cattle	Buffalo	Sheep	Goat	Horse	Domestic Pig	Cat	Dog	Spotted Deer	Sambar	Nilgai	Hanuman Langur	Tiger	Leopard	Pooled Mean±SE
Head	16.55 ± 0.65 ^{a}	19.01 ± 0.66 ^{c}	11.61± 0.23	19.87 ± 0.69 ^{a}	19.33 ± 1.08 ^{a}	17.14± 2.05 ^{b}	9.48 ± 0.76 ^{a}	12.96± 1.21 ^{a}	17.66 ± 1.63 ^{a}	26.19 ± 0.37 ^{b}	20.67± 0.63 ^{a}	17.98 ±0.82	18.20± 0.36 ^{bc}	12.58± 1.02 ^{a}	**17.09 ± 1.11**
Neck	17.76 ± 0.88 ^{bc}	18.88 ± 1.57 ^{c}	11.72± 0.30	21.54 ± 1.83 ^{b}	24.43 ± 2.67 ^{c}	21.52 ± 1.89 ^{c}	12.38± 1.07 ^{b}	18.31± 2.07 ^{e}	20.94 ± 0.98 ^{b}	25.58 ± 0.25 ^{ab}	25.31± 0.62 ^{c}	17.42 ±0.56	17.09± 0.62 ^{a}	13.66 ± 2.16 ^{b}	**19.04 ± 1.16**
Back	17.09 ± 0.84 ^{ab}	15.33 ± 0.18 ^{a}	11.35± 0.21	22.04 ± 1.82 ^{b}	21.79 ± 0.44 ^{b}	25.40 ± 2.43 ^{d}	17.03± 1.33 ^{d}	20.18 ± 1.88 ^{f}	22.21 ± 0.56 ^{c}	24.67 ± 0.21 ^{a}	23.29± 1.15 ^{b}	17.77 ±0.80	17.61± 0.48 ^{ab}	13.89 ± 1.99 ^{b}	**19.26 ± 1.08**
Abdomen	19.88 ± 1.01 ^{d}	16.89 ± 1.39 ^{b}	11.10± 0.16	20.38 ± 1.03 ^{a}	19.47 ± 0.77 ^{a}	21.00 ± 3.04 ^{c}	13.47± 2.05 ^{c}	16.76± 1.62 ^{d}	25.85 ± 1.50 ^{d}	35.09 ± 0.56 ^{e}	22.61± 0.87 ^{b}	18.36 ±1.12	18.19± 0.23 ^{bc}	17.50 ± 1.86 ^{d}	**19.75 ± 1.47**
Thigh	19.33 ± 1.57 ^{d}	21.03 ± 0.81 ^{d}	11.28± 0.19	22.23 ± 1.49 ^{b}	19.31 ± 1.07 ^{a}	18.02 ± 2.84 ^{b}	12.17± 0.85 ^{b}	15.49± 1.35 ^{c}	28.36 ± 1.55 ^{e}	31.92 ± 0.06 ^{d}	23.56± 1.22 ^{b}	17.56 ±1.07	17.53± 0.31 ^{ab}	16.09 ± 1.99 ^{c}	**19.56 ± 1.46**
Tail	18.55 ± 1.35 ^{cd}	19.75 ± 0.58 ^{c}	11.46 ± 0.15	22.14 ± 1.69 ^{b}	35.09 ± 3.45 ^{d}	15.60 ± 1.13 ^{a}	13.40± 0.66 ^{c}	14.35± 1.19 ^{b}	20.44 ± 0.64 ^{b}	28.10 ± 0.15 ^{c}	27.00± 1.24 ^{d}	17.87 ±0.35	19.20± 0.31 ^{c}	20.39 ± 1.10 ^{e}	**20.24 ± 1.63**
Pooled Mean±SE	**18.19** ± 0.46 ^{E}	**18.48** ± 0.49 ^{EF}	**11.42** ± 0.09 ^{A}	**21.37** ± 0.58 ^{G}	**23.24** ± 1.19 ^{H}	**19.78** ± 1.03 ^{F}	**12.99**± 0.59 ^{B}	**16.34**± 0.73 ^{CD}	**22.58** ± 0.76 ^{GH}	**28.59** ± 0.92 ^{I}	**23.74**± 0.50 ^{H}	**17.83** ^{DE} ±0.32	**17.97**± 0.19 ^{E}	**15.69** ± 0.80 ^{C}	

Means bearing same superscript in a column do not differ significantly
CD for Species is 1.54 and CD for Region is 1.01

The lowest cuticular scale height was noted at the distal part of hair from abdomen region in all the species under study except Sheep (Table 4.9). The highest cuticular scale height was noted in the proximal part of the hair in Buffalo, Cattle, Horse, Cat, Nilgai, Tiger Hanuman Langur and Leopard, at middle part of hair in Goat, Pig, Dog and Spotted Deer and at distal part in Sheep (Table 4.9).

The cuticular scale height at proximal part of hair in abdomen region was highest in Buffalo, Cattle, Horse, Cat, Nilgai, Tiger, Hanuman Langur, Leopard and Sambar, while it was highest at middle part of hair in Goat, Pig, Dog and Spotted Deer and at the distal part of hair in Sheep (Table 4.9).

The minimum cuticular scale height at the distal part of hair from the thigh region was found in all the species under study except Sheep, where it was minimum at proximal part of hair (Table 4.9). The maximum cuticular scale height was found at the proximal part of hair in thigh region in almost all species except Goat, Pig, Dog and Spotted Deer, where it was maximum in middle part of hair, while in Sheep it was maximum at distal part (Table 4.9). The cuticular scale height of the hair from thigh region, was lowest in distal part of hair in Tiger (2.48 ± 0.20 μm), while it was highest scale height was also noted at the distal part of hair in Sheep (10.78 ± 0.22 μm).

The lowest cuticular scale height in the distal part of hair of tail region was noted in almost all species except Goat, Sheep, and Sambar. The Sheep and Sambar had lowest cuticular scale height at the proximal part of the hair, while Goat had lowest cuticular scale height at the middle part of hair (Table 4.9). Highest cuticular scale height at the proximal part of hair of tail region in Buffalo, Goat, Cattle, Cat, Nilgai, Tiger, Hanuman Langur and Leopard, while it was highest at middle part of hair in Horse, Pig, Dog, Spotted Deer and Sambar. Only in case of Sheep the maximum cuticular scale height was measured at the distal part of hair (Table 4.9), amongst all the species. The lowest scale height was noted at the distal part of the hair from Tiger (2.19 ± 0.06 μm).

The overall mean cuticular scale height was significantly highest in Sheep amongst all parts of hair; body regions and the species while, it was significantly lowest in Nilgai (Table 4.9).

4.5.2. Cuticular Scale width

The mean and SE for cuticular scale width at proximal, part of the hair from all the six body regions and all the species under study were given in Table 4.10.

Table 4.11. Means and SE for scale width of hair (μ) at middle part of hair from various regions of body in different species.

Species→ Region ↓	Cattle	Buffalo	Sheep	Goat	Horse	Domestic Pig	Cat	Dog	Spotted Deer	Sambar	Nilgai	Hanuman Langur	Tiger	Leopard	Pooled Mean± SE
Head	17.23 ± 1.87 ^{b}	18.42 ± 1.29 ^{a}	17.75 ± 0.35	21.79 ± 1.29 ^{a}	19.94 ± 1.25 ^{a}	23.77 ± 1.67 ^{d}	11.98 ± 0.84 ^{a}	19.16± 2.45 ^{de}	19.21 ± 1.13 ^{a}	24.19 ^{a} ±0.19	19.92 ± 1.16 ^{a}	15.72 ± 0.75 ^{ab}	21.72 ± 0.97 ^{b}	14.70 ± 0.92 ^{a}	**18.96** ± 0.88
Neck	16.50 ± 0.97 ^{b}	20.57 ± 0.93 ^{b}	17.46 ± 0.36	23.87 ± 1.89 ^{cd}	24.48 ± 2.22 ^{b}	22.32 ± 1.84 ^{c}	13.38 ± 0.44 ^{b}	17.77 ± 1.97 ^{c}	21.15 ± 1.23 ^{b}	37.00 ^{d} ±0.33	23.39 ± 1.55 ^{b}	15.01 ± 0.94 ^{a}	21.63 ± 0.42 ^{b}	16.89 ± 0.55 ^{b}	**20.81** ± 1.49
Back	15.25 ± 0.46 ^{a}	18.67 ± 0.16 ^{a}	18.03 ± 0.54	24.45 ± 1.62 ^{d}	24.23 ± 1.14 ^{b}	23.61 ± 2.25 ^{d}	15.06 ± 0.21 ^{c}	19.51 ± 1.40 ^{e}	23.94 ± 1.28 ^{c}	27.70 ^{b} ±0.10	24.26 ± 0.95 ^{bc}	16.39 ± 0.99 ^{bc}	21.65 ± 0.40 ^{b}	17.41± 0.71 ^{b}	**20.73** ± 1.04
Abdomen	18.93 ± 1.12 ^{c}	18.68 ± 1.12 ^{a}	18.29 ± 0.26	23.29 ± 1.81 ^{bc}	22.90 ± 1.31 ^{b}	20.90 ± 2.84 ^{b}	14.83 ± 0.28 ^{c}	15.91 ± 1.38 ^{b}	25.67 ± 1.12 ^{d}	49.46 ^{e} ±0.67	20.60 ± 0.53 ^{a}	17.53 ± 1.06 ^{c}	19.32 ± 0.43 ^{b}	17.52 ± 0.97 ^{b}	**21.70** ± 2.19
Thigh	17.29 ± 1.67 ^{b}	21.74 ± 0.81 ^{c}	17.95 ± 0.47	26.25 ± 1.56 ^{e}	20.04 ± 0.54 ^{a}	20.00 ± 2.17 ^{ab}	13.36 ± 0.24 ^{b}	18.27 ± 1.26 ^{cd}	28.50 ± 1.04 ^{e}	29.59 ^{c} ±0.12	23.37 ± 1.01 ^{b}	17.21± 0.70 ^{c}	18.38 ± 0.67 ^{a}	16.88 ± 1.85 ^{b}	**20.63** ± 1.22
Tail	16.91 ± 0.98 ^{b}	19.77 ± 0.52 ^{b}	17.88 ± 0.34	22.46 ± 2.08 ^{ab}	37.68 ± 3.03 ^{c}	19.61 ± 1.94 ^{a}	15.15 ± 0.47 ^{c}	8.76 ± 0.57 ^{a}	24.23 ± 1.88 ^{c}	27.93 ^{b} ±0.07	24.88 ± 0.59 ^{c}	17.48 ± 0.57 ^{c}	22.37 ± 0.18 ^{b}	21.27 ± 1.05 ^{c}	**21.17** ± 1.72
Pooled Mean±SE	**17.02** ± 0.51 ^{B}	**19.64** ± 0.39 ^{C}	**17.89** ^{B} ±0.16	**23.68** ± 0.70 ^{F}	**24.88**± 1.22 ^{G}	**21.70** ± 0.86 ^{DE}	**13.96** ± 0.26 ^{A}	**16.56** ± 0.87 ^{B}	**23.78** ± 0.71 ^{F}	**32.64** ^{H} ±2.06	**22.74**± 0.50 ^{EF}	**16.56** ± 0.36 ^{B}	**20.85** ± 0.33 ^{CD}	**17.45** ± 0.53 ^{B}	

Means bearing same superscript in a column do not differ significantly
CD for Species is 1.45 and CD for Region is 0.95

The mean cuticular scale width of at the proximal part of the hair was found significantly lowest in Sheep (11.42 ± 0.09 μm) followed by Leopard (15.69 ± 0.80 μm), while significantly higher scale width was noted in Sambar (28.59 ± 0.92 μm) followed by Horse (23.24 ± 1.19 μm) as compared with all the fourteen species under study. Amongst all the species with respect to various body regions the cuticular scale width at proximal part of hair was noted lowest in hair from head region of Cat (9.48 ± 0.76 μm), while highest in the hair from abdomen region of Sambar (35.09 ± 0.56 μm).

The mean and SE for cuticular scale width at middle parts of the hair from all the regions of body and species under study were given in Table 4.11.

Table 4.12. Mean and SE for scale width of hair (μ) at distal part of hair from various regions of body in different species.

Species→ Region ↓	Cattle	Buffalo	Sheep	Goat	Horse	Domestic Pig	Cat	Dog	Spotted Deer	Sambar	Nilgai	Hanuman Langur	Tiger	Leopard	Pooled Mean± SE
Head	17.00 ± 1.12 ^{d}	21.36 ± 1.84 ^{bc}	10.30± 0.13	20.09 ± 2.00 ^{a}	22.31 ± 2.16 ^{a}	22.70 ± 1.97 ^{b}	16.57 ± 2.02 ^{bc}	18.81 ^{b} ±1.69	19.30 ± 0.92 ^{a}	25.51 ± 0.21 ^{b}	18.38 ± 0.92 ^{a}	16.18 ± 1.39 ^{c}	19.41 ± 0.63 ^{b}	17.20 ± 0.68 ^{b}	**18.94** ± 0.93
Neck	15.15 ± 1.15 ^{b}	20.84 ± 1.12 ^{ab}	10.86 ± 0.26	25.59 ± 3.09 ^{c}	24.57 ± 2.86 ^{b}	20.09 ± 1.50 ^{a}	14.60 ± 1.44 ^{a}	17.04 ^{a} ±1.19	20.88 ± 2.14 ^{b}	26.87 ± 0.23 ^{c}	21.13 ± 0.60 ^{c}	14.09 ± 0.68 ^{a}	23.25 ± 0.62 ^{c}	14.18 ± 0.89 ^{a}	**19.22** ± 1.27
Back	14.83 ± 0.80 ^{b}	22.42 ± 0.41 ^{d}	10.26 ± 0.12	24.55 ± 1.84 ^{c}	26.45 ± 0.71 ^{c}	24.33 ± 2.77 ^{c}	17.59 ± 0.71	18.29 ^{b} ±1.37	21.34 ± 1.84 ^{c}	29.46 ± 0.10 ^{d}	23.29 ± 2.20 ^{d}	15.46 ± 1.46 ^{bc}	25.28 ± 0.80 ^{d}	16.59 ± 0.30 ^{b}	**20.72** ± 1.38
Abdomen	13.33 ± 0.84 ^{a}	20.10 ± 0.65 ^{a}	10.37 ± 0.16	22.90 ± 1.96 ^{b}	23.30 ± 2.17 ^{a}	20.47 ± 2.12 ^{a}	17.15 ± 0.60 ^{bc}	18.10 ^{ab} ±2.50	24.04 ± 1.53 ^{d}	25.88 ± 0.18 ^{bc}	18.00 ± 0.75 ^{a}	15.04 ± 1.06 ^{ab}	16.55 ± 0.13 ^{a}	17.29 ± 0.64 ^{bc}	**18.75** ± 1.12
Thigh	15.25 ± 1.38 ^{b}	28.07 ± 1.98 ^{d}	10.23 ± 0.17	27.25 ± 2.53 ^{e}	22.58 ± 1.49 ^{a}	20.46 ± 1.52 ^{a}	16.06 ± 1.01 ^{b}	21.00 ^{c} ±0.92	26.23 ± 0.98 ^{e}	35.67 ± 4.39 ^{e}	19.67 ± 1.09 ^{b}	15.39 ± 0.15 ^{bc}	15.95 ± 1.29 ^{a}	17.53 ± 0.47 ^{bc}	**20.81** ± 1.72
Tail	16.81 ± 1.18 ^{b}	22.33 ± 0.84 ^{c}	10.18 ± 0.17	26.78 ± 2.92 ^{d}	35.31 ± 3.41 ^{d}	20.20 ± 2.26 ^{a}	18.04 ± 0.34	21.07 ^{c} ±2.12	20.18 ± 1.25 ^{ab}	24.24 ± 0.12 ^{a}	20.63 ± 1.16 ^{bc}	15.34 ± 0.84 ^{bc}	18.36 ± 0.91 ^{b}	18.34 ± 0.59 ^{c}	**20.56** ± 1.50
Pooled Mean± SE	**15.39** ± 0.46 ^{B}	**22.52** ± 0.70 ^{D}	**10.37** ± 0.08 ^{A}	**24.53** ± 1.01 ^{E}	**25.75** ± 1.16 ^{E}	**21.37** ± 0.83 ^{D}	**16.67** ± 0.48 ^{B}	**19.05** ^{C} ±0.70	**21.99** ± 0.70 ^{D}	**27.94** ± 1.27 ^{F}	**20.18**± 0.56 ^{C}	**15.25** ± 0.40 ^{B}	**19.80** ± 0.65 ^{C}	**16.85** ± 0.32 ^{B}	

Means bearing same superscript in a column do not differ significantly
CD for Species is 1.74 and CD for Region is 1.13

The significantly lowest mean cuticular scale width was noted in Cat (13.96 ± 0.26 μm) followed by Hanuman Langur (16.56 ± 0.36 μm) and Dog (16.56 ± 0.87 μm), while significantly higher scale width was noted in Sambar (32.64 ± 2.06 μm) followed by Horse (24.88 ± 1.22 μm) as compared with all the species under study. The minimum scale width was noted at the middle part of hair from head region in Cat (11.98 ± 0.84 μm), while maximum scale width was observed in hair from neck region of Sambar (37.00 ± 0.33 μm) among all the species.

The mean and SE for cuticular scale width at distal part of the hair from all the body regions and the species are given in Table 4.12.

Table 4.13. Means and SE for scale width (μ) at different parts of hair from various regions of body in different species.

Species→ Region ↓	Hair Parts	Cattle	Buffalo	Sheep	Goat	Horse	Domestic Pig	Cat	Dog	Spotted Deer	Sambar	Nilgai	Hanuman Langur	Tiger	Leopard	Pooled Mean± SE
Head	**Proximal**	16.55 ± 0.65	19.01 ± 0.66^{a}	11.61 ± 0.23^{b}	19.87 ± 0.69^{a}	19.33 ± 1.08^{a}	17.14 ± 2.05v	9.48 ± 0.76^{a}	12.96 ± 1.21^{a}	17.66 ± 1.63^{a}	26.19 ± 0.37^{b}	20.67 ± 0.63^{b}	17.98 ± 0.82^{b}	18.20 ± 0.36^{a}	12.58 ± 1.02^{a}	**17.09** ± 1.11
	Middle	17.23 ± 1.87	18.42 ± 1.29^{a}	17.75 ± 0.35^{c}	21.79 ± 1.29^{b}	19.94 ± 1.25^{a}	23.77 ± 1.67^{c}	11.98 ± 0.84^{b}	19.16 ± 2.45^{b}	19.21 ± 1.13^{b}	24.19 ± 0.19^{a}	19.92 ± 1.16^{b}	15.72 ± 0.75^{a}	21.72 ± 0.97^{c}	14.70 ± 0.92^{b}	**18.96** ± 0.88
	Distal	17.00 ± 1.12	21.36 ± 1.84^{b}	10.30 ± 0.13^{a}	20.09 ± 2.00^{ab}	22.31 ± 2.16^{b}	22.70 ± 1.97^{b}	16.57 ± 2.02^{c}	18.81 ± 1.69^{b}	19.30 ± 0.92^{b}	25.51 ± 0.21^{b}	18.38 ± 0.92^{a}	16.18 ± 1.39^{a}	19.41 ± 0.63^{b}	17.20 ± 0.68^{c}	**18.94** ± 0.93
	Mean ± SE	**16.93**± 0.50^{D}	**19.60**± **0.56**^{EF}	**13.22** ± 0.57^{AB}	**20.58** ± 0.57^{EF}	**20.53** ± 0.64^{EF}	**21.20** ± 0.88^{F}	**12.68** ± 0.72^{A}	**16.98** ± 0.86^{D}	**18.72** ± 0.50^{DE}	**25.30** ± 0.25^{G}	**19.66** ± 0.39^{EF}	**16.63** ± 0.43^{C}	**19.78** ± 0.37^{EF}	**14.83** ± 0.47^{BC}	
Neck	**Proximal**	17.76 ± 0.88^{c}	18.88 ± 1.57^{a}	11.72± 0.30^{a}	21.54 ± 1.83^{a}	24.43 ± 2.67	21.52 ± 1.89^{b}	12.38 ± 1.07^{a}	18.31 ± 2.07^{b}	20.94 ± 0.98	25.58 ± 0.25^{a}	25.31 ± 0.62^{c}	17.42 ± 0.56^{b}	17.09 ± 0.62^{a}	13.66 ± 2.16^{a}	**19.04** ± 1.16
	Middle	16.50 ± 0.97^{b}	20.57 ± 0.93^{b}	17.46 ± 0.36^{b}	23.87 ± 1.89^{b}	24.48 ± 2.22	22.32 ± 1.84^{b}	13.38 ± 0.44^{a}	17.77 ± 1.97^{ab}	21.15 ± 1.23	37.00 ± 0.33^{c}	23.39 ± 1.55^{b}	15.01 ± 0.94^{a}	21.63 ± 0.42^{b}	16.89 ± 0.55^{b}	**20.81** ± 1.49
	Distal	15.15 ± 1.15^{a}	20.84 ± 1.12^{b}	10.86 ± 0.26^{a}	25.59 ± 3.09^{c}	24.57 ± 2.86	20.09 ± 1.50^{a}	14.60 ± 1.44^{b}	17.04 ± 1.19^{a}	20.88 ± 2.14	26.87 ± 0.23^{b}	21.13 ± 0.60^{a}	14.09 ± 0.68^{a}	23.25 ± 0.62^{c}	14.18 ± 0.89^{a}	**19.22** ± 1.27
	Mean ± SE	**16.47** ± 0.43^{B}	**20.10** ± 0.50^{C}	**13.35** ± 0.52^{A}	**23.66** ± 0.94^{EF}	**24.49** ± 1.00^{F}	**21.31** ± 0.69^{CDE}	**13.45** ± 0.44^{A}	**17.71** ± 0.69^{B}	**20.99** ± 0.59^{CD}	**29.82** ± 1.28^{G}	**23.28** ± 0.49^{DEF}	**15.50** ± 0.37^{AB}	**20.66** ± 0.50^{C}	**14.91** ± 0.58^{A}	

Species→ Region ↓	Hair Parts	Cattle	Buffalo	Sheep	Goat	Horse	Domestic Pig	Cat	Dog	Spotted Deer	Sambar	Nilgai	Hanuman Langur	Tiger	Leopard	Pooled Mean± SE
Back	**Proximal**	17.09 ± 0.84^{b}	15.33 ± 0.18^{a}	11.35 ± 0.21^{b}	22.04 ± 1.82^{a}	21.79 ± 0.44^{a}	25.40 ± 2.43^{b}	17.03 ± 1.33^{b}	20.18 ± 1.88^{c}	22.21 ± 0.56^{a}	24.67 ± 0.21^{a}	23.29 ± 1.15^{a}	17.77 ± 0.80^{b}	17.61 ± 0.48^{a}	13.89 ± 1.99^{a}	**19.26** ± 1.08
	Middle	15.25 ± 0.46^{a}	18.67 ± 0.16^{b}	18.03 ± 0.54^{c}	24.45 ± 1.62^{b}	24.23 ± 1.14^{b}	23.61 ± 2.25^{a}	15.06 ± 0.21^{a}	19.51 ± 1.40^{b}	23.94 ± 1.28^{b}	27.70 ± 0.10^{b}	24.26 ± 0.95^{b}	16.39 ± 0.99^{a}	21.65 ± 0.40^{b}	17.41 ± 0.71^{b}	**20.73** ± 1.04
	Distal	14.83 ± 0.80^{a}	22.42 ± 0.41^{c}	10.26 ± 0.12^{a}	24.55 ± 1.84^{b}	26.45 ± 0.71^{c}	24.33 ± 2.77^{a}	17.59 ± 0.71^{b}	18.29 ± 1.37^{a}	21.34 ± 1.84^{a}	29.46 ± 0.10^{c}	23.29 ± 2.20^{a}	15.46 ± 1.46^{a}	25.28 ± 0.80^{c}	16.59 ± 0.30^{b}	**20.72** ± 1.38
	Mean ± SE	**15.72** ± 0.32^{B}	**18.81** ± 0.51^{C}	**13.21** ± 0.60^{A}	**23.68** ± 0.70^{E}	**24.15** ± 0.45^{E}	**24.45** ± 0.96^{E}	**16.56** ± 0.38^{B}	**19.33**± 0.62^{C}	**22.49** ± 0.54^{DE}	**27.27** ± 0.50^{F}	**23.61** ± 0.59^{E}	**16.54** ± 0.46^{B}	**21.52** ± 0.58^{D}	**15.96** ± 0.54^{B}	
Abdomen	**Proximal**	19.88 ± 1.01^{b}	16.89 ± 1.39^{b}	11.10± 0.16^{a}	20.38 ± 1.03^{a}	19.47 ± 0.77^{a}	21.00 ± 3.04^{b}	13.47 ± 2.05^{a}	16.76± 1.62^{a}	25.85 ± 1.50^{b}	35.09 ± 0.56^{b}	22.61 ± 0.87^{c}	18.36 ± 1.12^{c}	18.19 ± 0.23^{b}	17.50 ± 1.86	**19.75 ±1.47**
	Middle	18.93 ± 1.12^{b}	18.68 ± 1.12^{b}	18.29± 0.26^{b}	23.29 ± 1.81^{b}	22.90 ± 1.31^{b}	20.90 ± 2.84^{a}	14.83 ± 0.28^{b}	15.91± 1.38^{a}	25.67 ± 1.12^{b}	49.46 ± 0.67^{c}	20.60 ± 0.53^{b}	17.53 ± 1.06^{b}	19.32 ± 0.43^{c}	17.52 ± 0.97	**21.70 ± 2.19**
	Distal	13.33 ± 0.84^{a}	20.10 ± 1.65^{c}	10.37± 0.16^{a}	22.90 ± 1.96^{b}	23.30 ± 2.17^{b}	20.47 ± 2.12^{a}	17.15 ± 0.60^{c}	18.10± 2.50^{b}	24.04 ± 1.53^{a}	25.88 ± 0.18^{a}	18.00 ± 0.75^{a}	15.04 ± 1.06^{a}	16.55 ± 0.13^{a}	17.29 ± 0.64	**18.75 ± 1.12**
	Mean ± SE	**17.38 ± 0.63^{BC}**	**18.56 ± 0.58^{CD}**	**13.25± 0.62^{A}**	**22.19 ± 0.67^{E}**	**21.89 ± 0.66^{E}**	**20.79 ± 1.03^{DE}**	**15.15 ± 0.54^{AB}**	**16.92± 0.74^{BC}**	**25.18 ± 0.55^{F}**	**36.81 ± 2.44^{G}**	**20.40 ± 0.43^{DE}**	**16.98 ± 0.48^{BC}**	**18.02 ± 0.22^{C}**	**17.44 ± 0.49^{C}**	

Species→ Region ↓	Hair Parts	Cattle	Buffalo	Sheep	Goat	Horse	Domestic Pig	Cat	Dog	Spotted Deer	Sambar	Nilgai	Hanuman Langur	Tiger	Leopard	Pooled Mean± SE
Thigh	**Proximal**	19.33 ± 1.57^{c}	21.03 ± 0.81^{a}	11.28± 0.19^{a}	22.23 ± 1.49^{a}	19.31 ± 1.07^{a}	18.02 ± 2.84^{a}	12.17 ± 0.85^{a}	15.49± 1.35^{a}	28.36 ± 1.55^{b}	31.92 ± 0.06^{b}	23.56 ± 1.22^{b}	17.56 ± 1.07^{b}	17.53 ± 0.31^{b}	16.09 ± 1.99^{a}	**19.56 ± 1.46**
	Middle	17.29 ± 1.67^{b}	21.74 ± 0.81^{a}	17.95± 0.47^{b}	26.25 ± 1.56^{b}	20.04 ± 0.54^{a}	20.00 ± 2.17^{b}	13.36 ± 0.24^{b}	18.27± 1.26^{b}	28.50 ± 1.04^{b}	29.59 ± 0.12^{a}	23.37 ± 1.01^{b}	17.21 ± 0.70^{b}	18.38 ± 0.67^{b}	16.88 ± 1.85^{ab}	**20.63 ± 1.22**
	Distal	15.25 ± 1.38^{a}	28.07 ± 1.98^{b}	10.2± 0.17^{a}	27.25 ± 2.53^{b}	22.58 ± 1.49^{b}	20.46 ± 1.52^{b}	16.06 ± 1.01^{c}	21.00± 0.92^{c}	26.23 ± 0.98^{a}	35.67 ± 4.39^{c}	19.67 ± 1.09^{a}	15.39 ± 0.15^{a}	15.95 ± 1.29^{a}	17.53 ± 0.47^{b}	**20.81 ± 1.72**
	Mean ± SE	**17.29 ± 0.66^{BC}**	**23.61 ± 0.74^{FG}**	**13.16± 0.60^{A}**	**25.24 ± 0.82^{G}**	**20.64± 0.49^{DE}**	**19.49 ± 0.88^{CD}**	**13.86 ± 0.41^{A}**	**18.25± 0.60^{BC}**	**27.69 ± 0.50^{H}**	**32.39 ± 1.42^{I}**	**22.20 ± 0.53^{EF}**	**16.72 ± 0.33^{B}**	**17.29 ± 0.37^{BC}**	**16.83 ± 0.62^{B}**	
Tail	**Proximal**	18.55 ± 1.35	19.75 ± 0.58	11.46± 0.15	22.14 ± 1.69	35.09 ± 3.45	15.60 ± 1.13	13.40 ± 0.66	14.35 ± 1.19	20.44 ± 0.64	28.10 ± 0.15	27.00 ± 1.24	17.87 ± 0.35	19.20 ± 0.31	20.39 ± 1.10	**20.24 ± 1.63**
	Middle	16.91 ± 0.98	19.77 ± 0.52	17.88± 0.34	22.46 ± 2.08	37.68 ± 3.03	19.61 ± 1.94	15.15 ± 0.47	8.76 ± 0.57	24.23 ± 1.88	27.93 ± 0.07	24.88 ± 0.59	17.48 ± 0.57	22.37 ± 0.18	21.27 ± 1.05	**21.17 ± 1.72**
	Distal	16.81 ± 1.18	22.33 ± 0.84	10.18± 0.17	26.78 ± 2.92	35.31 ± 3.41	20.20 ± 2.26	18.04 ± 0.34	21.07 ± 2.12	20.18 ± 1.25	24.24 ± 0.12	20.63 ± 1.16	15.34 ± 0.84	18.36 ± 0.91	18.34 ± 0.59	**20.56 ± 1.50**
	Mean ± SE	**17.42 ± 0.47^{CDE}**	**20.62 ± 0.33^{FG}**	**13.17± 0.58^{A}**	**23.79 ± 0.95^{HI}**	**36.03^{K} ±1.28**	**18.47 ± 0.79^{DEF}**	**15.53 ± 0.38^{ABC}**	**14.73 ± 1.03^{AB}**	**21.62 ± 0.61^{GH}**	**26.75 ± 0.45^{J}**	**24.17 ± 0.60^{I}**	**16.90 ± 0.30^{BCD}**	**19.98 ± 0.37^{EFG}**	**20.00 ± 0.42^{EFG}**	

Means bearing same superscript for different region (for hair part) in a column do not differ significantly/Means bearing same superscript for different species (hair part of different region) in a row do not differ significantly/Head Region-CD for Species is 2.08 and CD for hair part is 0.96/Neck Region-CD for Species is 2.38 and CD for hair part is 1.10/Back Region-CD for Species is 2.05 and CD for hair part is 0.95/Abdomen Region-CD for Species is 2.27 and CD for hair part is 1.05/Thigh Region-CD for Species is 2.24 and CD for hair part is 1.04/Tail Region-CD for Species is 2.37

Significantly lowest cuticular scale width was noted in Sheep (10.37 ± 0.08 μm) followed by Hanuman Langur (15.25 ± 0.40 μm), while mean cuticular scale width was significantly higher in Sambar (27.94 ± 1.27 μm) followed by Horse (25.75 ± 1.16 μm) as compared to all the species considered for the present study. The cuticular scale width was found minimum in the distal part of tail region of Sheep (10.18 ± 0.17μm), while maximum cuticular scale width was noted at distal hair part at thigh region of Sambar (27.94 ± 1.27μm) among all the species and body regions.

No particular trend was observed regarding cuticular scale width of hair at proximal, middle and distal parts of the hair from all the six body regions amongst all the species.

The mean and SE for cuticular scale width at proximal, middle and distal parts of the hair from all the six body regions and all the species under study were given in Table 4.13.

The cuticular scale width of hair from head region was found significantly lower at the proximal part of hair in almost all the species except Buffalo, Sheep, Nilgai, Hanuman Langur and Sambar. The higher cuticular scale width was noted at the middle part of the hair in Goat, Cattle, Domestic Pig, Dog, Sheep and Tiger at the distal part in Buffalo, Horse, Cat, Spotted Deer and Leopard, while in Nilgai and Hanuman Langur at the proximal part (Table 4.13).

In the abdomen region the lowest cuticular scale width among three parts of hair in most of the species was observed at the distal part of hair except Buffalo, Goat, Horse, Dog and Cat. In Buffalo, Horse, Goat and Dog was noted at proximal part, while in Cat, it was observed at middle part of the hair (Table 4.13).

In the tail region, amongst all the wild animals the lowest cuticular scale width was observed at the distal part of the hair, while in domestic animals the lowest width was noted at the proximal part except Cattle, Dog and Sheep. Table 4.13.

The minimum cuticular scale width was noted in Sheep and maximum in Sambar at all body regions except in head region. At head region, lowest cuticular scale width was present in Cat while, highest width was noted in Sambar Table 4.13. In all the species with respect scale width in all the body regions and various parts of hair the distal hair part in Sheep had lowest scale width (Table 4.13).

4.5.3. Number of cuticular scales per mm length of hair

The mean and SE for number of scales per mm length of hair at proximal part of hair from different body regions in different species are given in Table 4.14.

Table 4.14. Mean and SE for number of cuticular scales per mm length of hair at proximal part of hair from various regions of body in different species.

Species→ Region ↓	Cattle	Buffalo	Sheep	Goat	Horse	Domes-tic Pig	Cat	Dog	Spotted Deer	Sambar	Nilgai	Hanu-man Langur	Tiger	Leop-ard	Pooled Mean± SE
Head	272.30± 12.97	274.39 ± 19.86	175.46± 4.54	230.51± 9.76	240.39± 10.29	371.14± 11.88	168.18± 3.04	227.42± 23.19	223.06± 9.82	217.27± 6.22	268.56± 13.62	226.79± 10.84	227.57± 6.00	251.72± 9.30	**241.05±** 12.59
Neck	242.94± 26.43	262.81± 5.74	183.84± 5.63	217.08± 12.49	209.17± 8.12	364.48± 8.66	174.03± 3.93	245.00± 17.40	233.84± 2.11	204.62± 3.96	279.52± 20.07	217.30± 10.41	235.22± 6.74	221.63± 23.18	**235.11±** 12.04
Back	267.53± 17.40	272.15± 4.31	181.50± 5.97	255.05± 9.34	202.66± 7.65	410.34± 18.83	179.48± 9.71	234.19± 23.28	241.54± 3.66	233.33± 4.63	277.58± 13.89	208.54 ± 14.05	265.21± 3.63	215.11± 24.52	**246.01±** 14.77
Abdomen	240.62± 11.84	253.59± 10.47	178.72± 4.44	248.08± 11.30	221.15± 10.03	429.27± 31.83	202.51± 22.34	247.29± 13.03	230.59± 2.84	159.84± 5.45	272.67± 14.94	198.36± 10.82	253.15± 18.16	267.08± 23.54	**243.07±** 16.25
Thigh	267.65± 9.90	255.11± 10.42	184.41± 8.32	231.45± 21.08	210.04± 11.05	357.08± 19.85	177.40± 6.41	260.69± 17.50	226.39± 10.60	191.08± 1.48	278.75± 12.23	202.83± 7.70	308.77± 9.43	257.40± 7.69	**243.50±** 13.06
Tail	252.74± 41.90	235.62± 80.26	191.24± 6.42	240.69± 11.71	224.12± 8.24	286.39± 25.36	176.59± 7.49	249.89± 19.81	244.79± 20.14	208.87± 3.18	314.31± 15.37	251.89± 11.99	238.43± 11.71	248.52± 8.75	**240.29±** 8.98
Pooled Mean±SE	**257.30±** 8.93 [E]	**258.94±** 4.72 [E]	**182.53±** 9.82 [A]	**237.14±** 5.42 [D]	**217.92±** 4.10 [BC]	**369.78±** 11.01 [F]	**179.70±** 4.51 [A]	**244.08±** 7.54 [DE]	**233.37±** 13.62 [CD]	**202.50±** 5.97 [B]	**281.90±** 6.29 [F]	**217.62±** 5.19 [BC]	**254.73±** 6.04 [E]	**243.58±** 7.46 [DE]	

Means bearing same superscript in a column do not differ significantly

CD for Species is 16.63

Table 4.15. Means and SE for number of cuticular scales per mm length of hair at middle part of hair from various regions of body in different species.

Species→ Region ↓	Cattle	Buffalo	Sheep	Goat	Horse	Domestic Pig	Cat	Dog	Spotted Deer	Sambar	Nilgai	Hanuman Langur	Tiger	Leopard	Pooled Mean±SE
Head	277.88 ± 18.73	291.26± 12.22	162.58± 6.55	273.80± 20.91	237.05± 4.57	430.12± 16.59	278.34 ±14.17	251.97± 9.92	229.38± 11.84	168.53± 5.76	299.62± 12.06	240.41 ±6.59	237.42± 8.34	260.83± 10.91	**259.94±** 16.38
Neck	252.98 ± 7.38	300.23± 12.33	171.76± 9.81	241.33± 21.18	216.72± 8.59	451.28± 12.72	261.55 ±23.18	204.39± 14.17	240.61± 6.78	215.36± 4.19	271.54± 4.11	221.92 ±8.02	258.74± 7.00	230.65± 7.52	**252.79±** 16.82
Back	288.93 ± 15.13	284.15± 14.86	185.75± 2.09	259.49± 12.00	242.66± 5.78	396.92± 11.60	269.85 ±14.80	203.50± 8.19	213.54± 7.13	206.22± 2.92	301.00± 19.21	245.08 ±9.95	295.09± 4.21	268.47± 15.43	**261.48±** 13.83
Abdomen	261.00 ± 16.75	284.63± 12.07	184.73± 4.24	256.18± 14.36	228.14± 7.66	421.02± 15.38	301.77 ±24.00	229.80± 13.02	229.45± 5.00	201.69± 3.28	321.92± 9.72	224.13 ±9.60	238.63± 6.94	244.99± 13.69	**259.15±** 15.32
Thigh	295.72 ± 13.90	262.96± 13.30	185.53± 5.25	235.25± 15.20	247.26± 8.20	429.11± 34.83	288.51 ±28.74	230.41± 17.64	229.24± 5.64	206.45± 7.24	280.97± 3.98	238.31 ±12.02	350.08± 8.41	251.94± 12.02	**266.55±** 16.00
Tail	246.20 ± 18.74	285.76± 10.62	177.37± 3.14	267.11± 17.19	231.16± 12.82	407.24± 25.31	333.55 ±10.33	203.11± 28.73	203.91± 17.52	208.86± 6.22	326.70± 18.67	265.05 ±12.96	289.57± 8.33	258.44± 8.20	**264.57±** 15.92
Pooled Mean± SE	**270.45** ± 6.66 ^{GH}	**284.83±** 5.14 ^{HI}	**177.95±** 2.61 ^{A}	**255.53±** 6.91 ^{FG}	**233.83±** 3.58 ^{CD}	**422.61 ±** 8.49 ^{K}	**288.93** ^{IJ} ±8.65	**220.53±** 7.03 ^{C}	**224.35±** 4.30 ^{CD}	**201.19±** 4.45 ^{B}	**300.29±** 5.90 ^{J}	**239.15** ^{DE} ±4.52	**278.25±** 7.18 ^{HI}	**252.55±** 4.86	

Means bearing same superscript in a column do not differ significantly

CD for Species is 15.75

The lowest number of cuticular scales per mm length of hair at the proximal part of the hair was found in Cat (177.70 ± 4.51) and Sheep (182.53 ± 9.82) where as highest number was noted in Domestic Pig (369.78 ± 11.01) as compared to all the species used for present work.

The middle part of Sheep hair showed minimum number of cuticular scales (177.95 ± 2.61) however, highest number of scales were noted in Domestic Pig (422.61 ± 8.49) during the present work. The mean and SE for number of cuticular scales per mm length of hair at middle part of hair is given in table 4.15.

Table 4.16. Means and SE for number of scales per mm length of hair at distal part of hair from various regions of body in different species.

Species→ Region ↓	Cattle	Buffalo	Sheep	Goat	Horse	Domestic Pig	Cat	Dog	Spotted Deer	Sambar	Nilgai	Hanuman Langur	Tiger	Leopard	Pooled Mean ±SE
Head	299.81± 13.78^{ab}	310.35 ± 16.00^{b}	158.87± 7.86^{a}	301.59± 14.14^{b}	252.18 ± 12.04^{a}	426.28 ± 7.71^{b}	389.49± 25.89^{a}	312.99 ± 11.79^{c}	246.13 ± 8.07	171.38 ± 1.76^{a}	359.61± 14.23^{a}	265.38 ± 9.68^{c}	363.85 ± 23.87^{b}	368.26 ± 25.54^{b}	**301.87** ± 21.12^{ab}
Neck	310.19± 27.27^{bc}	308.40 ± 17.45^{b}	166.36± 10.28^{ab}	264.28± 15.75^{a}	251.71 ± 17.35^{a}	399.15 ± 59.73^{a}	392.00± 21.28^{a}	256.65± 22.71^{a}	245.18 ± 9.89	264.21 ± 6.21^{c}	346.09± 19.20^{a}	224.52 ± 15.80^{a}	319.60 ± 14.02^{a}	334.19 ± 32.48^{a}	**291.61** ± 16.67^{a}
Back	321.66± 5.43^{c}	355.33 ± 7.24^{c}	176.89± 3.60^{b}	303.83± 14.72^{b}	276.24 ± 7.46^{b}	435.00 ± 14.74^{bc}	433.41± 6.90^{c}	285.93 ± 16.98^{b}	250.46 ± 10.08	279.48 ± 8.21^{d}	379.22± 23.52^{b}	255.85 ± 14.71^{bc}	387.52 ± 3.99^{d}	371.93 ±19.44	**322.34** ± 19.22^{c}
Abdomen	294.12± 11.68 a	304.05 ± 11.61^{b}	172.94± 5.88^{b}	292.55± 13.20^{b}	256.24 ± 15.22^{a}	445.70 ± 14.21^{d}	419.67± 13.48^{b}	289.95 ± 10.59^{b}	247.45 ± 9.12	291.03 ± 3.07^{d}	388.34± 30.59^{b}	251.98 ± 9.00^{b}	378.51 ± 15.45^{cd}	343.73 ± 18.35^{a}	**312.59** ± 19.25^{bc}
Thigh	322.51± 10.25^{c}	278.85 ± 9.15^{a}	174.56± 11.69	272.37± 18.89	261.59 ± 10.20^{a}	443.96 ± 41.71^{cd}	409.85± 16.27^{b}	267.57 ± 19.32^{a}	244.68 ± 10.23	308.49 ± 12.33^{e}	357.69± 13.36^{a}	256.43 ± 11.25^{bc}	368.77 ± 20.99^{bc}	359.55 ± 29.48^{b}	**309.06** ± 18.66^{bc}
Tail	310.98± 10.77^{bc}	306.84 ± 7.94^{b}	180.86± 7.33^{b}	292.88± 23.87^{b}	283.54 ± 17.42^{b}	456.61 ±27.12^{d}	439.24± 8.13^{c}	268.80 ± 17.46^{a}	253.57 ± 14.07	186.36 ± 4.91^{b}	351.13± 18.19^{a}	264.49 ± 14.18^{bc}	420.47 ± 15.83^{e}	343.68 ± 20.72^{a}	**311.39** ± 21.80^{bc}
Pooled Mean± SE	**309.88± 5.88^{D}**	**310.64** ± 5.99^{E}	**171.75**± 3.34^{A}	**287.92**± 6.94^{C}	**263.59** ± **5.61^{BC}**	**434.45** ± 12.84^{I}	**413.94**± 7.09^{H}	**280.31** ± 7.14^{C}	**247.91** ± 3.97^{B}	**250.16** ± 13.13^{B}	**363.68**± 8.23^{G}	**253.11** ± 5.33^{B}	**373.12** ± 8.17^{G}	**353.56** ± 9.70^{FG}	

Means bearing same superscript in a column do not differ significantly
CD for Species is 25.37 and CD for Region is 16.60

The mean number of cuticular scales per mm length of hair at distal part of hair was lowest in Sheep (171.75 ± 3.34), while it was highest in Domestic Pig (434.45 ± 12.84) as compared with all the species studied. The average number of cuticular scales per mm length of hair at distal part of hair in various regions of body of different species is given in table 4.16.

Table 4.17. Means and SE for number of cuticular scales per mm length of hair at different parts of hair from various regions of body in different species.

Species→ Region ↓	Hair Parts	Cattle	Buffalo	Sheep	Goat	Horse	Domestic Pig	Cat	Dog	Spotted Deer	Sambar	Nilgai	Hanuman Langur	Tiger	Leopard	Pooled Mean± SE
Head	**Proximal**	272.30 ± 12.97^{a}	274.39 ± 19.86^{a}	175.46± 4.54^{b}	230.51 ± 9.76^{a}	240.39 ± 10.29^{a}	371.14 ± 11.48^{a}	168.18 ± 3.04^{a}	227.42 ± 23.19^{a}	223.06 ± 9.82^{a}	217.27± 6.22^{b}	268.56± 13.62^{a}	226.79 ± 10.84^{a}	227.57 ± 6.00^{a}	251.72 ± 9.30^{a}	**241.05±** 12.59^{a}
	Middle	277.88 ± 18.373^{a}	291.26 ± 12.22^{b}	162.58± 6.55^{a}	273.80 ± 20.91^{b}	237.05 ± 4.57^{a}	430.12 ± 16.59^{b}	278.34 ± 14.17^{b}	251.97 ± 9.92^{b}	229.38 ± 11.84^{a}	168.53± 5.76^{a}	299.62± 12.06^{b}	240.41 ± 6.59^{b}	237.42 ± 8.34^{a}	260.83 ± 10.91^{a}	**259.94±** 16.38^{b}
	Distal	299.81 ± 13.78^{b}	310.35 ± 16.00^{c}	158.87± 7.86^{a}	301.59 ± 14.14^{c}	252.18 ± 12.14^{b}	426.28 ± 7.71^{b}	389.49 ± 25.89^{c}	312.99 ± 11.79^{c}	246.13 ± 8.07^{b}	171.38± 1.76^{a}	359.61 ± 14.23^{c}	265.38 ± 9.68^{c}	363.85 ± 23.87^{b}	368.26 ± 25.54^{b}	**301.87±** 20.12^{c}
	Mean ±SE	**283.33 ±** 8.82	**292.00 ±** 9.54^{DE}	**165.64±** 3.91^{A}	**268.64 ±** 11.05^{C}	**243.21 ±** 5.42^{BC}	**409.18 ±** 9.45^{E}	**278.67 ±** 23.80^{DC}	**264.13 ±** 12.33^{C}	**232.86 ±** 5.93^{B}	**185.73±** 8.65^{A}	**309.26 ±** 11.67^{E}	**244.19 ±** 6.32^{BC}	**276.28 ±** 17.11^{DC}	**293.60 ±** 15.78^{DE}	
Neck	**Proximal**	242.94 ± 26.43^{a}	262.81 ± 5.74^{a}	183.84± 5.63^{b}	217.08 ± 12.49^{a}	209.17 ± 8.12^{a}	364.48 ± 8.66^{a}	174.03 ± 3.93^{a}	245.00 ± 17.40^{b}	233.84 ± 2.11	204.62± 3.96^{a}	279.52 ± 20.07^{a}	217.30 ± 10.41	235.22 ± 6.74^{a}	221.63 ± 23.18^{a}	**235.11±** 12.04^{a}
	Middle	252.98 ± 7.38^{a}	300.23 ± 12.33^{b}	171.76± 9.81^{ab}	241.33 ± 21.18^{b}	216.72 ± 8.59^{a}	451.28 ± 12.72^{c}	261.55 ± 23.18^{b}	204.39 ± 14.17^{a}	240.61 ± 6.78	215.36± 4.19^{a}	271.54 ± 4.11^{a}	221.92 ± 8.02	258.74 ± 7.00^{b}	230.65 ± 7.52^{a}	**252.79±** 16.82^{b}
	Distal	310.19 ± 27.27^{b}	308.40 ± 17.45^{b}	166.36± 10.28^{a}	264.28 ± 15.75^{c}	251.71 ± 17.35^{b}	399.15 ± 59.73^{b}	392.00 ± 21.28^{c}	256.65 ± 22.71^{b}	245.18 ± 9.89	264.21± 6.21^{b}	346.09 ± 19.20^{b}	224.52 ± 15.80	319.60 ± 14.02^{c}	334.19 ± 32.48^{b}	**291.61±** 16.67^{c}
	Mean ±SE	**268.70 ±** 14.08^{DE}	**290.48 ±** 8.44^{EF}	**173.99±** 5.10^{A}	**240.90 ±** 10.27^{BCD}	**225.87 ±** 7.96^{B}	**404.97 ±** 21.16^{G}	**275.86±** 23.88^{EF}	**235.35 ±** 11.38^{BC}	**239.88 ±** 3.98^{BC}	**228.06±** 9.80^{B}	**299.05 ±** 11.96^{F}	**221.25 ±** 6.47^{B}	**271.19 ±** 10.14^{EF}	**262.15** ± 17.75^{CDE}	
Back	**Proximal**	267.53 ± 17.40^{a}	272.15 ± 4.31^{a}	181.50± 5.97	255.05 ± 9.34^{a}	202.66 ± 7.65^{a}	410.34 ± 18.83^{b}	179.48 ± 9.71^{a}	234.19 ± 23.28^{b}	241.54 ± 3.66^{b}	233.33± 4.63^{b}	277.58 ± 13.89^{a}	208.54 ± 14.05^{a}	265.21 ± 3.63^{a}	215.11 ± 24.52^{a}	**246.01±** 14.77^{a}
	Middle	288.93 ± 15.13^{b}	284.15 ± 14.86^{b}	185.75± 2.09	259.49 ± 12.00^{a}	242.66 ± 5.78^{b}	396.92 ± 11.60^{a}	269.85 ± 14.80^{b}	203.50 ± 8.19^{a}	213.54 ± 7.13^{a}	206.22± 2.93^{a}	301.00 ± 19.21^{b}	245.08 ± 9.95^{b}	295.09 ± 4.21^{b}	268.47 ± 15.43^{b}	**261.48±** 13.83^{b}
	Distal	321.66 ± 5.43^{c}	355.33 ± 7.24^{c}	176.89± 3.60	303.83 ± 14.72^{b}	276.24 ± 7.46^{c}	435.00 ± 14.74^{c}	433.41 ± 6.90^{c}	285.93 ± 16.98^{c}	250.46 ± 10.08^{b}	279.48± 8.21^{c}	379.22 ± 23.52^{c}	255.85 ± 14.71^{c}	387.52 ± 3.99^{c}	371.93 ± 19.44^{c}	**322.34±** 19.22^{c}
	Mean ±SE	**292.71±** 9.18^{CD}	**303.88 ±** 10.39^{DE}	**181.38±** 2.44^{A}	**272.79 ±** 8.51^{C}	**240.52 ±** 8.23^{B}	**414.08 ±** 9.16^{F}	**294.25 ±** 26.17^{D}	**241.20 ±** 12.49^{B}	**235.18 ±** 5.55^{B}	**239.68±** 11.43^{B}	**319.27 ±** 14.85^{E}	**236.49 ±** 8.62^{B}	**315.94 ±** 12.81^{E}	**285.17 ±** 19.20^{CD}	

Species→ Region ↓	Hair Parts	Cattle	Buffalo	Sheep	Goat	Horse	Domestic Pig	Cat	Dog	Spotted Deer	Sambar	Nilgai	Hanu-man Langur	Tiger	Leopard	Pooled Mean± SE
Abdomen	Proximal	240.62± 11.84^{a}	253.59 ± 10.47^{a}	178.72± 4.44^{ab}	248.08 ± 11.30^{a}	221.15 ± 10.03^{a}	429.27 ± 31.83^{a}	202.51 ± 22.34^{a}	247.29 ± 13.03^{b}	230.59 ± 2.84^{a}	159.84± 5.45^{a}	272.67± 14.94^{a}	198.36 ± 10.82^{a}	253.15± 18.16^{b}	267.08 ± 23.54^{b}	**243.07±** 16.25
	Middle	261.00 ± 16.75^{b}	284.63 ± 12.07^{b}	184.73^{b}± 4.24^{b}	256.18 ± 14.36^{b}	228.14 ± 7.66^{a}	421.02 ± 15.38^{a}	301.77 ± 24.00^{b}	229.80 ± 13.02^{a}	229.45 ± 5.00^{a}	201.69± 3.28^{b}	321.92± 9.72^{b}	224.13 ± 9.60^{b}	238.63± 6.94^{a}	244.99 ± 13.69^{a}	**259.15±** 15.32
	Distal	294.12 ± 11.68^{c}	304.05 ± 11.61^{c}	172.94 ± 5.88^{a}	292.55 ± 13.20^{c}	256.24 ± 15.22^{b}	445.70 ± 14.21^{b}	419.67 ± 13.48^{c}	289.95 ± 10.59^{c}	247.45 ± 9.12^{b}	291.03± 3.07^{c}	388.34± 30.59^{c}	251.98 ± 9.00^{c}	378.51± 15.45^{c}	343.73± 18.35^{c}	**312.59±** 19.25^{c}
	Mean ±SE	**265.25 ±** 9.12^{DE}	**280.76** ± 7.98 ^{EF}	**178.80 ±** 2.91^{A}	**265.60 ±** 8.47^{DE}	**235.18 ±** 7.20^{BC}	**431.99 ±** 12.19^{H}	**307.98±** 24.22^{FG}	**255.68 ±** 9.04^{CD}	**235.83 ±** 3.92^{BC}	**217.52±** 19.56^{B}	**327.64±** 15.97^{G}	**224.82 ±** 7.53^{B}	**290.10±** 17.10^{F}	**285.27±** 14.53^{EF}	
Thigh	Proximal	267.65 ± 9.90^{a}	255.11 ± 10.42^{a}	184.41 ± 8.32^{b}	231.45 ± 21.08^{a}	210.04 ± 11.05^{a}	357.08 ± 19.85^{a}	177.40 ± 6.41^{a}	260.69 ± 17.50^{b}	226.39 ± 10.60^{a}	191.08± 1.48^{a}	278.75± 12.23^{a}	202.83 ± 7.70^{a}	308.77± 9.43^{a}	257.40 ± 7.69^{a}	**243.50±** 13.06^{a}
	Middle	295.72 ± 13.90^{b}	262.96 ± 13.30^{a}	185.53 ± 5.25^{b}	235.25 ± 15.20^{a}	247.26 ± 8.20^{b}	429.11 ± 34.83^{b}	288.51± 28.74^{b}	230.41 ± 17.64^{a}	229.24 ± 5.64^{a}	206.45± 7.24^{b}	280.97± 3.98^{a}	238.31 ± 12.02^{b}	350.08± 8.41^{b}	251.94 ± 12.02^{a}	**266.55±** 16.00^{b}
	Distal	322.51 ± 10.25^{c}	278.85 ± 9.15^{b}	174.56 ± 11.69^{a}	272.37 ± 18.89^{b}	261.59 ± 10.20^{c}	443.96 ± 41.71^{c}	409.85± 16.27^{c}	267.57 ± 19.32^{b}	244.68 ± 10.23^{b}	308.49± 12.33^{c}	357.69± 13.36^{b}	256.43 ± 11.25^{c}	368.77± 20.99^{c}	359.55 ± 29.48^{b}	**318.10±** 21.01^{c}
	Mean ±SE	**295.29 ± 8.27**^{E}	**265.64** ± 6.48 ^{CD}	**181.50±** 4.93^{A}	**246.36 ±** 11.01^{BC}	**239.63 ±** 7.52^{BC}	**410.05 ±** 20.32^{G}	**291.92±** 25.32^{D}	**252.89 ±** 10.60^{BC}	**233.43 ±** 5.31^{B}	**235.34±** 19.33^{B}	**305.80±** 10.63^{E}	**232.53 ±** 7.85^{B}	**342.54±** 9.79^{F}	**289.63 ±** 15.79^{DE}	
Tail	Proximal	252.74 ± 41.90^{a}	235.62 ± 8.26^{a}	191.24 ± 6.42^{b}	240.69 ± 11.71^{a}	224.12 ± 8.24^{a}	286.39 ± 25.36^{a}	176.59 ±7.49^{a}	249.89 ± 19.81^{b}	244.79 ± 20.14^{b}	208.87± 3.18^{b}	314.31± 15.37^{a}	251.89 ± 11.99^{a}	238.43± 11.71^{a}	248.52 ± 8.75^{a}	**240.29±** 8.98^{a}
	Middle	246.20 ± 18.74^{a}	285.76 ± 10.62^{b}	177.37 ± 3.14^{a}	267.11 ± 17.79^{b}	231.16 ± 12.82^{a}	407.24 ± 25.31^{b}	333.55± 10.33^{b}	203.11 ± 28.73^{a}	203.91 ± 17.52^{a}	208.86± 6.22^{b}	326.70± 18.67^{b}	265.05 ± 12.96^{b}	289.57± 8.33^{b}	258.44 ± 8.20^{a}	**264.57±** 15.92^{b}
	Distal	310.98 ± 10.77^{b}	306.84 ± 7.94^{c}	180.86 ± 7.33^{ab}	292.88 ± 23.87^{c}	283.54 ± 17.42^{b}	456.61 ± 27.12^{c}	439.24^{c} ±8.13	268.80 ± 17.46^{c}	253.57± 14.07^{b}	186.36± 4.91^{a}	351.13± 18.19^{c}	264.49 ± 14.18^{b}	420.47± 15.83^{c}	343.68 ± 20.72^{b}	**311.39±** 21.80^{c}
	Mean ±SE	**269.97±** 16.36^{DEF}	**276.07** ± 8.74 ^{EF}	**183.16 ±** 3.51^{A}	**266.89** ± 11.27 ^{CDE}	**246.27 ±** 9.69^{BCD}	**383.41 ±** 22.33^{H}	**316.46±** 26.60^{G}	**240.60**^{BC} ±13.93	**234.09 ±** 10.81^{B}	**201.36±** 5.12^{A}	**330.71±** 10.18^{G}	**260.47±** 7..24^{BCD}	**316.16±** 19.76^{G}	**283.55 ±** 12.79^{F}	

Means bearing same superscript for different region (for hair part) in a column do not differ significantly/Means bearing same superscript for different species (hair part of different region) in a row do not differ significantly/Head Region-CD for Species is 22.17 and CD for hair part is 10.25/Neck Region-CD for Species is 28.95 and CD for hair part is 13.39/Back Region-CD for Species is 20.84 and CD for hair part is 9.64/Abdoman Region-CD for Species is 23.65 and CD for hair part is 10.94/Thigh Region-CD for Species is 26.32 and CD for hair part is 12.18/ Tail Region- CD for Species is 37.34 and CD for hair part is 12.65

From the above observations it can be observed that the number of cuticular scales were lowest in Sheep and highest in Domestic Pig at all the three parts of hair amongst all the body regions and species.

The mean and SE for number of cuticular scales per mm length of hair at different parts of hair in various body regions of different species are depicted in table 4.17. The lowest number of cuticular scales was found in Sheep while, highest number was noted in Domestic Pig in all the parts of hair, all body regions and amongst all the animal species considered for the present study.

4.5.4 Hair width

The mean and SE for hair width at proximal part of hair in various body regions and species are given in Table 4.18.

Table 4.18. Means and SE for hair width (µm) at proximal part of hair from various regions of body in different species.

Species→ Region ↓	Cattle	Buffalo	Sheep	Goat	Horse	Domestic Pig	Cat	Dog	Spotted Deer	Sambar	Nilgai	Hanuman Langur	Tiger	Leopard	Pooled Mean± SE
Head	21.34± 0.10 ^{a}	55.77± 0.51 ^{e}	41.99± 0.39 ^{b}	27.95± 1.55 ^{b}	37.45± 0.45 ^{e}	73.30± 0.15 ^{b}	11.71± 0.49 ^{a}	22.81± 0.11 ^{b}	37.69 ±0.24 ^{a}	64.53 ±0.49 ^{a}	34.75 ±0.49 ^{a}	30.44 ±0.18 ^{a}	53.79 ±0.44 ^{f}	37.32 ±0.28 ^{e}	**39.35 ±4.45 ^{a}**
Neck	31.87± 0.08 ^{d}	54.34± 0.22 ^{d}	59.52± 0.52 ^{e}	39.45± 0.55 ^{c}	35.36± 0.11 ^{d}	75.52± 0.18 ^{d}	17.25± 0.86 ^{c}	21.73± 0.19 ^{a}	42.62 ±0.25 ^{b}	74.64 ±0.29 ^{b}	93.19 ±0.54 ^{f}	41.50 ±0.27 ^{c}	38.95 ±0.12 ^{d}	31.07 ±0.08 ^{c}	**46.93 ±5.64 ^{d}**
Back	26.08± 2.31 ^{c}	32.34± 0.18 ^{a}	62.61± 3.11 ^{f}	42.82± 0.39 ^{d}	23.39± 0.17 ^{b}	92.41± 0.25 ^{f}	13.18 ^{b}± 0.27 ^{b}	22.42± 0.22 ^{b}	53.60 ±0.31 ^{c}	97.41 ±0.24 ^{e}	58.37 ±0.17 ^{c}	39.39 ±0.16 ^{b}	40.58 ±0.31 ^{e}	32.64 ±0.16 ^{d}	**45.52 ±6.50 ^{c}**
Abdomen	38.20± 5.44 ^{e}	49.24± 0.35 ^{c}	27.06± 0.25 ^{a}	28.47± 0.51 ^{b}	19.31± 0.32 ^{a}	77.03± 0.19 ^{e}	22.12± 0.18 ^{d}	24.22± 0.19 ^{c}	64.97 ±0.27 ^{d}	80.18 ±0.69 ^{c}	81.19 ±0.61 ^{e}	41.47 ±0.55 ^{c}	27.04 ±0.18 ^{a}	18.12 ±0.24 ^{a}	**42.76 ±6.08 ^{b}**
Thigh	23.46 ^{b}± 0.12	36.32± 0.39 ^{b}	45.66± 0.87 ^{d}	26.36 ± 0.41 ^{a}	32.75± 0.31 ^{c}	74.50± 0.28 ^{c}	23.88± 0.14 ^{e}	24.85± 0.14 ^{c}	75.11 ±1.06 ^{e}	83.22 ±0.17 ^{d}	40.83 ±0.21 ^{b}	48.74 ±0.19 ^{d}	31.99 ±0.17 ^{b}	26.57 ±0.20 ^{b}	**42.45 ±5.33 ^{b}**
Tail	38.94± 0.81 ^{f}	61.35± 0.71 ^{f}	43.15± 0.72 ^{c}	60.00± 0.30 ^{e}	101.57± 0.86 ^{f}	44.09± 0.54 ^{a}	38.41± 0.22 ^{f}	30.85± 0.52 ^{d}	82.90 ±0.31 ^{f}	97.16 ±0.70 ^{e}	74.16 ±0.14 ^{d}	59.05 ±0.21 ^{e}	36.48 ±0.07 ^{c}	26.52 ±0.19 ^{b}	**56.76 ±6.25 ^{e}**
Pooled Mean±SE	**29.98± 1.48 ^{D}**	**48.23± 1.78 ^{H}**	**46.67± 2.07 ^{H}**	**37.51± 2.01 ^{E}**	**41.64± 4.66 ^{F}**	**72.81± 2.43 ^{K}**	**21.09± 1.51 ^{A}**	**24.48± 0.52 ^{B}**	**59.48 ±2.78 ^{I}**	**82.86 ±2.87 ^{L}**	**63.75 ±3.57 ^{J}**	**43.43 ±1.49 ^{G}**	**38.14 ±1.41 ^{E}**	**28.70 ±1.02 ^{C}**	

Means bearing same superscript in a column do not differ significantly

CD for Species is 1.00 and CD for Region is 0.65

Table 4.19. Means and SE for hair width (μm) at middle part of hair from various regions of body in different species.

Species→ Region ↓	Cattle	Buffalo	Sheep	Goat	Horse	Domestic Pig	Cat	Dog	Spotted Deer	Sambar	Nilgai	Hanuman Langur	Tiger	Leopard	Pooled Mean± SE
Head	23.53 ± 0.10^{b}	56.92 ± 0.21^{b}	46.70 ± 0.32^{a}	32.28 ± 0.10^{b}	38.78 ± 0.58^{e}	75.53 ± 0.32^{e}	12.81 ± 0.19^{b}	44.32 ± 0.45^{f}	37.54 ± 0.21^{a}	65.29 ± 0.04^{b}	39.34 ± 0.30^{a}	26.70 ± 0.36^{a}	82.52 ± 0.28^{f}	47.14 ± 0.32^{d}	**44.96** ± 5.06^{b}
Neck	24.50 ± 0.26^{c}	62.48 ± 0.23^{d}	50.56 ± 0.27^{b}	45.05 ± 0.13^{d}	36.43 ± 0.17^{d}	72.53 ± 0.18^{d}	18.30 ± 0.18^{c}	18.22 ± 0.18^{a}	45.98 ± 0.13^{b}	83.48 ± 0.21^{e}	69.85 ± 0.21^{e}	39.51 ± 0.18^{b}	52.66 ± 0.36^{d}	54.30 ± 0.22^{f}	**48.13** ± 5.13^{d}
Back	27.77 ± 2.07^{d}	43.21 ± 0.32^{a}	60.16 ± 0.20^{e}	59.09 ± 0.09^{f}	30.12 ± 0.25^{c}	95.15 ± 0.13^{f}	12.53 ± 0.22^{a}	27.15 ± 0.27^{b}	54.59 ± 0.20^{c}	84.75 ± 0.07^{f}	61.35 ± 0.19^{c}	48.33 ± 0.31^{d}	55.38 ± 0.42^{e}	52.16 ± 0.50^{e}	**50.84** ± 5.75^{e}
Abdomen	28.28 ± 0.89^{e}	65.06 ± 0.28^{e}	53.35 ± 0.30^{c}	31.46 ± 0.24^{a}	22.33 ± 0.21^{a}	65.20 ± 1.09^{c}	24.81 ± 0.79^{e}	30.25 ± 0.18^{d}	76.69 ± 0.36^{f}	57.08 ± 0.27^{a}	63.20 ± 0.24^{d}	41.94 ± 0.31^{c}	25.05 ± 0.23^{a}	28.92 ± 0.27^{b}	**43.83** ± 4.85^{a}
Thigh	20.88 ± 0.24^{a}	61.89 ± 0.51^{c}	55.01^{c} ± 0.20^{c}	36.82 ± 0.36^{c}	28.75 ± 0.11^{b}	62.76 ± 0.15^{b}	19.97 ± 0.15^{d}	30.85 ± 0.15^{e}	76.16 ± 0.40^{e}	71.22 ± 0.46^{c}	43.90 ± 0.19^{b}	57.96 ± 0,33^{e}	40.34 ± 0.44^{c}	43.65 ± 0.21^{c}	**46.44** ± 4.66^{c}
Tail	39.22 ± 0.43^{f}	71.52 ± 0.38^{f}	66.51 ± 0.38^{f}	51.70 ± 0.14^{e}	89.65 ± 1.27^{f}	38.71 ± 0.19^{a}	31.51 ± 0.67^{f}	29.97 ± 0.16^{c}	73.36 ± 0.19^{d}	80.80 ± 0.13^{d}	63.34 ± 0.11^{d}	58.59 ± 0.12^{f}	36.05 ± 0.15^{b}	27.65 ± 0.23^{a}	**54.18** ± 5.29^{f}
Pooled Mean± SE	**27.36** ± 1.05^{B}	**60.18** ± 1.48^{J}	**55.38** ± 1.10^{H}	**42.73** ± 1.73^{E}	**41.01** ± 3.79^{D}	**68.31** ± 2.86^{L}	**19.99** ± 1.14^{A}	**30.13** ± 1.30^{C}	**60.72** ± 2.63^{K}	**73.77** ± 2.47^{M}	**56.83** ± 1.89^{I}	**45.50** ± 1.88^{F}	**48.67** ± 3.08^{G}	**42.30** ± 1.78^{E}	

Means bearing same superscript in a column do not differ significantly

CD for Species is 0.49 and CD for Region is 0.32

At the proximal part of hair the lowest hair width was noted in Cat (21.09 ± 1.51µm) followed by Dog (24.48 ± 0.52µm) while, highest hair width was found in Sambar (82.86 ± 2.87µm) as compared to all the species. It was also noted that the lower hair width amongst the body region was observed in hair of head region in almost all species except, Buffalo, Horse, Dog, Tiger and Leopard while, higher hair width was reported in the hair from tail region of almost all species except Pig, Sheep, Sambar, Tiger and Leopard Table 4.18. In Tiger and Leopard, lowest hair width of hair at proximal part was noted at abdomen region where as highest hair width was observed in head region Table 4.18. Amongst all the species and body regions, the lowest hair width at the proximal part of hair was noted in the hair of head region of Cat (11.71 ± 0.49µm) while, highest hair width was found in tail region of Horse (101.57 ± 0.86µm).

The mean and SE for hair width at middle part of hair in different body regions of various species of animal are presented in Table 4.19.

Table 4.20. Means and SE for hair width (μm) at distal part of hair from various regions of body in different species.

Species→ Region ↓	Cattle	Buffalo	Sheep	Goat	Horse	Domestic Pig	Cat	Dog	Spotted Deer	Sambar	Nilgai	Hanuman Langur	Tiger	Leopard	Pooled Mean ± SE
Head	22.20 ± 0.15 ^{b}	44.93 ± 0.21 ^{b}	55.54 ± 0.49 ^{d}	40.63 ± 0.51 ^{b}	36.86 ± 0.52 ^{d}	74.91 ± 0.45 ^{d}	12.24 ± 0.16 ^{a}	37.10 ± 0.13 ^{d}	38.51 ± 0.19 ^{b}	61.25 ± 0.34 ^{c}	30.81 ± 0.37 ^{a}	18.72 ± 0.28 ^{a}	44.95 ± 0.38 ^{e}	22.54 ± 0.35 ^{a}	**38.66** ± 4.45 ^{b}
Neck	19.52 ± 0.27 ^{a}	67.72 ± 0.23 ^{f}	58.14 ^{e} ±1.46	69.41 ± 0.15 ^{d}	40.92 ± 0.13 ^{e}	61.29 ± 0.33 ^{c}	13.39 ± 0.19 ^{b}	27.27 ± 0.55 ^{b}	49.98 ± 0.50 ^{d}	69.89 ± 0.04 ^{d}	58.78 ± 0.35 ^{f}	28.26 ± 0.25 ^{b}	43.48 ± 0.36 ^{d}	60.15 ± 0.35 ^{f}	**47.73** ± 4.93 ^{e}
Back	33.43 ± 2.99 ^{c}	33.23 ± 0.19 ^{a}	50.69 ^{c} ±0.09	43.29 ± 0.25 ^{c}	24.19 ± 0.42 ^{b}	90.10 ± 0.28 ^{e}	12.99 ± 0.22 ^{ab}	36.55 ± 0.34 ^{d}	26.77 ± 0.16 ^{a}	94.70 ± 0.20 ^{e}	40.31 ± 4.95 ^{b}	34.30 ± 0.32 ^{d}	51.12 ± 0.59 ^{f}	52.56 ± 3.43 ^{e}	**44.59** ± 5.93 ^{d}
Abdomen	21.62 ± 1.40 ^{b}	51.18 ± 0.15 ^{d}	55.67 ^{d} ±0.57	32.24 ± 0.37 ^{a}	21.80 ± 0.22 ^{a}	60.90 ± 0.15 ^{c}	29.66 ± 0.50 ^{e}	16.01 ± 0.26 ^{a}	26.38 ± 0.19 ^{a}	54.97 ± 0.19 ^{b}	51.46 ± 0.49 ^{d}	38.54 ± 0.25 ^{e}	31.79 ± 0.06 ^{b}	25.24 ± 0.33 ^{b}	**36.96** ± 3.86 ^{a}
Thigh	19.89 ± 0.81 ^{a}	48.21 ± 0.12 ^{e}	47.54 ^{a} ±0.39	41.55 ± 0.33 ^{b}	32.17 ± 0.88 ^{c}	57.58 ± 0.29 ^{b}	21.10 ± 0.07 ^{c}	31.73 ± 0.34 ^{c}	25.89 ± 1.28 ^{a}	61.64 ± 0.17 ^{c}	46.80 ± 0.46 ^{c}	32.60 ± 0.19 ^{c}	22.64 ± 0.38 ^{a}	36.57 ± 0.46 ^{d}	**37.57** ± 3.44 ^{a}
Tail	38.53 ± 1.59 ^{d}	62.81 ± 0.20 ^{e}	48.60 ^{b} ±0.17	31.52 ± 0.10 ^{a}	86.62 ± 0.19 ^{f}	35.87 ± 0.14 ^{a}	28.45 ± 0.15 ^{d}	37.02 ± 0.32 ^{d}	42.15 ± 1.31 ^{c}	27.54 ± 0.41 ^{a}	56.58 ± 0.43 ^{e}	42.25 ± 0.21 ^{f}	41.19 ± 0.28 ^{c}	27.40 ± 0.12 ^{c}	**43.32** ± 4.19 ^{c}
Pooled Mean± SE	**25.86** ± 1.37 ^{B}	**51.35** ± 1.93 ^{K}	**52.70** ^{L} ±0.72	**43.10** ± 2.13 ^{I}	**40.43** ± 3.67 ^{H}	**63.44** ± 2.80 ^{N}	**19.64** ± 1.23 ^{A}	**30.94** ± 1.29 ^{C}	**34.95** ± 1.59 ^{E}	**61.67** ± 4.82 ^{M}	**47.46** ± 1.80 ^{J}	**32.44** ± 1.28 ^{D}	**39.20** ± 1.59 ^{G}	**37.41** ± 2.47 ^{F}	

Means bearing same superscript in a column do not differ significantly
CD for Species is 1.03 and CD for Region is 0.67

The hair width at middle part of hair are found significantly lowest in Cat (19.99 ± 1.14μm), while highest hair width was noted in Sambar (73.77 ± 2.47μm) amongst all the species with respect to various body regions at middle part of hair lowest hair width was found in back region of Cat (12.53 ± 0.22μm) while, highest hair width was observed in back region of Domestic Pig (95.15 ± 0.13 μm). As observed in proximal part of hair of Tiger and Leopard, similar findings were noted in the present at the middle part of abdomen and higher at head region.

The mean and SE for hair width at distal part of hair in different body regions of various species of animal are given in Table 4.20. The mean hair width at distal part of hair was lowest in Cat (19.64 ± 1.23 μm) while, highest hair width was noted in Domestic Pig (63.44 ± 2.80 μm) followed by Sambar (61.67 ± 4.82 μm). Amongst all the species with respect to the various regions of body at distal part of hair, the lowest hair width was observed in head region of Cat (12.24 ± 0.16 μm), while highest hair width was observed in back region of Sambar (94.70 ± 0.20 μm).

Amongst all the species considered for the present study, the lowest mean hair width was observed in Cat at all the body regions except abdomen region. In abdomen region lowest hair width was noted in Domestic Pig. The highest mean diameter amongst all the species was recorded in Sambar at neck, back and thigh regions, while, at head and abdomen regions highest in Horse. Table 4.21.

Table 4.21. Means and SE for hair width (μm) at different parts of hair from various regions of body in different species.

Species→ Region ↓	Hair Parts	Cattle	Buffalo	Sheep	Goat	Horse	Domestic Pig	Cat	Dog	Spotted Deer	Sambar	Nilgai	Hanuman Langur	Tiger	Leopard	Pooled Mean± SE
Head	**Proximal**	21.34± 0.10^{a}	55.77 ± 0.51^{b}	41.99 ± 0.39^{a}	27.95 ± 1.55^{a}	37.45± 0.45^{b}	73.30± 0.15^{a}	11.71 ± 0.49^{a}	22.81± 0.11^{a}	37.69± 0.24^{a}	64.53± 0.49^{b}	34.75± 0.49^{b}	30.44± 0.18^{c}	53.79± 0.44^{a}	37.32 ± 0.28^{b}	**39.35** ± 4.45^{b}
	Middle	23.53± 0.10^{c}	56.92± 0.21^{c}	46.70± 0.32^{b}	32.28± 0.10^{b}	38.78± 0.58^{c}	75.53± 0.32^{c}	12.81± 0.19^{c}	44.32± 0.45^{c}	37.54± 0.21^{a}	65.29± 0.04^{c}	39.34± 0.30^{c}	26.70± 0.36^{b}	82.52± 0.28^{c}	47.14 ± 0.32^{c}	**44.96** ± 5.06^{c}
	Distal	22.20± 0.15^{b}	44.93± 0.21^{a}	55.54± 0.49^{c}	40.63± 0.51^{c}	36.86± 0.52^{a}	74.91± 0.45^{b}	12.24± 0.16^{b}	37.10± 0.13^{b}	38.51± 0.19^{b}	61.25± 0.34^{a}	30.81± 0.37^{a}	18.72± 0.28^{a}	44.95± 0.38^{b}	22.54 ± 0.35^{a}	**38.66**± 4.45^{a}
	Mean ±SE	**22.36** 0.23^{B}	**52.54**± 1.32^{I}	**48.08**± 1.38^{H}	**33.62**± 1.37^{D}	**37.70**± 0.34^{G}	**74.58**± 0.29^{L}	**12.25**± 0.20^{A}	**34.74**± 2.17^{E}	**37.91**± 0.16^{G}	**63.69**± 0.67^{K}	**34.97**± 0.87^{E}	**25.28**± 1.20^{C}	**60.42**± 3.89^{J}	**35.66** ± 2.46^{F}	
Neck	**Proximal**	31.87± 0.08^{c}	54.34± 0.22^{a}	59.52± 0.52^{c}	39.45± 0.55^{a}	35.36± 0.11^{a}	75.52± 0.18^{b}	17.25± 0.86^{b}	21.73± 0.19^{b}	42.62± 0.25^{a}	74.64± 0.29^{b}	93.19± 0.54^{c}	41.50± 0.27^{c}	38.95± 0.12^{a}	31.07 ± 0.08^{a}	**46.93**± 5.64^{a}
	Middle	24.50± 0.26^{b}	62.48± 0.38^{b}	50.56± 0.22^{a}	45.05± 0.13^{b}	36.43± 0.17^{b}	72.53± 0.44^{c}	18.30± 0.18^{c}	18.22± 0.18^{a}	45.98± 0.13^{b}	83.48± 0.21^{c}	69.85± 0.21^{b}	39.51± 0.18^{b}	52.66± 0.36^{c}	54.30 ± 0.22^{b}	**48.13**± 5.13^{c}
	Distal	19.52± 0.27^{a}	67.72± 0.23^{c}	58.14± 1.46^{b}	69.41± 0.15^{c}	40.92± 0.13^{c}	61.29± 0.33^{a}	13.39± 0.19^{a}	27.27± 0.55^{c}	49.98± 0.50^{c}	69.89± 0.04^{a}	58.78± 0.35^{a}	28.26± 0.25^{a}	43.48± 0.36^{b}	60.15 ± 0.35^{c}	**47.73**± 4.93^{b}
	Mean ±SE	**25.30**± 1.24^{C}	**61.51**± 1.34^{K}	**56.07**± 1.08^{J}	**51.30**± 3.16^{I}	**37.57**± 0.59^{E}	**69.78**± 1.50^{L}	**16.31**± 0.58^{A}	**22.4**± 0.92^{B}	**46.20**± 0.75^{G}	**76.00**± 2.00^{N}	**73.94**± 3.48^{M}	**36.42**± 1.42^{D}	**45.03**± 1.39^{F}	**48.51** ± 3.05 ^{H}	
Back	**Proximal**	26.08± 2.31^{a}	32.34± 0.18^{a}	62.61± 3.11^{c}	42.82± 0.39^{a}	23.39± 0.17^{a}	92.41± 0.25^{b}	13.18± 0.27	22.42± 0.22^{a}	53.60± 0.31^{b}	97.41 ± 0.24^{c}	58.37± 0.17^{b}	39.39± 0.16^{b}	40.58± 0.31^{a}	32.64 ± 0.16^{a}	**45.52**± 6.50^{b}
	Middle	27.77± 2.07^{b}	43.21± 0.32^{b}	60.16± 0.20^{b}	59.09± 0.09^{b}	30.12± 0.25^{b}	95.15± 0.13^{c}	12.53± 0.22	27.15± 0.27^{b}	54.59± 0.20^{c}	84.75 ± 0.07^{a}	61.35± 0.19^{c}	48.33± 0.31^{c}	55.38± 0.42^{c}	52.16 ± 0.50^{b}	**50.84**± 5.75^{c}
	Distal	33.43± 2.99^{c}	33.23± 0.19^{a}	50.69± 0.09^{a}	43.29± 0.25^{a}	24.19± 0.42^{a}	90.10± 0.28^{a}	12.99± 0.22	36.55± 0.34^{c}	26.77± 0.16^{a}	94.70 ± 0.20^{b}	40.31± 4.95^{a}	34.30± 0.32^{a}	51.12± 0.59^{b}	52.56 ± 3.43^{b}	**44.59**± 5.93^{a}
	Mean ±SE	**29.09**± 1.55^{CC}	**36.26**± 1.20^{D}	**57.82**± 1.58^{I}	**48.40**± 1.84^{G}	**25.90**± 0.75^{B}	**92.55**± 0.52^{J}	**12.90**± 0.15^{A}	**28.71**± 1.43^{C}	**44.99**± 3.13^{F}	**92.28**± 1.93^{J}	**53.34**± 2.74^{H}	**40.67**± 1.41^{E}	**49.03**± 1.53^{G}	**45.79** ± 2.50^{F}	

Species→ Region ↓	Hair Parts	Cattle	Buffalo	Sheep	Goat	Horse	Domestic Pig	Cat	Dog	Spotted Deer	Sambar	Nilgai	Hanu-man Langur	Tiger	Leopard	Pooled Mean± SE
Abdomen	**Proximal**	38.20± 5.44^{c}	49.24± 0.35^{a}	27.06± 0.25^{a}	28.47± 0.51^{a}	19.31± 0.32^{a}	77.03± 0.19^{c}	22.12± 0.18^{a}	24.22± 0.19^{b}	64.97± 0.27^{b}	80.18± 1.69^{c}	81.19± 0.61^{c}	41.47± 0.55^{b}	27.04± 0.18^{b}	18.12± 0.24^{a}	**42.76±** 6.08^{b}
	Middle	28.28± 0.89^{b}	65.06± 0.28^{c}	53.35± 0.30^{b}	31.46± 0.24^{b}	22.33± 0.21^{b}	65.20± 1.09^{b}	24.81± 0.79^{b}	30.25± 0.18^{c}	76.69 ± 0.36^{c}	57.08± 0.27^{b}	63.20± 0.24^{b}	41.94± 0.31^{b}	25.05± 0.23^{a}	28.92± 0.27^{c}	**43.83±** 4.85^{c}
	Distal	21.62± 1.40^{a}	51.18± 0.15^{b}	55.67± 0.57^{c}	32.24± 0.37^{c}	21.80± 0.22^{b}	60.90^{a}± 0.15^{a}	29.66± 0.50^{c}	16.01± 0.26^{a}	26.38± 0.19^{a}	54.97± 0.19^{a}	51.46± 0.49^{a}	38.54± 0.25^{a}	31.79± 0.06^{c}	25.24± 0.33^{b}	**36.96±** 3.86^{a}
	Mean ± SE	**29.37±** 2.43^{DE}	**55.16±** 1.71^{H}	**45.36±** 3.15^{G}	**30.72±** 0.45^{E}	**21.15±** 0.35^{A}	**67.71±** 1.69^{J}	**25.53±** 0.81^{C}	**23.49±** 1.42^{B}	**56.01±** 5.22^{H}	**64.07±** 4.10^{I}	**65.29±** 2.98^{I}	**40.65±** 0.42^{F}	**27.96±** 0.69^{D}	**24.09±** 1.10^{BC}	
Thigh	**Proximal**	23.46± 0.12^{c}	36.32 ± 0.39^{a}	45.66± 0.87^{a}	26.36± 0.41^{a}	32.75± 0.31^{b}	74.50± 0.28^{c}	23.88± 0.14^{c}	24.85± 0.14^{a}	75.11± 1.06^{b}	83.22± 0.17^{c}	40.83± 0.21^{a}	48.74± 0.19^{b}	31.99± 0.17^{b}	26.57± 0.20^{a}	**42.45±** 5.33^{b}
	Middle	20.88± 0.24^{b}	61.89± 0.51^{c}	55.01± 0.20^{c}	40.15± 3.51^{b}	28.75± 0.11^{a}	62.76± 0.15^{b}	19.97± 0.15^{a}	30.85± 0.15^{b}	76.16± 0.40^{c}	71.22± 0.46^{b}	43.90± 0.19^{b}	57.96± 0.33^{c}	40.34± 0.44^{c}	43.65± 0.21^{c}	**46.68±** 4.63^{c}
	Distal	19.89± 0.81^{a}	48.21± 0.12^{b}	47.54± 0.39^{b}	41.55± 0.33^{c}	32.17± 0.88^{b}	57.58± 0.29^{a}	21.10± 0.07^{b}	31.73± 0.34^{c}	25.89± 1.28^{a}	61.64± 0.17^{a}	46.80± 0.46^{c}	32.60± 0.19^{a}	22.64± 0.38^{a}	36.57± 0.46^{b}	**37.57±** 3.44^{a}
	Mean ± SE	**21.41±** 0.45^{A}	**48.81±** 2.54^{G}	**49.40±** 1.03^{G}	**36.02±** 2.00^{D}	**31.22±** 0.52^{C}	**64.95±** 1.72^{I}	**21.65±** 0.40^{A}	**29.15±** 0.75^{B}	**59.05±** 5.71^{H}	**72.03±** 3.13^{J}	**43.84±** 0.61^{E}	**46.43±** 2.55^{F}	**31.65±** 1.76^{C}	**35.60±** 1.71^{D}	
Tail	**Proximal**	38.94± 0.81^{b}	61.35± 0.71^{a}	43.15± 0.72^{a}	60.00± 0.30^{c}	101.57± 0.86^{c}	44.09± 0.54^{c}	38.41± 0.22^{c}	30.85± 0.52^{b}	82.90± 0.31^{c}	97.16± 0.70^{c}	74.16± 0.14^{c}	59.05± 0.21^{b}	36.48± 0.07^{a}	26.52^{a} ±0.19	**56.76±** 6.25^{c}
	Middle	39.22± 0.43^{b}	71.52± 0.38^{b}	66.51± 0.38^{c}	51.70± 0.14^{b}	89.65± 1.27^{b}	38.71± 0.19^{b}	31.51± 0.67^{b}	29.97± 0.16^{a}	73.36± 0.19^{b}	80.80± 0.13^{b}	63.34± 0.11^{b}	58.59± 0.12^{c}	36.05± 0.15^{a}	27.65± 0.23^{b}	**54.18±** 5.29^{b}
	Distal	38.53± 1.59^{a}	62.81± 0.20^{a}	48.60± 0.17^{b}	31.52± 0.10^{a}	86.62± 0.19^{a}	35.87± 0.14^{a}	28.45± 0.15^{a}	37.02± 0.37^{c}	42.15± 1.31^{a}	27.54± 0.41^{a}	56.58± 0.43^{a}	42.25± 0.21^{a}	41.19± 0.28^{b}	27.40± 0.12^{b}	**43.32±** 4.19^{a}
	Mean ± SE	**38.90±** 0.58^{D}	**65.23±** 1.12^{G}	**52.76±** 2.43^{F}	**47.74±** 2.90^{E}	**92.61±** 1.64^{J}	**39.56±** 0.85^{D}	**32.79±** 1.03^{B}	**32.61±** 0.79^{B}	**66.14±** 4.24^{H}	**68.50±** 10.51^{I}	**64.70±** 1.76^{G}	**53.29±** 1.90^{F}	**37.91±** 0.57^{C}	**27.19±** 0.15^{A}	

Means bearing same superscript for different region (for hair part) in a column do not differ significantly/Means bearing same superscript for different species (hair part of different region) in a row do not differ significantly/Head Region-CD for Species is 0.68 and CD for hair part is 0.31/Neck Region-CD for Species is 0.63 and CD for hair part is 0.29/ Back Region-CD for Species is 2.11 and CD for hair part is 0.97/Abdomen Region-CD for Species is 1.60 and CD for hair part is 0.74/Thigh Region-CD for Species is 1.16 and CD for hair part is 0.53/Tail Region- CD for Species is 0.86 and CD for hair part is 0.40

Table 4.22. Means and SE for medulla width (μm) at proximal part of hair at various region of body in different species.

Species→ Region ↓	Cattle	Sheep	Goat	Cat	Dog	Spotted Deer	Sambar	Nilgai	Tiger	Leopard	Pooled Mean ± SE
Head	3.17 ± 0.34^{a}	21.96 ± 0.62^{c}	6.75 ± 0.36^{b}	5.82± 0.33^{a}	7.92± 0.24^{b}	8.02 ± 0.56^{a}	49.37 ± 0.50^{a}	13.91 ± 0.26^{a}	22.06 ± 0.35^{e}	12.22 ± 0.34^{c}	**15.12 ± 4.32^{a}**
Neck	11.03 ± 0.46^{d}	49.57 ± 0.44^{f}	19.77 ± 0.50^{e}	12.27± 0.31^{c}	6.99± 0.17^{a}	18.47 ± 0.45^{b}	65.62 ± 0.08^{b}	55.16 ± 3.97^{f}	17.90 ± 0.38^{c}	15.40 ± 0.40^{e}	**27.22 ± 6.49^{c}**
Back	15.33 ± 2.87^{e}	48.65 ± 0.61^{e}	3.33 ± 0.13^{a}	8.53± 0.25^{b}	9.60± 0.25^{c}	41.52 ± 0.22^{c}	79.00 ± 0.27^{d}	31.02 ± 0.49^{c}	19.03 ± 0.26^{d}	18.78 ± 0.56^{f}	**27.48 ± 7.36^{c}**
Abdomen	20.13 ± 2.14^{f}	16.15 ± 0.33^{a}	8.43 ± 0.65^{c}	14.52± 0.13^{d}	7.72± 0.19^{b}	53.46 ± 0.34^{d}	74.91 ± 0.24^{c}	64.84 ± 0.57^{g}	5.29 ± 0.10^{a}	8.74 ± 0.22^{a}	**27.42 ± 8.35^{c}**
Thigh	7.35 ± 0.13^{c}	27.40 ± 1.19^{d}	9.02 ± 0.56^{d}	17.31± 0.18^{e}	8.24± 0.19^{b}	63.81 ± 1.36^{e}	74.95 ± 0.28^{c}	18.82 ± 0.40^{b}	11.12 ± 0.29^{b}	13.88 ± 0.21^{d}	**25.19 ± 7.65^{b}**
Tail	6.26 ± 0.67^{b}	20.30 ± 0.72^{b}	20.30 ± 0.47^{e}	29.83± 0.18^{f}	16.56± 0.65^{d}	63.84 ± 0.46^{e}	81.23 ± 0.62^{e}	43.23 ± 0.26^{e}	10.68 ± 0.25^{b}	9.54 ± 0.20^{b}	**30.18 ± 7.94^{d}**
Pooled Mean±SE	**10.54 ± 1.13^{B}**	**30.67 ± 2.29^{E}**	**11.27 ± 1.11^{B}**	**14.71± 1.31^{D}**	**9.50± 0.56^{A}**	**41.52 ± 3.65^{G}**	**70.85 ± 0.62^{H}**	**37.83 ± 3.08^{F}**	**14.34 ± 0.98^{D}**	**13.09 ± 0.59^{C}**	

Means bearing same superscript in a column do not differ significantly

CD for Species is 0.75 and CD for Region is 0.58

Table 4.23. Means and SE for medulla width (µm) at middle part of hair at various region of body in different species.

Species→ Region ↓	Cattle	Sheep	Goat	Cat	Dog	Spotted Deer	Sambar	Nilgai	Tiger	Leopard	Pooled Mean± SE
Head	3.66 ± 0.24^{a}	28.21± 0.36^{a}	8.96 ± 0.42^{b}	5.89 ± 0.22^{a}	28.26 ± 0.42^{e}	14.31 ± 0.22^{a}	50.97 ± 0.53^{b}	16.98 ± 0.70^{a}	51.97 ± 1.33^{f}	22.45 ± 0.38^{c}	**23.17** ± 5.43^{a}
Neck	5.00 ± 0.42^{b}	43.51 ± 1.51^{e}	22.21 ± 0.47^{e}	11.25 ± 0.18^{c}	6.65 ± 0.19^{a}	24.94 ± 0.37^{b}	74.20 ± 0.35^{f}	32.98 ± 0.38^{c}	30.63 ± 0.11^{e}	36.82 ± 0.22^{f}	**28.86** ± 6.49^{c}
Back	15.08 ± 2.15^{f}	47.13 ± 0.40^{f}	6.47 ± 0.25^{a}	6.87 ± 0.17^{b}	15.40 ± 0.30^{c}	40.06 ± 0.26^{c}	71.43 ± 0.14^{e}	37.35 ± 0.18^{d}	25.05 ± 0.25^{d}	34.64 ± 0.48^{e}	**29.95** ± 6.44^{d}
Abdomen	14.72 ± 1.73^{e}	40.77 ± 0.47^{d}	15.48 ± 0.46^{c}	18.00 ± 0.78^{e}	14.86 ± 0.22^{b}	64.88 ± 0.23^{e}	49.49 ± 0.28^{a}	45.07 ± 0.50^{f}	5.96 ± 0.22^{a}	13.21 ± 0.22^{b}	**28.24** ± 6.31^{b}
Thigh	8.49 ± 0.22^{d}	38.51 ± 0.44^{c}	19.94 ± 0.39^{d}	14.77 ± 0.42^{d}	14.88 ± 0.23^{b}	66.70 ± 0.57^{f}	64.49 ± 0.34^{c}	20.88 ± 0.26^{b}	18.33 ± 0.26^{c}	28.56 ± 0.24^{d}	**29.56** ± 6.54^{d}
Tail	5.84 ± 0.91^{c}	36.63 ± 0.34^{b}	28.83 ± 0.34^{f}	22.72 ± 1.11^{f}	16.70 ± 0.32^{d}	60.37 ± 0.60^{d}	67.62 ± 0.28^{d}	39.06 ± 0.30^{e}	13.77 ± 0.27^{b}	11.67 ± 0.34^{a}	**30.32** ± 6.56^{e}
Pooled Mean±SE	**8.80** ± 0.89^{A}	**39.13** ± 1.04^{G}	**16.98** ± 1.31^{D}	**13.25** ± 1.03^{B}	**16.13** ± 1.08^{C}	**45.21** ± 3.43^{H}	**63.0** ± 2.32^{I}	**32.05** ± 1.70^{F}	**24.28** ± 2.49^{E}	**24.56** ± 1.65^{E}	

Means bearing same superscript in a column do not differ significantly
CD for Species is 0.68 and CD for Region is 0.53

4.5.5. Medulla Width

During the present study it was noticed that the medulla was absent at distal part of hair from back region and proximal part of hair from tail region in Buffalo. In hair of Hanuman Langur, the medulla was not visible at proximal part of hair in head, neck, back and abdomen regions, distal part of hair in head and neck region and middle part of hair in back region. In Horse and Domestic Pig, the medulla was found absent at all the three parts of hair from tail region. Statistical analysis for medulla width and cortex width of hair in Buffalo, Domestic Pig, Horse and Hanuman Langur was not undertaken since medulla was absent in these species.

Amongst the ten species in which the medulla was present at the proximal part of hair in different body regions, it was noted that the medulla width was lowest in Dog (9.50 ± 0.56 μm) and highest medulla width was noted in Sambar (70.85 ± 0.62 μm). At the proximal part of hair amongst all the species and body region, the minimum medulla width was observed at proximal part of hair of head region in Cattle (3.17 ± 0.34 μm) while, maximum medulla width was noted at tail region in Sambar (81.23 ± 0.62 μm) Table 4.22.

At the middle part of hair minimum medulla width was observed in Cattle (8.80 ± 0.89 μm) while maximum medulla width was noted in Sambar (63.0 ± 2.32 μm). Amongst all the species and body regions, the lowest medulla width was found at head region of Cattle (3.66 ± 0.24 μm) while highest medulla width was noted at neck region in Sambar (74.20 ± 0.35 μm) at middle part of hair.

The mean and SE for medulla width at middle part of hair from various regions of body in different species are given in Table 4.23.

The medulla width was found lowest in Cattle (10.14 ± 1.08 μm) and highest in Sambar (47.45 ± 4.48 μm) at distal part of hair as compared with all the species under study. Amongst all the species and body regions the minimum medulla width was observed at head region of Cattle (3.22 ± 0.21 μm) whereas maximum medulla width was found at back region in Sambar (76.62 ± 0.54 μm).

The mean and SE for medulla width at distal part of hair from various regions of body in different species are given in Table 4.24.

Table 4.24. Means and SE for medulla width (µm) at distal part of hair at various regions of body in different species.

Species→ Region ↓	Cattle	Sheep	Goat	Cat	Dog	Spotted Deer	Sambar	Nilgai	Tiger	Leopard	Pooled Mean± SE
Head	3.22 ± 0.21^{a}	42.80 ± 0.34^{e}	10.19 ± 0.52^{c}	6.60 ± 0.18^{b}	19.46 ± 0.45^{d}	17.64 ± 0.24^{c}	43.62 ± 0.44^{b}	11.29 ± 0.16^{a}	22.12 ± 1.01^{d}	8.05 ± 0.10^{a}	$\mathbf{18.50 \pm 4.52^{b}}$
Neck	8.02 ± 0.61^{c}	49.10 ± 1.27^{f}	48.93 ± 0.34^{f}	6.92 ± 0.16^{b}	12.03 ± 0.40^{b}	31.64 ± 0.05^{e}	56.35 ± 0.25^{d}	22.63 ± 0.80^{b}	20.63 ± 0.28^{c}	40.37 ± 0.28^{f}	$\mathbf{29.66 \pm 5.78^{f}}$
Back	21.51 ± 2.32^{f}	40.01 ± 0.45^{c}	6.57 ± 0.19^{b}	5.40 ± 0.16^{a}	24.69 ± 0.42^{f}	10.84 ± 0.38^{a}	76.62 ± 0.54^{e}	24.99 ± 0.32^{c}	26.16 ± 0.56^{e}	34.17 ± 0.40^{e}	$\mathbf{27.09 \pm 6.56^{e}}$
Abdomen	12.56 ± 0.42^{e}	41.50 ± 0.79^{d}	17.25 ± 0.23^{d}	22.31 ± 0.63^{e}	7.26 ± 0.16^{a}	12.88 ± 0.42^{b}	43.69 ± 0.26^{b}	28.06 ± 0.25^{e}	10.59 ± 0.46^{b}	12.22 ± 0.41^{c}	$\mathbf{20.83 \pm 4.10^{c}}$
Thigh	9.90 ± 0.83^{d}	36.66 ± 0.49^{b}	21.45 ± 0.45^{e}	15.68 ± 0.29^{c}	18.30 ± 0.47^{c}	17.34 ± 0.91^{c}	49.86 ± 0.22^{c}	26.15 ± 0.79^{d}	6.70 ± 0.20^{a}	21.15 ± 0.57^{d}	$\mathbf{22.32 \pm 4.03^{d}}$
Tail	5.65 ± 1.10^{b}	27.51 ± 0.21^{a}	4.67 ± 0.34^{a}	17.31 ± 0.29^{d}	21.55 ± 0.28^{e}	24.31 ± 0.14^{d}	14.57 ± 0.20^{a}	34.82 ± 0.31^{f}	10.78 ± 0.11^{b}	10.19 ± 0.16^{b}	$\mathbf{17.13 \pm 3.12^{a}}$
Pooled Mean±SE	$\mathbf{10.14 \pm 1.08^{A}}$	$\mathbf{39.59 \pm 1.14^{I}}$	$\mathbf{18.18 \pm 2.53^{E}}$	$\mathbf{12.37 \pm 1.09^{B}}$	$\mathbf{17.21 \pm 1.00^{D}}$	$\mathbf{19.11 \pm 1.20^{F}}$	$\mathbf{47.45 \pm 4.48^{J}}$	$\mathbf{24.66 \pm 1.21^{H}}$	$\mathbf{16.1 \pm 1.22^{C}}$	$\mathbf{21.02 \pm 2.09^{G}}$	

Means bearing same superscript in a column do not differ significantly

CD for Species is 0.65 and CD for Region is 0.50

Table 4.25. Means and SE for medulla width (µm) at different part of hair in various region of body in different species.

Species→ Region ↓	Hair Parts	Cattle	Buffalo	Sheep	Goat	Horse	Domestic Pig	Cat	Dog	Spotted Deer	Sambar	Nilgai	Hanuman Langur	Tiger	Leopard	Pooled Mean± SE
Head	**Proximal**	3.17± 0.34^{a}	3.13± 0.25^{b}	21.96± 0.62^{a}	6.75± 0.36^{a}	14.45± 0.48^{c}	3.29± 0.24^{a}	5.82± 0.33^{a}	7.92± 0.24^{a}	8.02± 0.56^{a}	49.37± 0.50^{b}	13.91± 0.26^{b}	-	22.06± 0.35^{a}	12.22± 0.34^{b}	**13.24± 3.37^{a}**
	Middle	3.66± 0.24^{b}	3.11± 0.37^{b}	28.21± 0.36^{b}	8.96± 0.42^{b}	9.06± 0.67^{b}	5.15± 0.40^{c}	5.89± 0.22^{a}	28.26± 0.42^{c}	14.31± 0.22^{b}	50.97± 0.53^{c}	16.98± 0.70^{c}	-	51.97± 1.33^{b}	22.45± 0.38^{c}	**19.15± 4.46^{c}**
	Distal	3.22± 0.21^{a}	2.03± 0.19^{a}	42.80± 0.34^{c}	10.19± 0.52^{c}	7.01± 0.50^{a}	3.94± 0.17^{b}	6.60± 0.18^{b}	19.46± 0.45^{b}	17.64± 0.24^{c}	43.62± 0.44^{a}	11.29± 0.16^{a}	-	22.12± 1.01^{a}	8.05± 0.10^{a}	**15.23± 3.70^{b}**
	Mean ±SE	**3.35± 0.16^{B}**	**2.75± 0.20^{A}**	**30.99± 2.13^{J}**	**8.63± 0.42^{E}**	**10.18± 0.82^{F}**	**4.13± 4.13^{C}**	**6.10± 0.16^{D}**	**18.54± 18.54^{I}**	**13.32± 0.99^{G}**	**47.99± 1.17^{L}**	**14.06± 0.61^{H}**	-	**32.05± 3.46^{K}**	**14.24± 1.48^{H}**	
Neck	**Proximal**	11.03± 0.46^{c}	7.27± 0.56^{a}	49.57± 0.44^{c}	19.77± 0.50^{a}	12.25± 0.60^{b}	3.32± 0.37^{a}	12.27± 0.31^{c}	6.99± 0.17^{a}	18.47± 0.45^{a}	65.62± 0.08^{b}	55.16± 0.64^{c}	-	17.90± 0.38^{a}	15.40± 0.40^{a}	**22.69± 5.21^{a}**
	Middle	5.00± 0.42^{a}	15.78± 0.50^{c}	43.51± 1.51^{a}	22.21± 0.47^{b}	7.53± 0.54^{a}	3.94± 0.23^{b}	11.25± 0.18^{b}	6.65± 0.19^{a}	24.94± 0.37^{b}	74.20± 0.35^{c}	32.98± 0.38^{b}	-	30.63± 0.11^{c}	36.82± 0.22^{b}	**24.27± 5.11^{b}**
	Distal	8.02± 0.61^{b}	14.85± 0.22^{b}	49.10± 1.27^{b}	48.93± 0.34^{c}	7.20± 0.46^{a}	5.45± 0.56^{c}	6.92± 0.16^{a}	12.03± 0.40^{b}	31.64± 0.05^{c}	56.35± 0.25^{a}	22.63± 0.80^{a}	-	20.63± 0.28^{b}	40.37± 0.28^{c}	**24.93± 4.70^{c}**
	Mean ±SE	**8.02± 0.66^{B}**	**12.63± 0.96^{E}**	**47.39± 0.92^{J}**	**30.30± 3.21^{H}**	**8.99± 0.63^{C}**	**4.24± 0.31^{A}**	**10.15± 0.58^{D}**	**8.56± 0.61^{BC}**	**25.02± 1.32^{G}**	**65.39± 2.58^{K}**	**36.92± 3.31^{I}**	-	**23.05± 1.34^{F}**	**30.86± 2.68^{H}**	
Back	**Proximal**	15.33± 2.87^{a}	-	48.65± 0.61^{c}	3.33± 0.13^{a}	14.60± 0.31^{c}	40.34± 0.60^{b}	8.53± 0.25^{c}	9.60± 0.25^{a}	41.52± 0.22^{c}	79.00± 0.27^{c}	31.02± 0.49^{b}	-	19.03± 0.26^{a}	18.78± 0.56^{a}	**27.48± 6.01^{b}**
	Middle	15.08± 2.15^{a}	-	47.13± 0.40^{b}	6.47± 0.25^{b}	13.09± 0.20^{b}	48.10± 0.73^{c}	6.87± 0.17^{b}	15.40± 0.30^{b}	40.06± 0.26^{b}	71.43± 0.14^{a}	37.35± 0.18^{c}	-	25.05± 0.25^{b}	34.64± 0.48^{b}	**30.06± 5.50^{c}**
	Distal	21.51± 2.32^{b}	-	40.01± 0.45^{a}	6.57± 0.19^{b}	7.33± 0.44^{a}	26.85± 2.50^{a}	5.40± 0.16^{a}	24.69± 0.42^{c}	10.84± 0.38^{a}	76.62± 0.54^{b}	24.99± 0.32^{a}	-	26.16± 0.56^{c}	34.17± 0.40^{b}	**25.43± 5.42^{a}**
	Mean ±SE	**17.31± 1.52^{D}**	-	**45.26± 0.95^{I}**	**5.46± 0.38^{A}**	**11.67± 0.78^{C}**	**38.43± 2.29^{H}**	**6.93± 0.33^{B}**	**16.56± 1.52^{D}**	**30.81± 3.43^{G}**	**75.68± 1.15^{J}**	**31.12± 1.24^{G}**	-	**23.41± 0.79^{E}**	**29.20± 1.81^{F}**	

Species→ Region ↓	Hair Parts	Cattle	Buffalo	Sheep	Goat	Horse	Domestic Pig	Cat	Dog	Spotted Deer	Sambar	Nilgai	Hanuman Langur	Tiger	Leopard	Pooled Mean± SE
Abdomen	**Proximal**	20.13± 2.14^{c}	18.74± 0.61^{a}	16.15± 0.33^{a}	8.43± 0.65a	6.83± 0.51^{a}	4.40± 0.20	14.52± 0.13^{a}	7.72± 0.19^{a}	53.46± 0.34^{b}	74.91± 0.24^{c}	64.84± 0.57^{c}	-	5.29± 0.10^{a}	8.74± 0.22^{a}	**23.40± 6.24^{b}**
	Middle	14.72± 1.73^{b}	39.17± 0.35^{c}	40.77± 0.47^{b}	15.48± 0.46^{b}	12.64± 0.31^{b}	4.49± 0.25	18.00± 0.78^{b}	14.86± 0.22^{b}	64.88± 0.23^{c}	49.49± 0.28^{b}	45.07± 0.50^{b}	-	5.96± 0.22^{b}	13.21± 0.22^{c}	**26.06± 4.95^{c}**
	Distal	12.56± 0.42^{a}	30.10± 0.43^{b}	41.50± 0.79^{c}	17.25± 0.23^{c}	14.53± 0.18^{c}	4.74± 0.11	22.31± 0.63^{c}	7.26± 0.16^{a}	12.88± 0.42^{a}	43.69± 0.26^{a}	28.06± 0.25^{a}	-	10.59± 0.46^{c}	12.22± 0.41^{b}	**19.82± 3.22^{a}**
	Mean ±SE	**15.80± 1.16^{F}**	**29.34± 2.04^{H}**	**32.81± 2.87^{I}**	**13.72± 0.96^{E}**	**11.33± 0.82^{D}**	**4.54± 0.11^{A}**	**18.28± 0.83^{G}**	**9.95± 0.85^{C}**	**43.74± 5.41^{J}**	**56.03± 4.80^{L}**	**45.99± 3.65^{K}**	-	**7.28± 0.59^{B}**	**11.39± 0.49^{D}**	
Thigh	**Proximal**	7.35± 0.13^{a}	17.92± 0.49^{b}	27.40± 1.19^{a}	9.02± 0.56^{a}	19.82± 0.21^{c}	4.32± 0.19^{a}	17.31± 0.18^{c}	8.24± 0.19^{a}	63.81± 1.36^{b}	74.95± 0.28^{b}	18.82± 0.40^{a}	4.82± 0.37^{b}	11.12± 0.29^{b}	13.88± 0.21^{a}	**21.34± 5.53^{b}**
	Middle	8.49± 0.22^{b}	32.04± 0.59^{c}	38.51± 0.44^{c}	19.94± 0.39^{b}	17.06± 0.21^{b}	4.87± 0.34^{b}	14.77± 0.42^{a}	14.88± 0.23^{b}	66.70± 0.57^{c}	64.49± 0.34^{c}	20.88± 0.26^{b}	8.44± 0.43^{c}	18.33± 0.26^{c}	28.56± 0.24^{c}	**25.57± 4.97^{c}**
	Distal	9.90± 0.83^{c}	10.76± 0.99a	36.66± 0.49^{b}	21.45± 0.45^{c}	8.16± 0.53^{a}	9.35± 0.48^{c}	15.68± 0.29^{b}	18.30± 0.47^{c}	17.34± 0.91^{a}	49.86± 0.22^{a}	26.15± 0.79^{c}	4.37± 0.13^{a}	6.70± 0.20^{a}	21.15± 0.57^{b}	**18.27± 3.24^{a}**
	Mean ±SE	**8.58± 0.37^{B}**	**20.24± 2.18^{H}**	**34.19± 1.25^{J}**	**16.81± 1.37^{G}**	**15.01± 1.22^{E}**	**6.18± 0.58^{A}**	**15.92± 0.31^{F}**	**13.81± 1.03^{D}**	**49.28± 5.51^{K}**	**63.10± 3.64^{L}**	**21.95± 0.80^{I}**	**5.88± 0.48^{A}**	**12.05± 1.17^{C}**	**21.20± 1.47^{I}**	
Tail	**Proximal**	6.26± 0.67^{b}	-	20.30± 0.72^{a}	20.30± 0.47^{b}	-	-	29.83± 0.26^{c}	16.56± 0.65^{a}	63.84± 0.46^{c}	81.23± 0.62^{c}	43.23± 0.26^{c}	5.76± 0.31^{a}	10.68± 0.25^{a}	9.54± 0.20^{a}	**27.96± 7.16^{b}**
	Middle	5.84± 0.91^{a}	-	36.63± 0.34^{c}	28.83± 0.34^{c}	-	-	22.72± 1.11^{b}	16.70± 0.32^{a}	60.37± 0.60^{b}	67.62± 0.28^{b}	39.06± 0.30^{b}	9.24± 0.49^{b}	13.77± 0.27^{b}	11.67± 0.34^{c}	**28.41± 5.95^{c}**
	Distal	5.65± 1.10^{a}	-	27.51± 0.21^{b}	4.67± 0.34^{a}	-	-	17.31± 0.29^{a}	21.55± 0.28^{b}	24.31± 0.14^{a}	14.57± 0.20^{a}	34.82± 0.31^{a}	5.78± 0.55^{a}	10.78± 0.11^{a}	10.19± 0.16^{b}	**16.10± 2.86^{a}**
	Mean ±SE	**5.91± 0.50^{A}**	-	**28.15± 1.64^{G}**	**17.94± 2.44^{E}**	-	-	**23.29± 1.30^{F}**	**18.27± 0.61^{E}**	**49.51± 4.34^{I}**	**54.47± 10.17^{J}**	**39.04± 0.85^{H}**	**6.93± 0.47^{B}**	**11.74± 0.37^{D}**	**10.47± 0.25^{C}**	

Means bearing same superscript for different region (for hair part) in a column do not differ significantly/Means bearing same superscript for different species (hair part of different region) in a row do not differ significantly/Head Region-CD for Species is 0.73 and CD for hair part is 0.35/Neck Region-CD for Species is 0.82 and CD for hair part is 0.39/Back Region-CD for Species is 1.45 and CD for hair part is 0.72/Abdomen Region-CD for Species is 0.94 and CD for hair part is 0.45/Thigh Region-CD for Species is 0.86 and CD for hair part is 0.40/Tail Region- CD for Species is 0.81 and CD for hair a part is 0.42

During the present study, the maximum medulla width was observed in Sambar at all the body regions, whereas minimum medulla width was observed in neck and abdomen region of Domestic Pig while, at head region in Buffalo, back region in Goat, thigh region in Hanuman Langur and tail region in Cattle Table 4.26. It was also observed that the lower medulla width mostly found at the proximal part of hair in almost in all the regions except head and tail. In head and tail regions, the lowest medulla width was notated at distal part of hair. The maximum medulla width was noted at the proximal part of hair from abdomen, thigh and tail region whereas at middle part of hair from head and neck region and distal part of hair from back region Table 4.25.

4.5.6. Cortex width

At the proximal part of hair, the width of cortex was found maximum in Goat (12.52 ± 0.79 μm) and Nilgai (12.48 ± 0.65 μm), while minimum cortex width was observed in Cat (2.75 ± 0.27 μm) as compared with all the species and body regions. Amongst all the species at the proximal part of hair from the neck region of Cat (1.97 ± 0.04 μm) showed lowest cortex width while highest cortex width was noted at back region of Goat (19.03 ± 0.20 μm). The mean and SE for cortex width of proximal part of hair from various regions of body and all the species are given in Table 4.26.

Table 4.26. Means and SE for cortex width of hair (μm) at proximal end in various region of body in different species.

Species→ Region ↓	Cattle	Sheep	Goat	Cat	Dog	Spotted Deer	Sambar	Nilgai	Tiger	Leopard	Pooled Mean±SE
Head	8.68± 0.20^{c}	9.02± 0.14^{e}	10.52± 0.69^{c}	2.20± 0.15^{b}	6.57± 0.06^{ab}	15.18± 0.32^{e}	7.09± 0.10^{e}	11.51± 0.25^{c}	15.37± 0.25^{e}	12.39± 0.15^{f}	**9.85± 1.28^{d}**
Neck	10.06± 0.16^{d}	4.65± 0.24^{a}	8.85± 0.26^{a}	1.97± 0.04^{a}	7.19± 0.13^{c}	11.23± 0.13^{d}	4.70± 0.11^{c}	19.02± 0.38^{f}	9.62± 0.15^{a}	7.41± 0.24^{d}	**8.47± 1.48^{c}**
Back	5.13± 0.59^{a}	7.63± 0.46^{c}	19.03± 0.20^{e}	2.15± 0.08^{ab}	6.38± 0.16^{a}	5.59± 0.20^{b}	9.75± 0.10^{f}	12.55± 0.36^{d}	10.73± 0.16^{c}	7.03± 0.23^{c}	**8.60± 1.49^{c}**
Abdomen	6.08± 0.75^{b}	5.18± 0.32^{b}	9.35± 0.68^{b}	3.62± 0.14^{d}	7.99± 0.10^{d}	4.84± 0.13^{a}	3.43± 0.04^{a}	7.28± 0.14^{a}	10.56± 0.10^{c}	4.43± 0.11^{a}	**6.28± 0.77^{a}**
Thigh	8.50± 0.14^{c}	8.74± 0.15^{d}	8.60± 0.24^{a}	2.78± 0.23^{c}	7.94± 0.23^{d}	5.06± 0.25^{a}	3.88± 0.08^{b}	9.60± 0.37^{b}	10.02± 0.24^{b}	6.03± 0.13^{b}	**7.11± 0.79^{b}**
Tail	16.37± 0.49^{e}	9.87± 0.16^{f}	18.75± 0.43^{d}	3.77± 0.05^{d}	6.78± 0.25^{b}	8.73± 0.11^{c}	5.99± 0.01^{d}	14.93± 0.11^{e}	11.89± 0.23^{d}	8.89± 0.19^{e}	**10.60± 1.53^{e}**
Pooled Mean±SE	**9.14± 0.64^{G}**	**7.51± 0.39^{D}**	**12.52± 0.79^{I}**	**2.75± 0.27^{A}**	**7.14± 0.12^{C}**	**8.44± 0.64^{F}**	**5.81± 0.52^{B}**	**12.48± 0.65^{I}**	**11.36± 0.33^{H}**	**7.70± 0.43^{E}**	

Means bearing same superscript in a column do not differ significantly

CD for Species is 0.32 and CD for Region is 0.25

Table 4.27. Means and SE for cortex width (µm) at distal part of hair in various regions of body in different species.

Species→ Region ↓	Cattle	Sheep	Goat	Cat	Dog	Spotted Deer	Sambar	Nilgai	Tiger	Leopard	Pooled Mean±SE
Head	10.01± 0.14^{d}	8.18± 0.15^{d}	12.21± 0.25^{e}	3.01± 0.11^{bc}	6.09± 0.21^{c}	11.92± 0.16^{f}	7.20± 0.11^{e}	11.82± 0.33^{d}	15.22± 0.25^{d}	11.31± 0.16^{c}	**9.70± 1.13^{d}**
Neck	9.60± 0.40^{c}	4.40± 0.36^{a}	11.79± 0.33^{d}	3.04± 0.13^{bc}	4.54± 0.09^{a}	10.23± 0.21^{e}	4.12± 0.16^{b}	17.39± 0.40^{e}	10.16± 0.37^{a}	7.91± 0.20^{b}	**8.32± 1.41^{b}**
Back	6.05± 0.88^{b}	6.11± 0.20^{b}	22.95± 0.20^{f}	3.21± 0.22^{c}	5.52± 0.06^{b}	6.75± 0.24^{d}	6.64± 0.24^{d}	11.81± 0.14^{d}	15.63± 0.13^{e}	7.94± 0.33^{b}	**9.26± 1.89^{c}**
Abdomen	6.28± 0.76^{b}	6.00± 0.16^{b}	8.13± 0.07^{b}	2.88± 0.10^{b}	8.19± 0.22^{e}	5.56± 0.25^{c}	3.33± 0.19^{a}	8.69± 0.33^{a}	10.51± 0.22^{b}	7.36± 0.19^{a}	**6.69± 0.75^{a}**
Thigh	5.42± 0.19^{a}	7.72± 0.17^{c}	7.89± 0.20^{a}	2.39± 0.10^{a}	7.55± 0.13^{d}	4.79± 0.20^{a}	3.16± 0.07^{a}	10.70± 0.12^{b}	10.98± 0.26^{c}	7.49± 0.12^{a}	**6.81± 0.91^{a}**
Tail	15.39± 0.18^{e}	15.68± 0.50^{e}	9.29± 0.83^{c}	4.10± 0.39^{d}	6.02± 0.24^{c}	5.20± 0.22^{b}	5.62± 0.02^{c}	11.36± 0.33^{c}	10.20± 0.15^{a}	7.31± 0.26^{a}	**9.02± 1.31^{c}**
Pooled Mean±SE	**8.79± 0.61^{F}**	**8.02± 0.62^{E}**	**12.04± 0.88^{H}**	**3.11± 0.21^{A}**	**6.32± 0.22^{C}**	**7.41± 0.46^{D}**	**5.01± 0.39^{B}**	**11.96± 0.46^{G}**	**12.11± 0.41^{H}**	**8.22± 0.25^{E}**	

Means bearing same superscript in a column do not differ significantly
CD for Species is 0.34 and CD for Region is 0.26

The cortex width at middle part of hair was observed lowest in Cat (3.11 ± 0.21 μm) and highest cortex width and highest cortex width was noted in Nilgai (12.11 ± 0.41 μm) and Goat (12.04 ± 0.88 μm) as compared with all the species considered for the present study. Amongst all the species with respect to body region at middle part of hair, the lowest cortex width was noted at the high region of Cat (2.39 ± 0.10 μm) while highest cortex width was observed at back region of Goat (22.95 ± 0.20 μm).). The mean and SE for cortex width at middle part of hair in various regions of body and all the species are presented in Table 4.27.

During the present study it was observed that the lowest cortex width at distal part of hair in Cat (3.30 ± 0.21 μm), while highest width in Goat (11.94 ± 0.88 μm) as compared with all species. From the present observations it can be concluded that the Cat hair had minimum cortex width, while of Goat hair had maximum cortex width from all the parts of hair amongst all the species considered for the present study Table 4.28.

Table 4.28. Means and SE for cortex width (μm) at distal part of hair in various regions of body in different species.

Species→ Region ↓	Cattle	Sheep	Goat	Cat	Dog	Spotted Deer	Sambar	Nilgai	Tiger	Leopard	Pooled Mean±SE
Head	8.99± 0.22^{d}	5.95± 0.15^{c}	15.52± 0.42^{e}	2.16± 0.14^{a}	8.93± 0.17^{e}	10.10± 0.28^{f}	6.75± 0.51^{d}	10.68± 0.18^{c}	10.37± 0.29^{c}	5.85± 0.22	**8.53± 1.13^{d}**
Neck	4.59± 0.10^{a}	4.59± 0.16^{a}	10.61± 0.26^{c}	2.83± 0.12^{b}	7.41± 0.09^{cd}	8.16± 0.26^{d}	5.70± 0.17^{b}	15.70± 0.17^{d}	10.62± 0.16^{c}	9.48± 0.18	**7.97± 1.41^{c}**
Back	5.99± 0.64^{c}	4.62± 0.20^{a}	18.50± 0.31^{f}	2.88± 0.18^{b}	4.85± 0.27^{b}	8.49± 0.25^{e}	7.59± 0.16^{e}	9.90± 0.25^{a}	12.53± 0.31^{d}	10.28± 0.23	**8.56± 1.89^{d}**
Abdomen	4.40± 0.72^{a}	6.58± 0.34^{d}	6.11± 0.30^{a}	3.85± 0.15^{c}	3.70± 0.12^{a}	4.78± 0.17^{b}	5.10± 0.08^{a}	9.74± 0.09^{a}	8.82± 0.35^{b}	6.34± 0.13	**5.94± 0.75^{a}**
Thigh	4.87± 0.34^{b}	5.15± 0.20^{b}	9.96± 0.30^{b}	2.68± 0.14^{b}	7.27± 0.15^{c}	3.34± 0.29^{a}	6.31± 0.07^{c}	10.18± 0.22^{b}	7.97± 0.28^{a}	8.35± 0.14	**6.61± 0.91^{b}**
Tail	15.51± 1.23^{e}	8.61± 0.25^{e}	10.96± 0.22^{d}	5.39± 0.45^{d}	7.67± 0.27^{d}	7.11± 0.23^{c}	6.09± 0.09^{c}	9.70± 0.25^{a}	14.47± 0.18^{e}	8.52± 0.10	**9.40± 1.31^{e}**
Pooled Mean±SE	**7.39± 0.61^{F}**	**5.92± 0.62^{B}**	**11.94± 0.88^{I}**	**3.30± 0.21^{A}**	**6.64± 0.22^{D}**	**7.00± 0.46^{E}**	**6.25± 0.39^{C}**	**10.98± 0.46^{H}**	**10.80± 0.41^{H}**	**8.14± 0.25^{G}**	

Means bearing same superscript in a column do not differ significantly

CD for Species is 0.35 and CD for Region is 0.37

Table 4.29. Means and SE for cortex width (µm) at different parts of hair in various regions of body in different species.

Species→ Region ↓	Hair Parts	Cattle	Buffalo	Sheep	Goat	Horse	Domestic Pig	Cat	Dog	Spotted Deer	Sambar	Nilgai	Hanuman Langur	Tiger	Leopard	Pooled Mean± SE
Head	**Proximal**	8.68 ± 0.20^{a}	25.88 ± 0.46^{b}	9.02 ± 0.14^{c}	10.52 ± 0.69^{a}	11.40 ± 0.41^{a}	33.15 ± 0.30^{a}	2.20 ± 0.15^{a}	6.57± 0.06^{b}	15.18± 0.32^{c}	7.09± 0.10^{b}	11.51± 0.25^{b}	-	15.37± 0.25^{b}	12.39± 0.15^{b}	**13.00± 2.20^{b}**
	Middle	10.01 ± 0.14^{c}	30.23 ± 0.44^{c}	8.18 ± 0.15^{b}	12.21 ± 0.25^{b}	12.25 ± 0.47^{b}	35.66 ± 0.44^{b}	3.01± 0.11^{b}	6.09± 0.21^{a}	11.92± 0.16^{b}	7.20± 0.11^{b}	11.82± 0.33^{b}	-	15.22± 0.25^{b}	11.31± 0.16^{b}	**13.47± 2.47^{c}**
	Distal	8.99 ± 0.22^{b}	19.22 ± 0.45^{a}	5.95 ± 0.15^{a}	15.52 ± 0.42^{c}	14.89 ± 0.35^{c}	36.38 ± 0.33^{c}	2.16± 0.14^{a}	8.93± 0.17^{c}	10.10± 0.28^{a}	6.75± 0.51^{a}	10.68± 0.18^{a}	-	10.37± 0.29^{a}	5.85± 0.22^{a}	**11.98± 2.30^{a}**
	Mean ±SE	**9.22 ± 0.17^{D}**	**25.11 ± 1.13^{I}**	**7.72 ± 0.32^{C}**	**12.75 ± 0.57^{G}**	**12.85 ± 0.42^{G}**	**35.06 ± 0.39^{J}**	**2.46± 0.12^{A}**	**7.19± 0.31^{B}**	**12.40± 0.53^{G}**	**7.01± 0.23^{B}**	**11.33± 0.18^{F}**	-	**13.65± 0.58^{H}**	**9.85± 0.70^{E}**	
Neck	**Proximal**	10.06 ± 0.16^{c}	22.20 ± 0.24^{b}	4.65 ± 0.24^{b}	8.85 ± 0.26^{a}	11.32 ± 0.48^{a}	33.28 ± 0.41^{c}	1.97± 0.04^{a}	7.19± 0.13^{b}	11.23± 0.13^{c}	4.70± 0.11^{b}	19.02± 0.38^{c}	-	9.62± 0.15^{a}	7.41± 0.24^{a}	**11.66± 0.22^{b}**
	Middle	9.60 ± 0.40^{b}	21.34 ± 0.44^{a}	4.40 ± 0.36^{a}	11.79 ± 0.33^{c}	13.33 ± 0.24^{b}	31.96 ± 0.35^{b}	3.04± 0.13^{c}	4.54± 0.09^{a}	10.23± 0.21^{b}	4.12± 0.16^{a}	17.39± 0.40^{b}	-	10.16± 0.37^{b}	7.91± 0.20^{b}	**11.52± 2.09^{b}**
	Distal	4.59 ± 0.10^{a}	24.03 ± 0.33^{c}	4.59 ± 0.16^{ab}	10.61 ± 0.26^{b}	15.13 ± 0.28^{c}	25.10 ± 0.48^{a}	2.83± 0.12^{b}	7.41± 0.09^{b}	8.16± 0.26^{a}	5.70± 0.17^{c}	15.70± 0.17^{a}	-	10.62± 0.16^{c}	9.48± 0.18^{c}	**11.07± 1.83^{a}**
	Mean ±SE	**8.09 ± 0.62^{C}**	**22.52 ± 0.33^{I}**	**4.55 ± 0.15^{B}**	**10.41 ± 0.33^{F}**	**13.26 ± 0.42^{G}**	**30.11 ± 0.90^{J}**	**2.62± 0.13^{A}**	**6.38± 0.32^{C}**	**9.87± 0.33^{E}**	**4.84± 0.25^{B}**	**17.37± 0.37^{H}**	-	**10.13± 0.17^{EF}**	**8.27± 0.24**	
Back	**Proximal**	5.13 ± 0.59^{a}	-	7.63 ± 0.46^{c}	19.03 ± 0.20^{b}	3.93 ± 0.15^{a}	22.92 ± 0.45^{b}	2.15± 0.08^{a}	6.38± 0.16^{c}	5.59± 0.20^{a}	9.75± 0.10^{c}	12.55± 0.36^{c}	-	10.73± 0.16^{a}	7.03± 0.23^{a}	**9.40± 1.71^{a}**
	Middle	6.05 ± 0.88^{b}	-	6.11 ± 0.20^{b}	22.95 ± 0.20^{c}	8.38 ± 0.19^{b}	19.11 ± 0.44^{a}	3.21± 0.22^{c}	5.52± 0.06^{b}	6.75± 0.24^{b}	6.64± 0.24^{a}	11.81± 0.14^{b}	-	15.63± 0.13^{c}	7.94± 0.33^{b}	**10.01± 1.69^{b}**
	Distal	5.99 ± 0.54^{b}	-	4.62 ± 0.20^{a}	18.50 ± 0.31^{a}	8.08 ± 0.47^{b}	30.48 ± 1.25^{c}	2.88± 0.18^{b}	4.85± 0.27^{a}	8.49± 0.25^{c}	7.59± 0.16^{b}	9.90± 0.25^{a}	-	12.53± 0.31^{b}	10.28± 0.23^{c}	**10.35± 2.09^{c}**
	Mean ±SE	**5.73 ± 0.39^{BC}**	-	**6.12 ± 0.34^{C}**	**20.16 ± 0.50^{I}**	**6.79 ± 0.52^{D}**	**24.17 ± 1.23^{J}**	**2.75± 0.14^{A}**	**5.58± 0.18^{B}**	**6.94± 0.32^{D}**	**7.99± 0.48^{E}**	**11.42± 0.31^{G}**	-	**12.96± 0.50^{H}**	**8.42± 0.36^{F}**	

Species→ Region ↓	Hair Parts	Cattle	Buffalo	Sheep	Goat	Horse	Domestic Pig	Cat	Dog	Spotted Deer	Sambar	Nilgai	Hanuman Langur	Tiger	Leopard	Pooled Mean± SE
Abdomen	Proximal	6.08± 0.75^{b}	14.84± 0.37^{c}	5.18± 0.32^{a}	9.35± 0.68^{c}	5.46± 0.15^{c}	34.79± 0.22^{c}	3.62± 0.14^{b}	7.99± 0.10^{b}	4.84± 0.13^{a}	3.43± 0.04^{a}	7.28± 0.14^{a}	-	10.56± 0.10^{b}	4.43± 0.11^{a}	**9.07± 2.15^{b}**
	Middle	6.28± 0.76^{b}	12.12± 0.22^{b}	6.00± 0.16^{b}	8.13± 0.07^{b}	5.18± 0.18^{b}	28.95± 0.34^{b}	2.88± 0.10^{a}	8.19± 0.22^{c}	5.56± 0.25^{b}	3.33± 0.19^{a}	8.69± 0.33^{b}	-	10.51± 0.22^{b}	7.36± 0.19^{c}	**8.71± 1.70^{c}**
	Distal	4.40± 0.72^{a}	9.75± 0.19^{a}	6.58± 0.34^{b}	6.11± 0.30^{a}	4.47± 0.10^{a}	23.69± 0.08^{a}	3.85± 0.15^{b}	3.70± 0.12^{a}	4.78± 0.17^{a}	5.10± 0.08^{b}	9.74± 0.09^{c}	-	8.82± 0.35^{a}	6.34± 0.13^{b}	**7.49± 1.36^{a}**
	Mean ±SE	**5.59± 0.45^{D}**	**12.24± 0.53^{I}**	**5.92± 0.21^{D}**	**7.86± 0.40^{F}**	**5.04± 0.13^{C}**	**29.14± 1.11^{J}**	**3.45 ± 0.12^{a}**	**6.62± 0.51^{E}**	**5.06± 0.13^{C}**	**3.95± 0.30^{B}**	**8.57± 0.27^{G}**	-	**9.96± 0.24^{H}**	**6.04± 0.31^{D}**	
Thigh	Proximal	8.50± 0.14^{c}	6.72± 0.21^{a}	8.74± 0.15^{c}	8.60± 0.24^{b}	7.17± 0.23^{b}	31.64± 0.21^{c}	2.78± 0.23^{b}	7.94± 0.23^{c}	5.06± 0.25^{c}	3.88± 0.08^{b}	9.60± 0.37^{a}	19.96± 0.37^{b}	10.02± 0.24^{b}	6.03± 0.13^{a}	**9.76± 1.92^{ab}**
	Middle	5.42± 0.19^{b}	13.57± 0.45^{b}	7.72± 0.17^{b}	7.89± 0.20^{a}	5.26± 0.27^{a}	27.04± 0.33^{b}	2.39± 0.10^{a}	7.55± 0.13^{b}	4.79± 0.20^{b}	3.16± 0.07^{a}	10.70± 0.12^{c}	24.58± 0.34^{c}	10.98± 0.26^{c}	7.49± 0.12^{b}	**9.90± 1.91^{b}**
	Distal	4.87± 0.34^{a}	17.86± 0.63^{c}	5.15± 0.20^{a}	9.96± 0.30^{c}	13.19± 0.37^{c}	23.53± 0.27^{a}	2.68± 0.14^{b}	7.27± 0.15^{a}	3.34± 0.29^{a}	6.31± 0.07^{c}	10.18± 0.22^{b}	13.88± 0.55^{a}	7.97± 0.28^{a}	8.35± 0.14^{c}	**9.61± 1.50^{a}**
	Mean ±SE	**6.27± 0.41^{C}**	**12.72± 1.14^{H}**	**7.20± 0.38^{D}**	**8.82± 0.25^{E}**	**8.54± 0.84^{E}**	**27.40± 0.82^{J}**	**2.62± 0.10^{A}**	**7.59± 0.12^{D}**	**4.40± 0.23^{B}**	**4.45± 0.48^{B}**	**10.16± 0.18^{G}**	**19.47± 1.09^{I}**	**9.66± 0.33^{F}**	**7.29± 0.24^{D}**	
Tail	Proximal	16.37± 0.49^{b}	-	9.87± 0.16^{b}	18.75± 0.43^{c}	-	-	3.77± 0.05^{a}	6.78± 0.25^{b}	8.73± 0.11^{c}	5.99± 0.01^{b}	14.93± 0.11^{c}	25.16± 0.28^{c}	11.89± 0.23^{b}	8.89± 0.19^{c}	**11.92± 1.82^{b}**
	Middle	15.39± 0.18^{a}	-	15.68± 0.50^{c}	9.29± 0.83^{a}	-	-	4.10± 0.39^{b}	6.02± 0.24^{a}	5.20± 0.22^{a}	5.62± 0.02^{a}	11.36± 0.33^{b}	21.51± 0.24^{b}	10.20± 0.15^{a}	7.31± 0.26^{a}	**10.15± 1.56^{a}**
	Distal	15.51± 1.23^{a}	-	8.61± 0.25^{a}	10.96± 0.22^{b}	-	-	5.39± 0.45^{c}	7.67± 0.27^{c}	7.11± 0.23^{b}	6.09± 0.09^{b}	9.70± 0.25^{a}	16.56± 0.40^{a}	14.47± 0.18^{c}	8.52± 0.10^{b}	**10.05± 1.11^{a}**
	Mean ±SE	**15.75± 0.43^{H}**	-	**11.39± 0.77^{E}**	**13.00± 1.04^{G}**	-	-	**4.42± 0.25^{A}**	**6.82± 0.21^{C}**	**7.01± 0.36^{C}**	**5.90± 0.08^{B}**	**12.00± 0.55^{EF}**	**21.08± 0.87^{I}**	**12.19± 0.44^{F}**	**8.24± 0.19^{D}**	

Means bearing same superscript for different region (for hair part) in a column do not differ significantly/Means bearing same superscript for different species (hair part of different region) in a row do not differ significantly/Head Region-CD for Species is0.47 and CD for hair part is 0.23/Neck Region-CD for Species is 0.43and CD for hair part is 0.21/Back Region-CD for Species is 0.61 and CD for hair part is 0.30/Abdomen Region-CD for Species is 0.49 and CD for hair part is 0.23/Thigh Region-CD for Species is 0.44 and CD for hair part is 0.20/Tail Region-CD for Species is 0.62 and CD for haira part is 0.32

At the distal part of hair, amongst all the species with respect to various body regions, minimum cortex width was noted at head region of Cat (2.16 ± 0.14 μm) while maximum cortex width was found at back region of Goat (18.50 ± 0.31 μm).

Mean and SE for cortex width at different parts of hair at various body regions of different species are depicted in Table 4.29. Amongst all the species, the lowest cortex width was noted at all the body regions in Cat Table 4.29, whereas the highest cortex width was observed in the hair of pig at all the body regions except tail region where medulla was not seen Table 4.29. At tail region the maximum cortex width was seen in Hanuman Langur (21.08 ± 0.87 μm).

Table. 4.30. Means and SE for maximum hair diameter (µm) in various region of body in different species.

Species→ Region ↓	Cattle	Buffalo	Sheep	Goat	Horse	Domestic Pig	Cat	Dog	Spotted Deer	Sambar	Nilgai	Hanuman Langur	Tiger	Leopard	Pooled Mean ± SE
Head	42.43 ± 3.76^{d}	64.62 ± 3.75^{a}	40.32± 2.64^{a}	47.82± 1.09^{a}	44.37± 0.94^{c}	118.16± 7.26^{d}	23.50± 2.38^{d}	23.26± 0.82^{a}	33.15 ± 1.28^{a}	62.95 ± 0.94^{a}	58.81 ± 1.38^{a}	42.35 ± 4.09^{bc}	50.34 ± 3.68^{c}	37.72 ± 4.00^{a}	**49.27±** 6.06^{a}
Neck	29.40 ± 3.61^{a}	88.07 ± 2.16^{e}	52.82± 3.81^{c}	78.87± 4.14^{f}	34.06± 1.93^{b}	61.17± 2.47^{a}	17.05± 1.25^{b}	21.44± 1.67^{a}	35.96 ± 1.96^{b}	95.92 ± 1.57^{f}	108.36± 4.66^{e}	40.05 ± 2.78^{ab}	51.49 ± 3.21^{c}	40.72 ± 1.15^{b}	**53.96±** 7.38^{b}
Back	48.78 ± 2.87^{e}	83.60 ± 2.43^{d}	38.70± 2.28^{a}	72.11± 3.18^{e}	44.46± 1.43^{c}	74.57± 2.23^{b}	43.03± 1.77^{e}	23.32± 2.14^{a}	38.16 ± 2.13^{b}	77.12 ± 9.62^{c}	86.34 ± 4.95^{d}	43.53 ± 2.91^{c}	38.72 ± 1.27^{b}	35.33 ± 1.58^{a}	**53.41±** 5.33^{b}
Abdomen	37.68 ± 0.85^{c}	77.83 ± 4.17^{c}	70.46± 7.72^{e}	57.44± 1.21^{c}	29.04± 1.29^{a}	75.40± 3.92^{b}	11.67± 0.66^{a}	36.85± 3.33^{c}	44.63 ± 0.51^{d}	91.65 ± 4.42^{e}	76.07 ± 2.06^{b}	40.09 ± 0.93^{ab}	58.95 ± 4.59^{d}	47.66 ± 2.14^{d}	**53.96±** 5.77^{b}
Thigh	32.22 ± 2.40^{b}	62.54 ± 2.41^{a}	48.60± 2.58^{b}	54.33± 0.76^{b}	45.96± 2.18^{c}	105.43± 3.38^{c}	24.59± 3.89^{d}	39.23± 5.32^{c}	42.03 ± 1.37^{c}	83.56 ± 0.38^{d}	83.40 ± 2.67^{c}	41.40 ± 1.06^{abc}	38.07 ± 1.10^{b}	46.49 ± 3.02^{d}	**53.42±** 5.85^{b}
Tail	48.58 ± 1.16^{e}	74.51 ± 2.86^{b}	57.78± 8.18^{d}	63.73± 4.21^{d}	97.92± 1.66^{d}	74.22± 8.03^{b}	20.65± 1.07^{c}	27.74 ± 2.97^{b}	53.68 ± 3.76^{e}	72.33 ± 4.17^{b}	57.87 ± 0.69^{a}	39.28 ± 0.91^{a}	35.36 ± 0.54^{a}	43.73 ± 1.91^{c}	**54.81±** 5.40^{b}
Pooled Mean±SE	**39.85 ±** 1.62^{C}	**75.20 ±** 1.95^{G}	**51.45±** 2.67^{E}	**62.38±** 2.08^{F}	**49.30±** 3.87^{E}	**84.83±** 3.89^{I}	**23.42±** 1.84^{A}	**28.64±** 1.65^{B}	**41.27 ±** 1.38^{C}	**80.59 ±** 3.56^{H}	**78.47 ±** 3.16^{GH}	**41.12 ±** 0.96^{C}	**45.49±** 1.81^{D}	**41.94 ±** 1.21^{CD}	

Means bearing same superscript in a column do not differ significantly

CD for Species is 3.69 and CD for Region is 2.41

4.5.7 Diameter of hair in cross section

The Means and SE for maximum and minimum diameter of hair from the cross section in various regions of body in different species are summarized in Table 4.30 and 4.31 respectively. The lowest maximum diameter of hair was found in Cat (23.42 ± 1.84 µm), while highest maximum diameter of hair was found in Pig (84.83 ± 3.89 µm). Amongst all species with respect to all the body regions, the lowest maximum hair diameter was observed in the abdomen region of Cat (11.67 ± 0.66 µm), while highest maximum hair diameter was seen in the head region of Domestic Pig (118.16 ± 7.26 µm).

The present study indicated that the minimum diameter of hair was in the range of (18.64 ± 1.50 µm) found in Cat and (77.75 ± 2.91 µm) in Domestic Pig. Amongst all the species and all the regions of body, the lowest minimum diameter of hair was noted in the abdomen region of Cat (9.66 ± 0.22 µm), while highest minimum diameter of hair was reported in head region of Domestic Pig (104.10 ± 4.13 µm).

4.5.8 Diameter of medulla in cross section

The lowest diameter of medulla was found in the hair of Domestic Pig (9.47 ± 0.50 µm), while highest diameter of medulla was observed in Sambar (49.43 ± 3.53 µm).

Amongst all the species and body regions, the lowest diameter of medulla was recorded in hair of neck region of Domestic Pig (6.81 ± 0.80 µm) and highest diameter of medulla was observed in the hair of neck region of Sambar (65.16 ± 2.95 µm). The Means and SE for diameter of medulla in cross section of hair from various body regions in different species are presented in Table 4.32.

Table. 4.31. Means and SE for minimum hair diameter (μm) in various region of body in different species.

Species→ Region ↓	Buffalo	Goat	Cattle	Horse	Domestic Pig	Dog	Cat	Sheep	Spotted Deer	Nilgai	Tiger	Hanuman Langur	Leopard	Sambar	Pooled Mean ± SE
Head	45.30 ± 4.48^{b}	41.45 ± 1.47^{b}	28.78± 1.56^{c}	34.21 ± 1.43^{d}	104.10 ± 4.13^{f}	19.58± 0.60^{a}	16.30 ± 1.98^{c}	26.69 ± 1.55^{d}	28.34 ± 1.42^{a}	50.06 ± 1.94^{b}	42.97 ± 4.55^{d}	32.55 ± 1.79^{a}	29.74 ± 3.28^{b}	45.52 ± 1.41^{b}	**38.97**± 5.48^{a}
Neck	60.74 ± 2.92^{e}	58.73 ± 0.80^{e}	20.40± 2.56^{a}	29.43 ± 2.15^{b}	58.30 ± 1.77^{a}	18.21± 1.30^{a}	13.76 ± 0.83^{b}	15.07 ± 1.35^{b}	26.69 ± 1.98^{a}	97.68 ± 5.37^{e}	42.88 ± 1.26^{d}	36.90 ± 1.03^{b}	34.84 ± 2.16^{d}	59.01 ± 2.76^{e}	**40.90**± 6.10^{b}
Back	55.35 ± 1.65^{d}	49.34 ± 1.47^{d}	29.09± 2.29^{cd}	32.28 ± 1.49^{c}	69.84 ± 2.35^{c}	21.37± 1.61^{b}	35.08 ± 1.39^{e}	26.84 ± 1.24^{d}	33.06 ± 2.43^{b}	78.89 ± 5.05^{d}	29.53 ± 0.88^{a}	36.54 ± 2.06^{b}	31.22 ± 1.88^{bc}	42.08 ± 7.29^{a}	**40.75**± 4.34^{b}
Abdomen	53.44 ± 1.74^{c}	38.55 ± 2.04^{a}	30.57± 0.90^{d}	27.06 ± 1.24^{a}	67.35 ± 4.65^{b}	30.53± 3.06^{c}	9.66 ± 0.22^{a}	24.11 ± 2.99^{c}	34.29 ± 2.01^{b}	59.67 ± 3.38^{c}	38.37 ± 0.72^{c}	39.75 ± 1.56^{c}	38.78 ± 1.74^{e}	46.57 ± 1.68^{b}	**38.48**± 3.84^{a}
Thigh	43.13 ± 2.99^{a}	44.45 ± 0.48^{c}	22.15± 1.17^{b}	45.96 ± 2.18^{e}	90.75 ± 4.90^{e}	21.53± 3.14^{b}	20.66 ± 2.85^{d}	10.82 ± 1.13^{a}	38.29 ± 1.18^{c}	48.96 ± 1.24^{b}	33.37 ± 0.98^{b}	40.23 ± 1.98^{c}	32.01 ± 1.48^{c}	50.13 ± 4.72^{c}	**38.75**± 4.95^{a}
Tail	60.60 ± 3.12^{e}	49.18 ± 1.20^{d}	47.01± 0.95^{e}	89.15 ± 0.77^{f}	76.15 ± 1.01^{d}	22.54± 2.28^{b}	16.40 ± 1.17^{c}	13.98 ± 1.04^{b}	39.68 ± 1.39^{c}	41.52 ± 0.71^{a}	33.22 ± 0.60^{b}	42.00 ± 2.20^{d}	14.12 ± 0.94^{a}	56.75 ± 3.45^{d}	**43.02**± 5.84^{c}
Pooled Mean±SE	**53.09** ± 1.61^{F}	**46.95** ± 1.22^{G}	**29.67**± 1.59^{C}	**43.01** ± 3.68^{F}	**77.75** ± 2.91^{H}	**22.29**± 1.07^{B}	**18.64** ± 1.50^{A}	**19.58** ± 1.27^{A}	**33.39** ± 1.05^{D}	**62.80** ± 3.55^{G}	**36.72** ± 1.15^{E}	**38.00** ± 0.86^{E}	**30.12** ± 1.52^{C}	**50.01** ± 2.47^{E}	

Means bearing same superscript in a column do not differ significantly

CD for Species is 2.67 and CD for Region is 1.75

Table. 4.32. Means and SE for medulla diameter (μm) of hair in various region of body in different species.

Species→ Region ↓	Cattle	Buffalo	Sheep	Goat	Horse	Domestic Pig	Cat	Dog	Spotted Deer	Sambar	Nilgai	Tiger	Leopard	Pooled Mean ± SE
Head	17.56± 2.77^{c}	26.72± 0.55^{a}	20.14± 1.30^{a}	19.93± 0.97^{b}	20.68± 1.27^{c}	11.95± 1.21^{c}	10.75± 1.15^{b}	9.42± 0.44^{a}	18.94± 1.02^{a}	30.22± 6.44^{a}	21.83± 1.60^{a}	23.21± 2.69^{e}	19.70± 2.36^{c}	**19.31± 1.53^{b}**
Neck	7.43± 1.26^{a}	36.41± 2.90^{c}	33.95± 2.26^{b}	30.45± 1.73^{d}	10.55± 0.65^{a}	6.81± 0.80^{a}	9.12± 0.89^{b}	9.75± 1.29^{a}	20.33± 1.96^{a}	65.16± 2.95^{e}	57.01± 3.12^{f}	22.24± 1.56^{e}	17.91± 1.67^{b}	**25.16± 4.85^{c}**
Back	22.87± 1.08^{e}	37.26± 2.63^{c}	21.08± 2.47^{a}	37.50± 3.45^{e}	21.37± 0.95^{c}	9.40± 0.95^{b}	21.19± 1.99^{c}	9.70± 0.92^{a}	25.53± 1.84^{b}	45.74± 8.48^{b}	37.15± 4.88^{c}	14.94± 1.31^{c}	20.05± 2.71^{c}	**24.91± 2.90^{c}**
Abdomen	20.03± 0.41^{d}	50.49± 2.46^{d}	39.75± 3.37^{c}	30.63± 1.09^{d}	15.72± 1.28^{b}	8.92± 1.01^{b}	5.44± 0.36^{a}	14.70± 1.41^{b}	30.10± 2.04^{d}	54.59± 3.44^{d}	39.50± 1.75^{d}	19.02± 1.20^{d}	22.13± 1.30^{d}	**27.00± 3.96^{d}**
Thigh	13.23± 2.03^{b}	32.49± 2.29^{b}	33.46± 2.66^{b}	24.81± 1.34^{c}	31.45± 2.28^{d}	10.33± 1.72^{bc}	12.61± 2.55^{c}	15.69± 2.03^{b}	28.44± 1.54^{c}	51.24± 2.73^{c}	44.43± 1.38^{e}	12.66± 1.28^{b}	24.42± 1.50^{e}	**25.79± 3.30^{cd}**
Tail	16.74± 2.47^{c}	-	39.18± 5.39^{c}	10.98± 0.66^{a}	-	9.41± 0.69^{b}	5.79± 0.58^{a}	9.84± 0.94^{a}	30.82± 2.74^{d}	49.64± 3.44^{b}	26.80± 0.99^{b}	7.79± 0.52^{a}	0.87± 0.08^{a}	**15.99± 1.08^{a}**
Pooled Mean±SE	**16.31± 1.09^{B}**	**30.56± 2.73^{D}**	**31.26± 1.79^{D}**	**25.72± 1.59^{C}**	**16.63± 1.72^{B}**	**9.47± 0.50^{A}**	**10.82± 1.05^{A}**	**11.52± 0.65^{A}**	**25.69± 1.07^{C}**	**49.43± 3.53^{F}**	**37.79± 2.19^{E}**	**16.64± 1.09^{B}**	**17.51± 1.47^{B}**	

Means bearing same superscript in a column do not differ significantly

CD for Species is 2.42 and CD for Region is 1.64

4.5.9. Cortex width in cross section

The cortex width of hair was observed significantly lowest in Cat (4.28 ± 0.34 μm) and was significantly highest in Domestic Pig (33.42 ± 1.40 μm). Amongst all the regions of body and all the species, cortex width of hair was lowest in the hair from abdomen region of Cat (2.34 ± 0.18 μm), while highest cortex width was seen in the hair from head region of Domestic Pig (45.01 ± 2.87 μm). The Means and SE for cortex width (μm) of hair in various region of body in different species are summarized in Table 4.33.

4.5.10. Cuticle width in cross section

The cuticular width of hair was noted lowest in Cat (0.76 ± 0.07 μm), while cuticle width was found highest in Domestic Pig (2.78 ± 0.21 μm). Amongst all species with respect to all the regions of body minimum cuticle width was observed in the hair from abdomen region of Cat (0.44 ± 0.04 μm) and highest cuticular width was reported in the hair from thigh region of Domestic Pig (4.03 ± 0.80 μm). The Means and SE for cuticle width in cross section of hair from various regions of body in different species are given in Table 4.34.

4.5.11. Medulla index

The medulla index is referred as ratio between medulla width and hair width. It was always considered for the differentiation of animal hairs from human hair during legal cases by Kshirsagar et al. (2009). The medulla index was calculated as per the formula given by Kitpipit and Thanakiatkrai (2013). The medulla index could not be calculated in few regions of some species due to absence of medulla viz. back region and tail region in Buffalo, tail region in Horse and Domestic Pig, head, neck, back and abdomen regions in Hanuman Langur. The lowest medulla index was noted in the hair from head region of Buffalo as 5.49 followed by abdomen region of Domestic Pig as 5.83 and highest medulla index was noted in the hair from abdomen region of Sambar as 93.43. The values regarding medulla index of all the body regions in all the species under study are given in Table 4.35. Medulla index in Tiger ranged between33.31 to 62.98. These observations were in agreement with the findings reported by Kitpipit and Thanakiatkrai (2013). Who reported medulla index in Tiger hair as (35.2 ± 11.3). They also mentioned that the, medulla index in Cat as (67.8 ± 7.4), In Domestic Dog as (57.5 ± 9.4). Similar medulla index were noted during the present study in Cat ranged from 51.52 to 77.67 and in Dog it were ranged from 44.11 to 67.55. These findings of the present study were also akin with the findings reported by Sato et al. (2006). In Hanuman Langur the medulla index was found varied from 14.55 to 15.65 amongst all the body regions. These findings were not in agreement with the findings reported by Sarkar et al. (2011) regarding medulla. They observed that the medulla index were ranged from 0.72 to 0.75 in Hanuman Langur.

4.5.12. Hair Index

The hair index is a ratio between minimum and maximum diameter of hair in cross section. It was calculated as per the formula given by Kitpipit and Thanakiatkrai (2013). The lowest hair index was observed in thigh region of Sheep as 22.26 while, highest hair index was noted in tail region of Hanuman Langur as 106.91. The hair index of all the fourteen animal species of six different body regions are mentioned in Table 4.36. The hair index of Tiger ranged from 65.09 to 93.94 irrespective of body regions. Similar range of hair index (64.0 – 100.0) was noted by Kitpipit and Thanakiatkrai (2013) in Tiger hair. During the present study it was observed that the hair index varied from 54.87 to 91.65 in Dog and 69.37 to 84.02 in Cat amongst all the body regions. Similar hair index were noted by Sato et al. (2006) in Dog. They also found that the hair index varied from 41.0 to 100.0 in Domestic Dog, while findings regarding hair index of Cat hair (35.9 to 100.0) do not corroborate with the findings of the present study.

4.5.13. Cuticle Index

The cuticular index was found lowest in the hair of abdomen region of Sambar and highest was observed in the hair from back region of Domestic Pig. The values of cuticle index are given in Table 4.37. It ranged from 1.35 to 2.50 in hair of Tiger irrespective of body regions. Similar range of cuticle index was observed by Kitpipit and Thanakiatkrai (2013) in Tiger hair (2.4 to 3.4).

Table. 4.33. Means and SE for cortex width (µm) of hair in various region of body in different species.

Species→ Region ↓	Cattle	Buffalo	Sheep	Goat	Horse	Domes-tic Pig	Dog	Cat	Spotted Deer	Nilgai	Tiger	Leopard	Sambar	Pooled Mean ± SE
Head	7.57± 0.84^{b}	15.15± 1.06^{b}	8.04± 0.79^{f}	9.97± 0.61^{a}	9.54± 0.72^{c}	45.01± 2.87^{f}	6.72± 0.43^{c}	4.08± 0.77^{b}	5.23± 0.27^{a}	13.73± 1.36^{b}	10.75± 1.21^{b}	6.66± 0.76^{a}	7.54± 0.60^{c}	**11.54± 2.71^{c}**
Neck	6.99± 1.01^{ab}	19.78± 1.51^{e}	5.02± 0.39^{c}	17.04± 0.35^{d}	10.46± 0.88^{d}	24.36± 0.84^{a}	5.28± 0.39^{a}	2.73± 0.30^{a}	4.91± 0.29^{a}	21.46± 2.55^{c}	12.11± 0.46^{c}	9.77± 0.16^{d}	6.80± 0.47^{b}	**11.29± 1.83^{bc}**
Back	7.75± 0.44^{c}	17.91± 0.55^{d}	7.18± 0.47^{e}	11.28± 0.99^{c}	8.74± 0.56^{b}	27.18± 1.23^{b}	5.98± 0.83^{b}	7.42± 0.49^{d}	5.13± 0.30^{a}	22.12± 1.08^{c}	9.36± 0.31^{a}	7.35± 0.36^{b}	6.02± 0.29^{a}	**11.03± 1.78^{bc}**
Abdomen	9.90± 0.42^{d}	15.94± 1.05^{c}	6.42± 0.75^{d}	10.32± 0.26^{b}	5.78± 0.26^{a}	29.41± 1.90^{c}	9.64± 0.94^{d}	2.34± 0.18^{a}	5.44± 0.32^{a}	13.84± 1.46^{b}	12.31± 0.73^{c}	10.43± 0.67^{e}	7.61± 0.09^{cd}	**10.72± 1.72^{b}**
Thigh	6.44± 0.36^{a}	11.66± 0.52^{a}	2.58± 0.08^{a}	11.70± 0.78^{c}	9.29± 0.53^{bc}	41.16± 1.04^{e}	5.70± 0.92^{ab}	4.26± 0.66^{bc}	6.40± 0.33^{b}	11.21± 1.11^{a}	11.22± 0.31^{b}	8.38± 0.60^{c}	8.26± 0.83^{d}	**10.64± 2.47^{ab}**
Tail	13.13± 1.38^{e}	-	3.75± 0.52^{b}	21.04± 0.63^{e}	-	33.38± 0.87^{d}	7.23± 0.87^{c}	4.87± 0.25^{c}	7.88± 0.52^{c}	10.94± 0.59^{a}	11.91± 0.40^{c}	9.65± 0.82^{d}	6.23± 0.31^{ab}	**10.00± 2.31^{a}**
Pooled Mean±SE	**8.63± 0.50^{E}**	**13.41± 1.15^{G}**	**5.50± 0.38^{B}**	**13.56± 0.73^{G}**	**7.30± 0.64^{D}**	**33.42± 1.40^{H}**	**6.76± 0.38^{CD}**	**4.28± 0.34^{A}**	**5.83± 0.22^{BC}**	**15.55± 0.95^{G}**	**11.28± 0.30^{F}**	**8.71± 0.32^{E}**	**7.08± 0.31^{D}**	

Means bearing same superscript in a column do not differ significantly
CD for Species is 1.00 and CD for Region is 0.68

Table. 4.34. Means and SE for cuticle width (μm) of hair in various region of body in different species.

Species→ Region ↓	Cattle	Buffalo	Sheep	Goat	Horse	Domestic Pig	Cat	Dog	Spotted Deer	Sambar	Nilgai	Hanuman Langur	Tiger	Leopard	Pooled Mean ± SE
Head	1.15± 0.17	1.11± 0.17	1.46± 0.15	1.43± 0.06	1.11± 0.06	2.14± 0.22	0.90± 0.07	0.85± 0.05	0.80± 0.06	1.00± 0.04	1.43± 0.11	0.97± 0.11	1.12± 0.08	0.76± 0.03	**1.16±** 0.09
Neck	0.92± 0.11	1.56± 0.28	0.75± 0.11	2.17± 0.33	0.81± 0.07	1.67± 0.19	0.52± 0.04	0.71± 0.06	0.87± 0.09	1.19± 0.13	1.38± 0.11	0.81± 0.10	0.99± 0.13	1.20± 0.19	**1.11±** 0.12
Back	1.21± 0.20	1.16± 0.12	1.04± 0.13	1.17± 0.16	1.00± 0.12	3.31± 0.46	1.48± 0.17	0.78± 0.07	0.60± 0.05	0.81± 0.18	1.27± 0.16	0.71± 0.09	0.82± 0.07	0.91± 0.07	**1.16±** 0.17
Abdomen	0.98± 0.12	0.99± 0.10	0.89± 0.11	1.31± 0.12	0.77± 0.05	3.09± 0.27	0.44± 0.04	1.09± 0.08	0.97± 0.13	0.71± 0.12	1.11± 0.13	1.07± 0.10	0.79± 0.08	1.18± 0.10	**1.10±** 0.16
Thigh	0.65± 0.05	1.38± 0.20	0.54± 0.04	1.66± 0.11	1.25± 0.09	4.03± 0.80	0.59± 0.06	0.69± 0.11	0.56± 0.06	1.10± 0.18	1.18± 0.18	1.08± 0.16	0.95± 0.08	0.89± 0.07	**1.18±** 0.23
Tail	1.58± 0.14	1.41± 0.09	1.11± 0.13	1.90± 0.21	2.03± 0.38	2.42± 0.38	0.63± 0.08	0.69± 0.16	0.89± 0.09	0.87± 0.10	1.05± 0.11	0.86± 0.10	0.85± 0.14	0.87± 0.08	**1.23±** 0.14
Pooled Mean±SE	**1.08±** 0.07^{CDE}	**1.27±** 0.07^{E}	**0.97±** 0.07^{BCD}	**1.61±** 0.09^{F}	**1.16±** 0.10^{DE}	**2.78±** 0.21^{G}	**0.76±** 0.07^{A}	**0.80±** 0.04^{AB}	**0.78±** 0.04^{AB}	**0.95±** 0.08^{ABC}	**1.24±** 0.06^{E}	**0.92±** 0.05^{ABC}	**0.92±** 0.04^{ABC}	**0.97±** 0.05^{BCD}	

Means bearing same superscript in a column do not differ significantly

CD for Species is 0.20

Table. 4.35. Medullary Index of hair shaft at various regions of body in different species

Species→ Region ↓	Cattle	Buffalo	Sheep	Goat	Horse	Domestic Pig	Cat	Dog	Spotted Deer	Sambar	Nilgai	Hanuman Langur	Tiger	Leopard
Head	15.54	5.49	77.05	25.09	37.26	6.81	51.52	63.76	45.80	78.07	43.16	-	62.98	47.62
Neck	34.60	23.31	83.28	70.49	29.93	7.21	67.06	44.11	49.90	88.89	59.19	-	58.16	67.10
Back	64.34	-	77.71	11.11	48.47	50.55	64.71	67.55	76.06	78.66	60.87	-	47.24	65.90
Abdomen	52.70	60.20	74.55	53.52	65.06	5.83	75.21	49.12	84.59	93.43	79.86	-	33.31	45.69
Thigh	42.20	51.76	70.00	51.64	60.52	12.55	72.48	48.24	87.58	90.05	55.88	14.55	45.44	65.44
Tail	15.95	-	55.08	48.06	-	-	77.67	58.22	77.01	83.60	58.30	15.65	33.44	42.22

Table. 4.36. Hair Index of various regions of body in different species

Species→ Region ↓	Cattle	Buffalo	Sheep	Goat	Horse	Domestic Pig	Cat	Dog	Spotted Deer	Sambar	Nilgai	Hanuman Langur	Tiger	Leopard
Head	67.83	70.09	66.20	86.68	77.10	88.1	69.37	84.17	85.50	72.31	85.12	76.88	85.35	78.83
Neck	69.38	68.96	28.52	74.46	86.41	95.30	80.71	84.95	74.23	61.52	90.14	92.13	83.27	85.55
Back	59.64	66.21	69.35	68.43	72.50	93.67	81.51	91.65	86.64	54.57	91.37	83.94	76.27	88.39
Abdomen	81.14	68.66	34.21	67.13	93.16	89.32	82.72	82.85	76.82	50.81	78.45	99.16	65.09	81.37
Thigh	68.74	68.96	22.26	81.82	100.00	86.08	84.02	54.87	91.11	59.99	58.70	97.18	87.65	68.87
Tail	96.77	81.33	24.19	77.18	91.04	102.59	79.41	81.24	73.93	78.47	71.75	106.91	93.94	32.29

Table. 4.37. Cuticle Index of hair shaft at various regions of body in different species

Species→ Region ↓	Cattle	Buffalo	Sheep	Goat	Horse	Domestic Pig	Cat	Dog	Spotted Deer	Sambar	Nilgai	Hanuman Langur	Tiger	Leopard
Head	2.71	1.71	3.61	2.99	2.50	1.81	3.85	3.65	2.42	1.59	2.43	2.29	2.22	2.01
Neck	3.13	1.78	1.43	2.75	2.37	2.74	3.07	3.33	2.41	1.24	1.28	2.02	1.91	2.95
Back	2.48	1.39	2.69	1.62	2.26	4.44	3.45	3.35	1.58	1.05	1.47	1.62	2.12	2.56
Abdomen	2.59	1.27	1.26	2.28	2.65	4.10	3.81	2.97	2.18	0.78	1.46	2.66	1.35	2.47
Thigh	2.01	2.21	1.11	3.05	2.73	3.82	2.42	1.75	1.34	1.31	1.41	2.62	2.50	1.92
Tail	3.26	1.89	1.92	2.99	2.08	3.26	3.04	2.49	1.65	1.20	1.82	2.19	2.40	2.00

4.6 Scanning electron microscopy

The hair samples of all the species under study were subjected for Scanning Electron Microscopy (SEM). The proximal parts of hair from back region of all species were considered for SEM as a representative of dorsal guard hair. The cuticular scales of dorsal guard hair were observed at 600X and 1000X magnification.

The scanning electron microscopic findings of the present study revealed that the cuticular scales of dorsal guard hair amongst all the species under study showed free cuticular scale margins, directed towards the tip or distal part of the hair shaft. These findings are in concurrence with the observations of Rajaram and Menon (1986) in Porcupine.

The cuticular scales of guard hair of Cattle (*Bos indicus*) were transversely placed and were slightly oblique at some places. The cuticular scales were arranged in regular wave pattern with crenate free borders. It was observed that the cuticular scales were distantly placed (Fig. 4.199.). Generally one to two scales were observed in a single row throughout the length of hair shaft.

The dorsal guard hair of Buffalo (*Bubalus bubalis*) showed transversally placed cuticular scales with regular wave pattern. The individual cuticular scales showed crenate margins and the distance between scale margins was observed as distant (Fig. 4.198.). Two to three scales were noted in each row.

The Sheep (*Ovis aries*) hair showed intermediate position of cuticular scales. The scales were arranged in regular wave pattern with smooth scale margins and distant, scale margin distance. It was also seen that the number of scales per row was two to three (Fig. 4.199.).

It was observed that the cuticular scales were transverse in position in Goat (*Capra hircus*). They were arranged in irregular wave pattern with rippled border. The distance between scale margin was observed as near (Fig. 4.200). The number of scales per row varies from three to four scales.

The cuticular scale position in dorsal guard hair of Horse (*Equus cabellus*) was noticed as intermediate with irregular wave pattern. The individual scales were observed with the smooth to slightly rippled scale margins and the distance between scale margins was distant (Fig. 4.201).

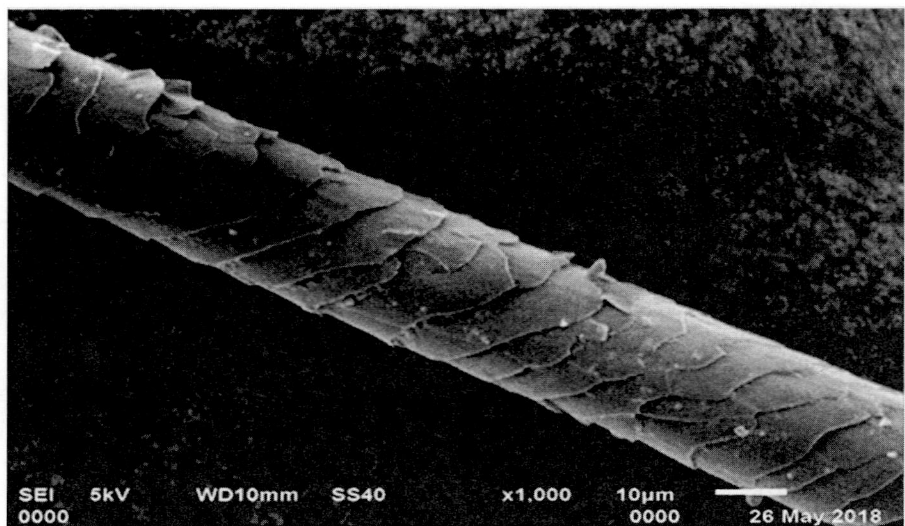

Fig. 4.197. Scanning electron photomicrograph showing cuticular scale pattern of dorsal guard hair of Cattle (*Bos indicus*) (1,000X)

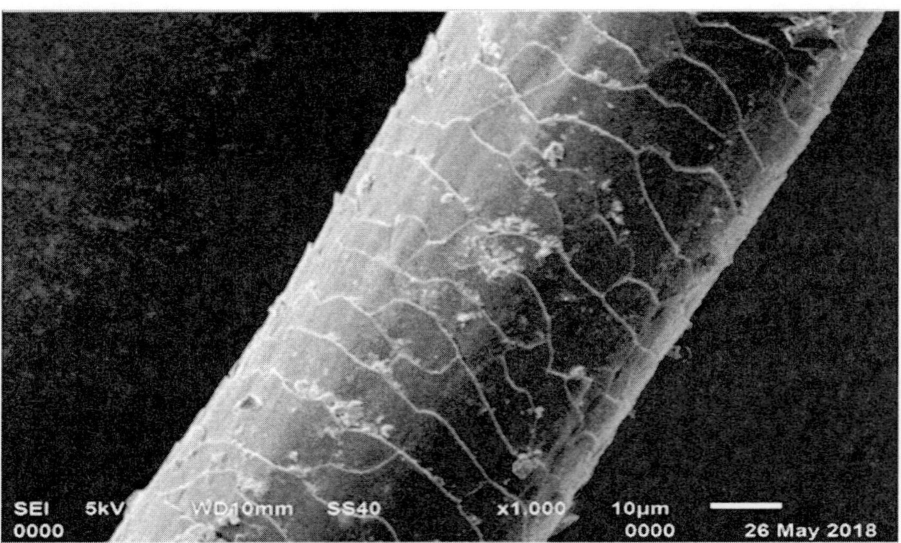

Fig. 4.198. Scanning electron photomicrograph showing cuticular scale pattern of dorsal guard hair of Buffalo (*Bubalus bubalis*) (1,000X)

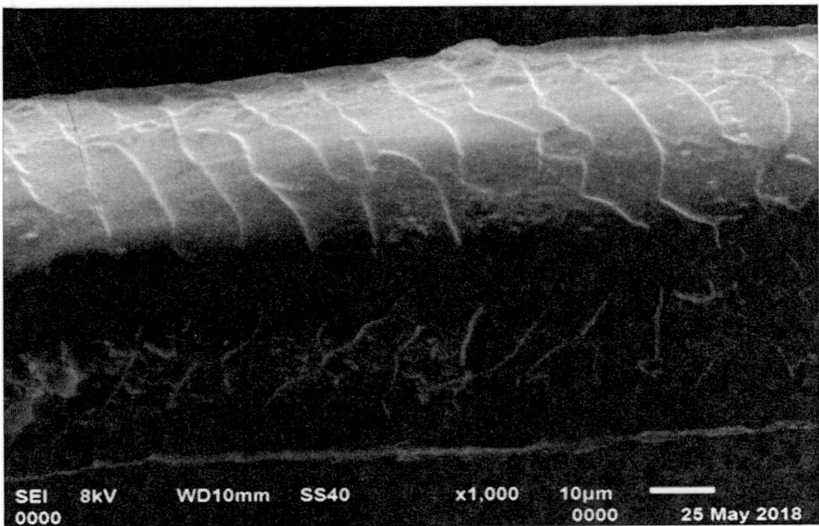

Fig. 4.199. Scanning electron photomicrograph showing cuticular scale pattern of dorsal guard hair of Sheep (*Ovis aries*) (1,000X)

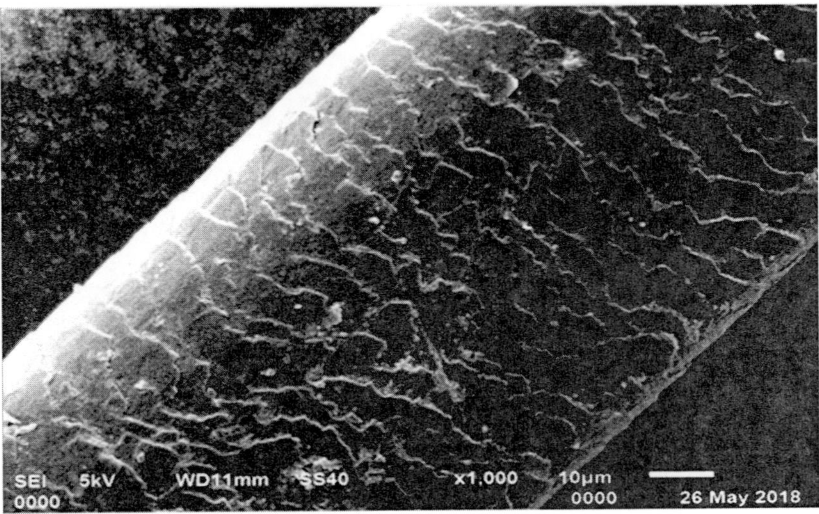

Fig. 4.200. Scanning electron photomicrograph showing cuticular scale pattern of dorsal guard hair of Goat (*Capra hircus*) (1,000X)

The cuticular characteristics in Domestic Pig (*Sus scrofa domesticus*) showed transversal scale position and had regular wave scale pattern. The structure of scale margins was found rippled and distance between scale margins was close. The number of scales per row was observed as six to seven (Fig. 4.202.).

Scanning electron microscopically the dorsal guard hair of Cat (*Felis catus domesticus*) showed that the scales were placed transversely. It was also noticed that the scales were arranged in irregular wave pattern. The scales were found with slightly rippled scale margins with near scale margin distance. It was also observed that five to six scales were present in each row (Fig. 4.203).

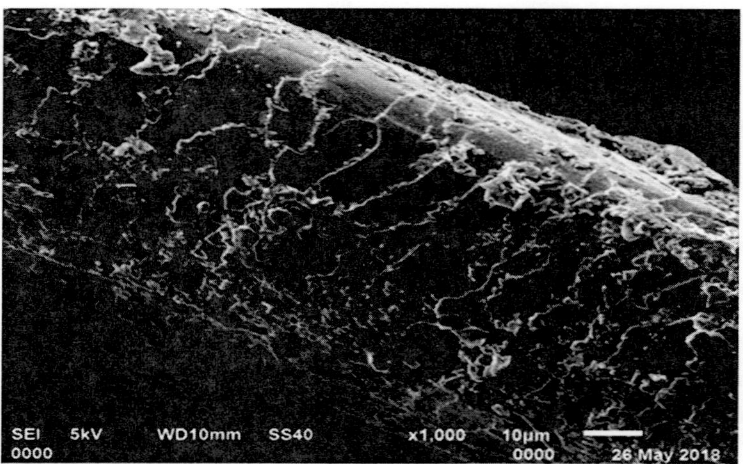

Fig. 4.201. Scanning electron photomicrograph showing cuticular scale pattern of dorsal guard hair of Horse (*Equus caballus*) (1,000X)

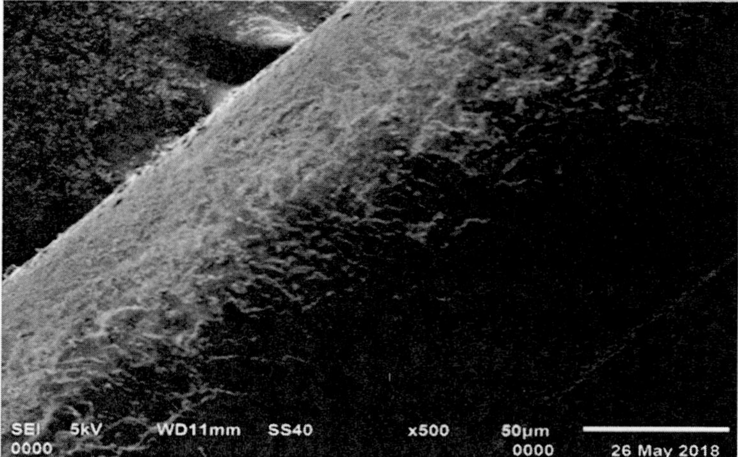

Fig. 4.202. Scanning electron photomicrograph showing cuticular scale pattern of dorsal guard hair of Domestic Pig (*Sus crofa domesticus*) (500X)

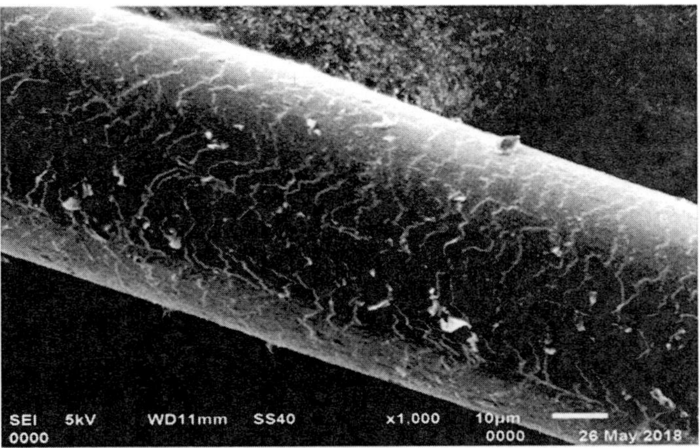

Fig. 4.203. Scanning electron photomicrograph showing cuticular scale pattern of dorsal guard hair of Cat (*Felis catus*) (1,000X)

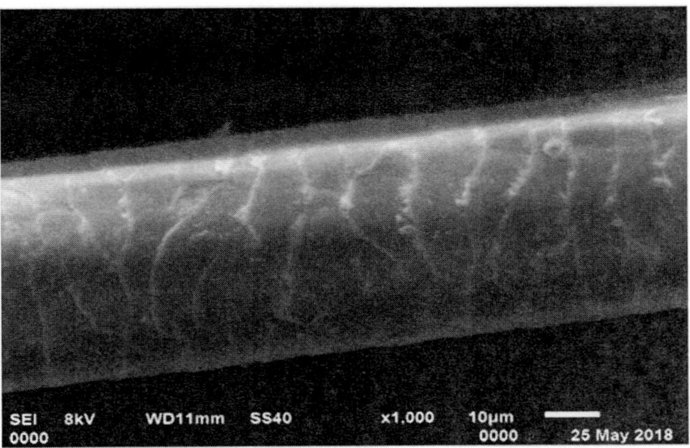

Fig. 4.204. Scanning electron photomicrograph showing cuticular scale pattern of dorsal guard hair of Dog (*Canis familiaris*) (1,000X)

The cuticular scales Dog (*Canis lupus familiaris*) were transversely placed and arranged in regular wave pattern. Each scale showed centrally elevated area with rippled margin and peripheral sloppy area with smooth margin. The distance between scale margins was noted as distant (Fig. 4.204). Two to three scales were observed in each row of the cuticle.

The was found that the cuticular scales of proximal part of hair in Spotted Deer (*Axis axis*) hair were transversally arranged. The cuticular scale pattern was observed as regular wave with smooth scale margins. The distance between scale margins was distant. It was also observed that one to two scales were arranged in every row (Fig. 4.205). The observations of the present study regarding transversely arranged cuticular scale is in agreement with the findings of Verma *et al.* (2016) in Spotted Deer. They further reported that that the cuticular scales in middle part of hair had crenate margin and distance between margins was near. These variations in the observations regarding scale margin and distance between margins may be attributed to the variations in the part of hair (proximal or middle part of hair). it need to be mentioned that during present study the scanning electron microscopy was performed on proximal par of hair whereas Verma *et al.* (2016) observed scale pattern at middle part of hair.

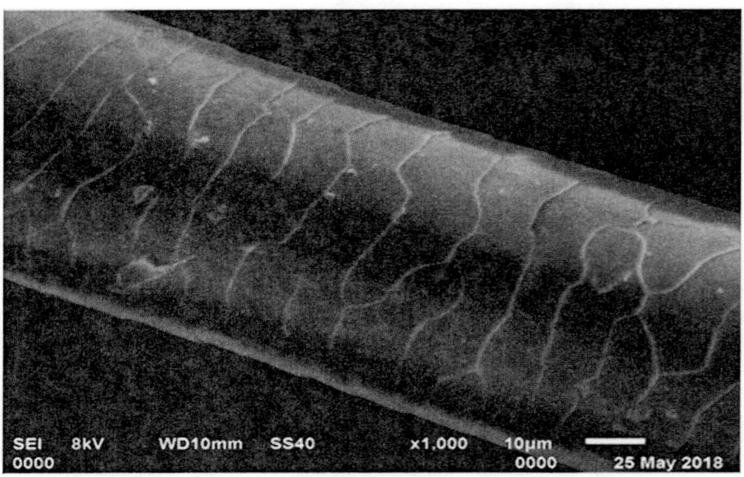

Fig. 4.205. Scanning electron photomicrograph showing cuticular scale pattern of dorsal guard hair of Spotted deer (*Axis axis*) (1,000X)

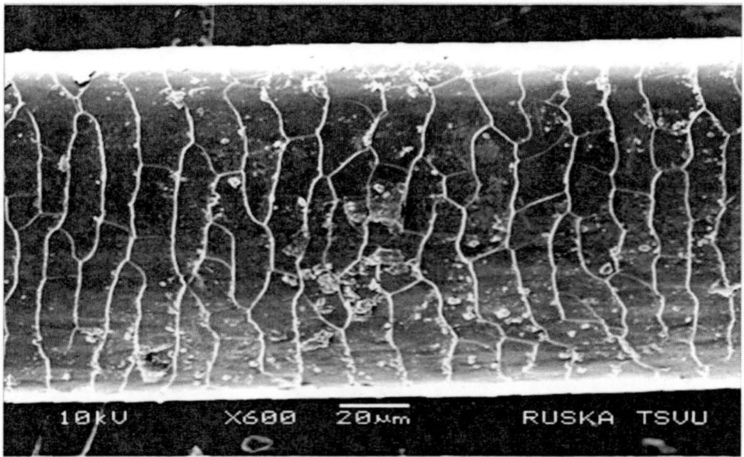

Fig. 4.206: Scanning electron photomicrograph showing cuticular scale pattern of dorsal guard hair of Sambar (*Rusa unicolor*) (600X)

IThe cuticular scales were arranged in regular wave pattern in Sambar (*Rusa unicolor*). The scales were transversely placed and were having smooth margins. It was also observed that the distance between margins of two successive scale margins was distant (Fig. 4.206). It was noted that the around 4 to 5 scales were present in the single row.

The cuticular scales of dorsal guard hair were transversely placed in Nilgai (*Bosellaphus tragocamellus*). The scales were arranged in regular wave pattern with rippled scale margin. It was also found that the distance between scale margins as near (Fig. 4.207). These observations are in partial consonance with the findings of the Kamalakanna (2017d) who stated that the cuticular scale pattern was regular wave with smooth margins and had near scale distance.

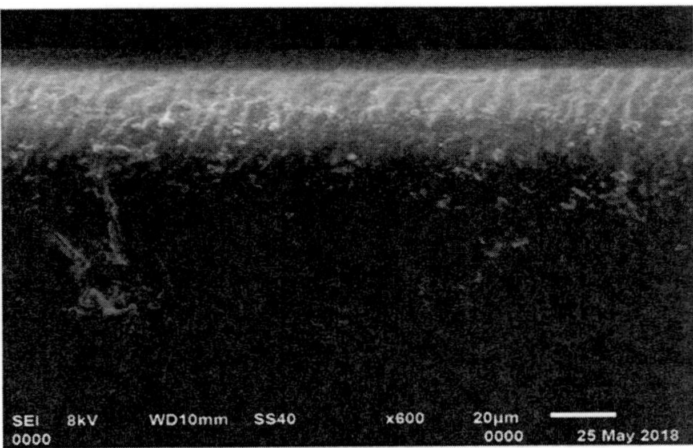

Fig. 4.207: Scanning electron photomicrograph showing cuticular scale pattern of dorsal guard hair of Nilgai (*Boselapphus tragocamelus*) (600X)

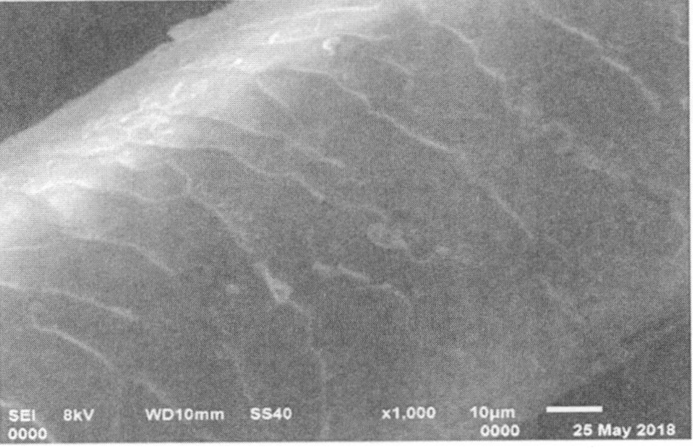

Fig. 4.208. Scanning electron photomicrograph showing cuticular scale pattern of dorsal guard hair of Hanuman Langur (*Semnopithecus entellus*) (1,000X)

The cuticular scale characteristics of dorsal guard hair of Hanuman Langur (*Samnopithecus entellus*) under electron microscope were as, scale position – transversal, scale pattern – irregular wave, scale margins – slightly rippled, distance between scale margins – distant. It was observed that two to three scales were present in a row (Fig. 4.208).

It was found that the cuticular scales were transversely placed in the Tiger (*Panthera tigris*) hair and were arranged in regular wave pattern. The cuticular scale margins were observed as rippled with near scale margin distance (Fig. 4.209).

The cuticular scales in Leopard (*Panthera pardus*) are placed transversal in relation to the longitudinal direction of hair. The scales were arranged in irregular wave pattern with rippled scale margins and close, distance between scale margins. Three to four scales were present in each row of the cuticular scales (Fig. 4.210).

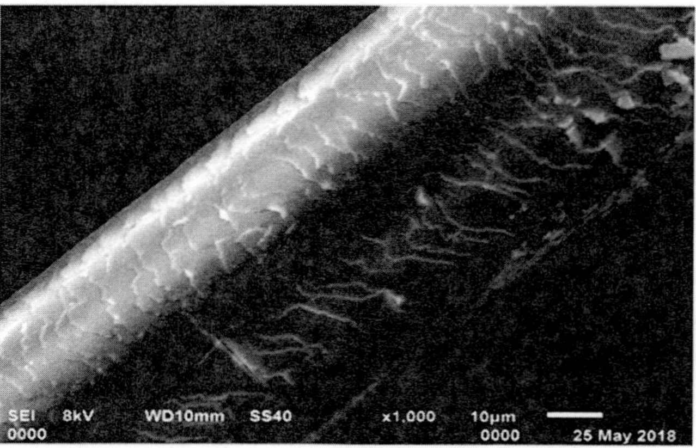

Fig. 4.209: Scanning electron photomicrograph showing cuticular scale pattern of dorsal guard hair of Tiger (*Panthera pardus*) (1,000X)

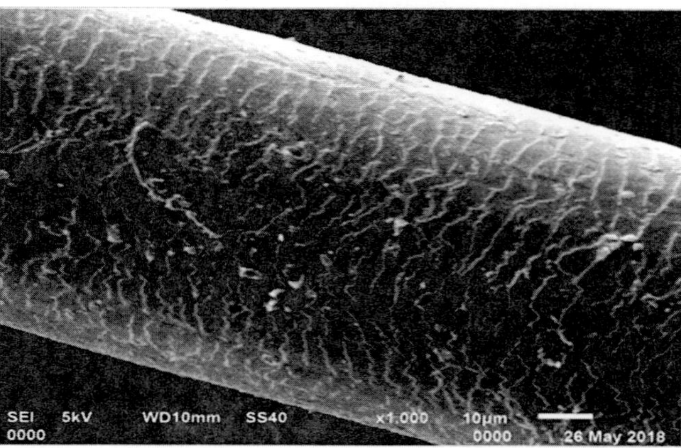

Fig. 4.210: Scanning electron photomicrograph showing cuticular scale pattern of dorsal guard hair of Leopard (*Panthera pardus*) (1,000X)

The observations of the above study indicated that the cuticular scales were transversely arranged in all the species considered for the present work. Regular wave pattern with crenate margin and distant scale margin distance was seen in Cattle and Buffalo. Regular wave pattern with rippled scale margin and near scale margin distance was noted in Domestic Pig, Tiger and Nilgai while, regular wave pattern with smooth margin and distant scale margin distance was found in Sheep and Sambar. The irregular wave pattern with slightly rippled or rippled margin and near or close scale margin distance was observed in Goat, Cat, Hanuman Langur and Leopard. The scales with smooth margin, arranged in irregular wave pattern with distant scale margin distance were noticed in Horse hair. A unique feature of cuticular scale margin with central elevation and slightly rippled structure and peripheral smooth structure with distant scale margin distance was observed in hair of Dog.

4.7. Molecular characterization of hair

Molecular characterization for identification of species from hair was performed using conventional polymerase chain reaction. The product was amplified by using universal primer and amplified *cytochrome b* (398bp) and *12s rRNA* (400bp) gene segments as per the details given earlier.

The *12s rRNA* and cytochrome b (Cyt b) located in the mitochondrial genome have highly conserved nucleotide sequences which are known to be species specific and are used for species identification. All extracted DNA samples under study were amplified using *12srRNA* primers and cytochrome b gene primers as detailed by Sahajpal (2010). The amplified products were submitted for sequencing (Eurofins, Bangalore). The sequences (forward and reverse) were aligned continges were prepared. The trimmed sequences were subjected to BLAST analysis and further submitted to NCBI for getting Accession number.

Gene Bank Data base sequence with *12s rRNA* Gene used for comparison

Sr. No.	GenBank Accession number	Species Name
1.	MN447115	Leopard (*Panthera pardus*)
2.	MN447116	Nilgai (*Boselaphus tragocamelus*)
3.	MN447117	Spotted Deer (*Axis axis*)
4.	MN447118	Dog (*Canis lupus familiaris*)
5.	MN447119	Hanuman Langur (*Semnopthecus entellus*)
6.	MN447120	Tiger (*Panthera tigris*)
7.	MN447121	Cat (*Felis catus domesticus*)
8.	MN447122	Domestic Pig (*Sus scrofa domesticus*)
9.	MN447123	Cattle (*Bos indicus*)

A total of 14 different amplicons of various species were submitted for sequencing of which nine PCR amplicons revealed proper results as detailed in above table.

The findings of microscopy were matched with the findings of BLAST analysis and all the nine samples gave proper species specific matching. The remaining isolated showed error in BLAST results.

Gene Bank Data base sequence with *Cytochrome b* Gene used for comparison

Sr. No.	GenBank Accession number	Species Name
1.	MN462610	Nilgai (*Boselaphus tragocamelus*)
2.	MN462611	Sambar (*Rusa unicolor*)
3.	MN462612	Spotted Deer (*Axis axis*)
4.	MN462613	Dog (*Canis lupus familiaris*)
5.	MN462614	Cat (*Felis catus domesticus*)
6.	MN462615	Buffalo (*Bubalus bubalis*)

A total of 14 DNA samples were used for *Cytochrome b* PCR assay. Of total 14 PCR amplicons, six amplicons showed species specific matching after BLAST analysis as detailed in above table.

Hsieha *et al.* (2003) and Tarditi *et al.*, (2010) used *Cytochrome b* gene for confirmation of Cat hair and Rhinoceros horn samples. Sahajpal and Goyal (2010) used *12s rRNA* and *Cytochrome b* genes for confirmation of hair of Indian Civet Cat and proved cent percent matching with *Viverricula indica*..

Branicki *et al* (2003) used *Cytochrome b* gene for validation and accessing the possibility of applying sequence analysis of region coding *Cytochrome b* as a method of species identification in the field of forensic science.

Similarly, in our study also all *12s rRNA* and *Cytochrome b* confirmed the respective species under study. But when compared in between the *12s rRNA* and *Cytochrome b* BLAST analysis *12s rRNA* proved better for identification of species.

As discussed above, most of the scientists confirmed different samples such as horns tissues and hair samples for confirmation of specific species and further also concluded that *Cytochrome b* can be used in the field of forensic genetics. Microscopic examination of hair samples has a limited scope in forensic investigation because of limited availability of sample amount as an evidence. Whereas molecular characterization of hair samples using *Cytochrome b* and *12srRNA* can be a useful tool for confirmation of animal species.

Phylogenetic tree of *Cytochrome b* gene

The *Cytochrome b* gene sequences were used to develop phylogenetic tree based on maternal lineage for different species.

Neighbour joining tree revealed two distinct clusters. The one with the Cattle and Hanuman Langur and other with all other animal species. Cluster one includes two sub cluster of Cattle (*Bos indicus*) with that of the sequence available in the NCBI for Murrah Buffalo (*Bubalus bubalis*) and another sub cluster had Hanuman Langur (*Semnopithecus entellus*) and *Homo sapiens* together on one node showing the higher genetic similarity between Hanuman Langur and *Homo sapiens* for *Cytochrome b* gene sequences.

The second major cluster cover all other species which were included in the present study. The animals of the same species, both the studied and published sequences taken from NCBI as reference were clustered together in the neighbour joining tree. The Domestic Pig was placed as external branch revealed high genetic differentiation of *Sus scrofa domesticus* with rest of the species. The Leopard and Tiger showed genetic similarities and the sequences were also placed as external branches of second cluster showing high variability in the *Cytochrome b* gene of these two species with rest of the species. The ruminants, feline and canine were further differentiated from other species as they were placed on two different nodes in the neighbour joining tree. Under the ruminant's cluster, buffalo were distantly related with other wild ruminant species like Spotted Deer, Sambar and Nilgai and domestic ruminants such as Goat and Sheep were found having much closer relationship as they were plotted on different nodes but neighbouring to each other. The neighbor joining tree based on mitochondrial Cytochrome gene sequences given in Fig. 4.213.

Phylogenetic tree of *12s rRNA* gene

The phylogenetic tree constructed based on *12s rRNA* gene for species differentiation given in Fig. 4.212. The animals of the same species, both the studied sequences and published sequences taken from NCBI as reference were clustered together in the neighbor joining tree. Neighbour joining tree revealed two distinct clusters, one of the ruminant stomach animals and other one was of simple stomach animals and the Domestic pig i e. omnivore animal was placed in between two clusters. Lastly the small separate cluster of Hanuman Langur showing the distinctness of this species with all other animals based on *12s rRNA* gene diversity.

The major cluster cover all other species grouped in two distinct sub clusters differentiating ruminants and simple stomach separately. However, the domestic pig was placed as external branch with the group of ruminant animals. The domestic ruminants Cattle, Buffalo and Goat were clustered together being genetically similar species while Spotted Deer and Sambar were cluster together having high similarity for the *12s rRNA* gene sequence.

Under the second subcluster having all non ruminant species Horses were shown distant relationship with other species while as expected the wild species (leopard and tiger) clustered together and Cats were found having high genetic similarity with them based on the 12S rRNA sequences, while Dogs were plotted on another branch of the same node.

The results of the phylogenetic tree were similar with the findings of the trichological studies. During trichlogical studies the characteristic features of ruminants and simple stomach animals were similar but distinct from the wild ruminants. The domestic carnivores showed some trichological similarities but were distinct from the wild carnivores. Hence, trichological feature of various animals could be used for the species identifications where genomic facilities are limited.

L_1 - Blank

L_2 - 100 bp DNA ladder

(*12s rRNA gene*)

L_3 - Horse

L_4 - Goat

L_5 - Buffalo

L_6 - Sambar

L_7 - Domestic Pig

L_8 - Sheep

(*Cytochrome b gene*)

L_9 - Sambar

L_{10} - Hanuman Langur

L_{11} - Tiger

L_{12} - Buffalo

L_{13} - Domestic Pig

L_{14} - Horse

L_{15} - Cattle

L_{16} - Sheep

Fig. 4.211: Amplification of *12s rRNA* and *cytochrome b* gene of hair sample of various species.

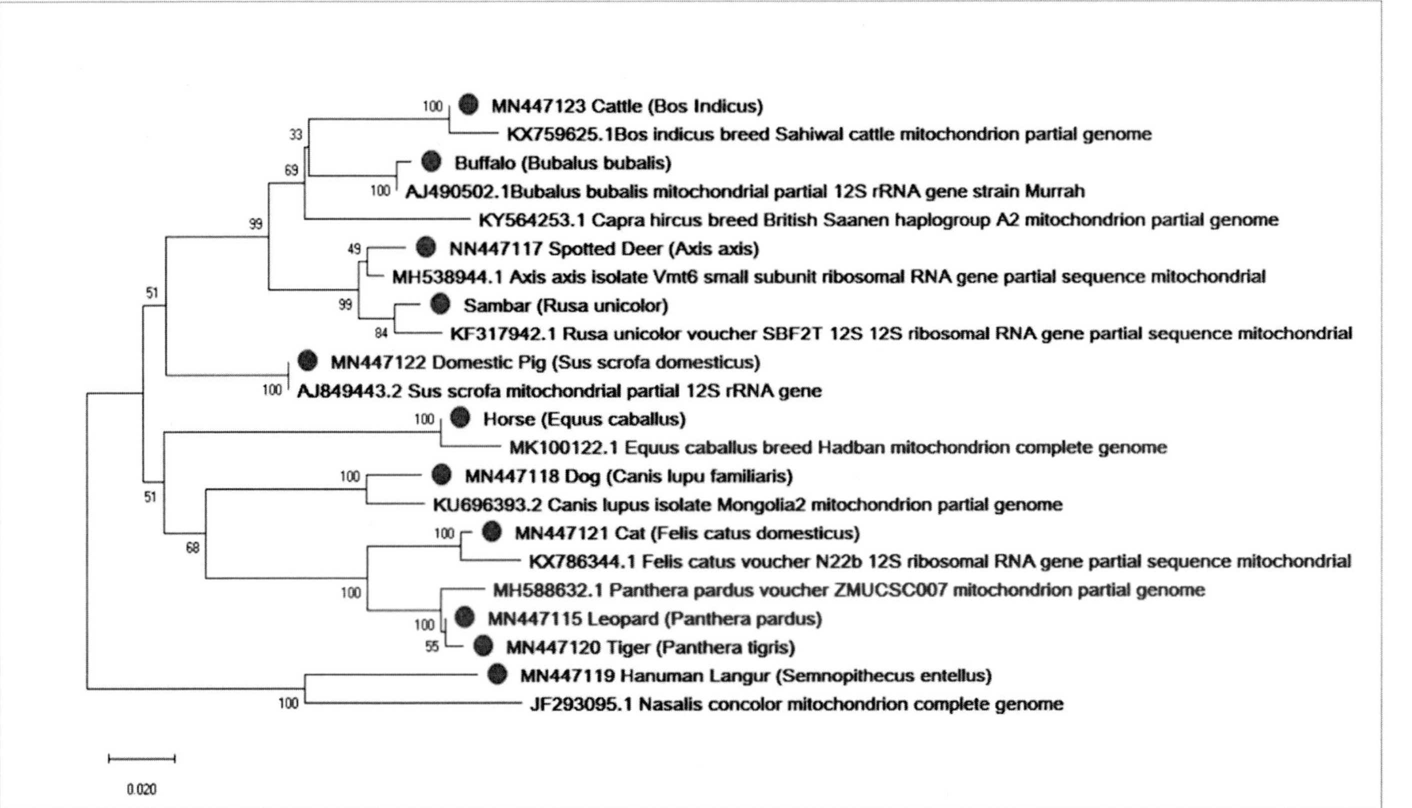

Fig. 4.212: Neighbour joining tree based on mitochondrial *12s rRNA* gene sequence

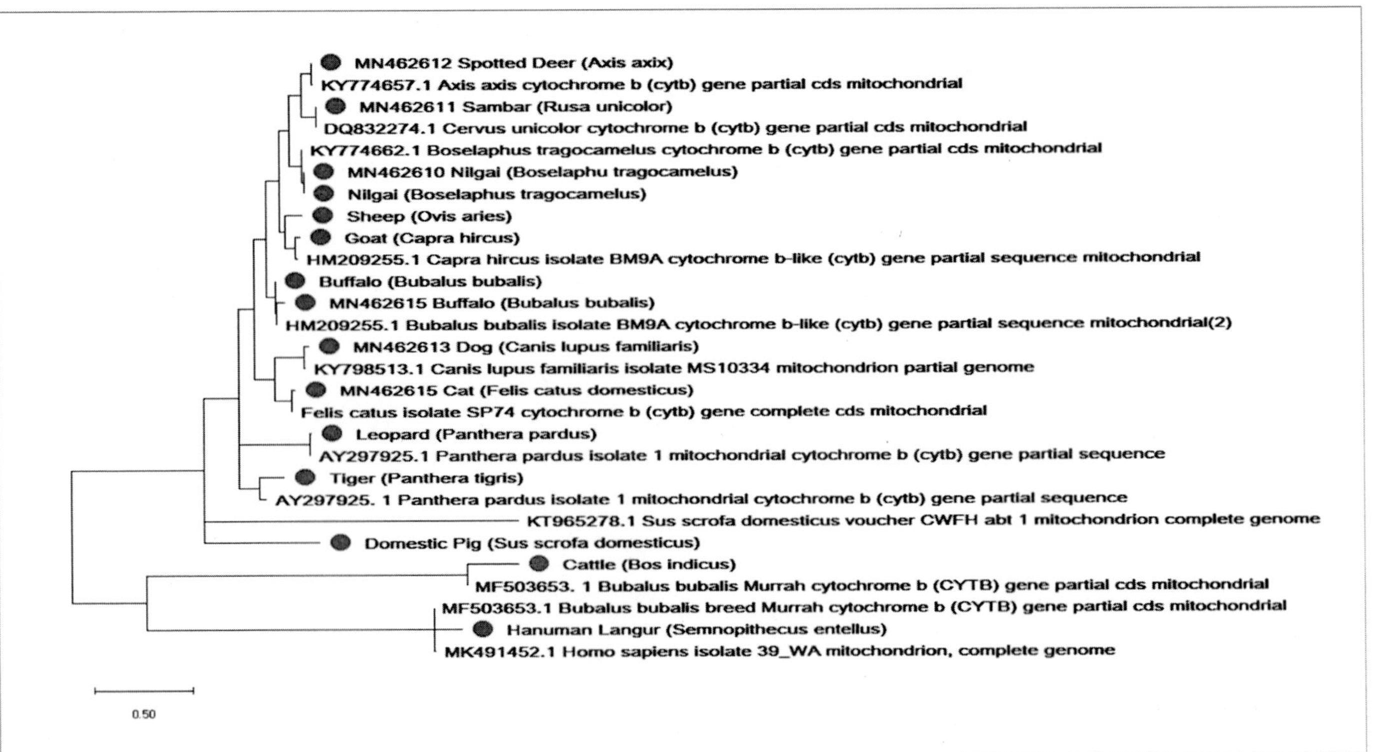

Fig. 4.213: Neighbour joining tree based on mitochondrial cytochrome b gene sequence

Summary and Conclusion

The present trichological work was carried out on guard hair of eight domestic and six wild animal species to study physical, microscopic, scanning electron microscopic and molecular parameters in view to identify the species of animal from hair.

The physical characteristics of hair like colour, number of bands and hair profile showed huge variations in the same individual from region to region particularly in the domestic animal species. Due to presence of metallic shine in the hair from wild herbivores, they could be distinguished grossly from that of the domestic herbivore.

The hair of Domestic Pig could be grossly identified because of its hard consistency, roughness and splits at the tip of hair. The hair from all the six body regions were split into more than two parts at the tip of hair shaft.

The findings of the present study revealed that the shortest hair were found in the head region of most of the species under study except Cattle, Horse, Pig and Tiger. In Cattle, Pig and Tiger, the shortest hair were found in the thigh region and in Horse at the back region. The longest hair were observed in the tail region of Cattle, Buffalo, Goat, Horse, Spotted Deer, Nilgai, Sambar, Tiger and Leopard. The maximum total length irrespective of body region amongst all the species was observed in Sheep and minimum in Tiger.

The length of hair in different body regions of different species of animals, during present study indicated that the effect of species, region and species verses region interaction could be used as a significant source for identification of animal species.

A significant variation in hair diameter at proximal, middle and distal parts of the hair in different regions and species was noted during present study. The hair diameter in all the regions reduced towards the distal part of the hair in all the species. The hair from the back region of Spotted Deer had nearly uniform diameter throughout the length of hair shaft.

The present study showed transversal cuticular scale position in relation to the longitudinal direction of hair in all the species under study except Dog, where the scales were longitudinally positioned. It was also observed that the cuticular scale height was less than the cuticular scale width in all the parts of hair in all body regions of all species, except Dog, where the scale width was less than scale height at the middle part of hair of tail region. The observations of the present study indicated that the cuticular structure varied along the length of hair shaft amongst all the species and body regions.

The four different cuticular scale patterns were observed in Tiger hair during the present study viz., streaked, regular as well as irregular wave and single chevron patterns.

In the hair of all body regions and all the domestic as well as wild animal species, the medulla was absent at the tip of hair. The medulla tapers from the proximal part of hair towards the distal part of hair in all the species and body regions.

The tip and root of hair from spotted Deer (*Axis axis*) in all the body regions showed unique microscopic structures. Wine glass shaped medulla at the root of hair was a characteristic feature in animals of cervidae family. The hair tip was observed as broken or bifurcated at all the body regions. The hair in all the six body regions of Sambar (*Rusa unicolor*) showed unique broad / wide form of medulla at the base / root of hair. Polygonal medulla cells were found in the medulla, which were filled with the slate black coloured pigments. The hair from all the body regions of Nilgai (Bosellaphus tragocamellus) showed wine glass shaped medulla at the base of hair.

A very unique structural feature i e. cortical fusi were noted in the cortex Hanuman Langur (*Samnopithecus entellus*) at all the parts of hair amongst all the body regions.

The dumb-bell shaped cross section with large sized medulla was observed in hair from neck, abdomen, thigh and tail regions of Sheep (*Ovis aries*). The medulla cells were present in the form of strands, which were arranged in a web attern in the hair of Spotted Deer (*Axis axis*) and Sambar (*Rusa unicolor*) in all the body regions. The cigar shaped cross section with a large sized medulla, which occupied almost two third of hair diameter was noted in the hair of Sambar.

The overall mean cuticular scale height was significantly highest in Sheep amongst all parts of hair; body regions and the species while, it was significantly lowest in Nilgai. The minimum cuticular scale width was noted in Sheep and maximum in Sambar at all body regions except in head region. At head region, lowest cuticular scale width was present in Cat while, highest width was noted in Sambar. The lowest number of cuticular scales was found in Sheep while, highest number was noted in Domestic Pig in all the parts of hair, all body regions and amongst all the animal species considered for the present study.

Amongst all the species considered for the present study, the lowest mean hair width was observed in Cat in all the body regions except abdomen region. In abdomen region, lowest hair width was noted in Domestic Pig. The highest mean hair width amongst all the species was recorded in Sambar at neck, back and thigh regions, while, it was highest in head and abdomen regions in Horse.

During the present study, the maximum width of medulla was observed in Sambar at all the body regions, whereas minimum medulla width was observed in neck and abdomen region in Domestic Pig, It was also observed that the lowest width of medulla was found at the proximal part of hair in almost all the regions except head and tail. In head and tail regions, the lowest width of medulla was noted at distal part of hair. The maximum width of medulla was noted at the proximal part of hair from abdomen, thigh and tail region at middle part of hair from head and neck region and distal part of hair from back region.

Amongst all the species, the lowest width of cortex was noted in all the body regions in Cat, whereas the highest width of cortex was observed in the hair of Domestic Pig in all the body regions, except tail region, where medulla was not seen. At tail region, the maximum width of cortex was seen in Hanuman Langur.

The present study indicated that the minimum diameter of hair was in the range of 18.64 ± 1.50 µm (Cat) to 77.75 ± 2.91 µm in (Domestic Pig). Amongst all the species and all the regions of body, the lowest "minimum diameter" of hair was noted in the abdomen region of Cat (9.66 ± 0.22 µm), while highest "minimum diameter" of hair was reported in head region of Domestic Pig (104.10.13 µm).

Amongst all the species and body regions, the lowest diameter of medulla was recorded in hair of neck region in Domestic Pig and highest diameter of medulla was observed in the hair of neck region of Sambar.

Amongst all the regions of body and all the species, cortex width of hair in cross section was lowest in the hair from abdomen region of Cat, while highest cortex width was seen in the hair from head region of Domestic Pig.

The lowest medulla index was noted in the hair from head region of Buffalo followed by abdomen region of Domestic Pig and highest medulla index was noted in the hair from abdomen region of Sambar.

The lowest hair index was observed in the hair of Sheep from thigh region while, highest hair index was noted in the hair from tail region of Hanuman Langur. The cuticular index was found lowest in the hair of abdomen region of Sambar and highest was observed in the hair from back region of Domestic Pig.The scanning electron microscopic findings of the present study revealed that the cuticular scales of dorsal guard hair amongst all the species under study showed free cuticular scale margins, directed towards the tip or distal part of the hair shaft. It was also revealed that the cuticular scale patterns were different in all the domestic as well as wild animals.

Bibliography

Agren, T. (1995) Fur in birka : An examination of hair residue on penannular brooches. Laborativ Arkeologi 8 : 50-58.

Anwar, M. B., M. S. Nadeem, M. A. Beg, A. R. Kayani and G. Muhammad (2012) A photographic key for the identifaction of mammalian hairs of prey species in snow leopard (*Panthera uncia*) habitats of gilgit-baltistan province of Pakistan. Pakistan J. Zool. 44(3) : 737-743.

Aparna, R. and S. K. Yadav (2013) Role of hair as evidence in investigation : A forensic approach. International Journal of Scientific and Engineering Research. 4(11) :1779-1784.

Baddi, S., R. V. Prasad, K. V. Jamuna, S. M. Byregowda, S. Rao and V. Ramkrishna (2014) Morphological studies on the hair of Sloth Bear (*Melursus ursinus*) under different microscopes. Journal of Wildlife Research. 2(1) : 01-03.

Bahuguna, A. (2010) Trichotaxonomy of Indian species of genus Ratufa Gray (Mammalia : Rodentia : Sciuridae). Rec. zool. Surv. India. 110 (Part-3) : 37-57.

Bahuguna, A. and S. K. Mukherjee (2000) Use of SEM to recognise Tibetan antelope (Chiru) hair and blending in wool products Science & Justice *40:*177-182.

Bekaert B., M.H.D. Larmuseau , M. P.M. Vanhove, A. Opdekamp and R. Decorte (2012) Automated DNA extraction of single dog hairs without roots for mitochondrial DNA analysis. Forensic Science International: Genetics 6:277–281

Bhat, M. A., A. B. Shrivastav, A. A. Dar and S. W. Bari (2014) Studies on hair of some wild animals for species identification as an aid to wildlife forensics. IJAVMS. 8(1) : 2-5.

Branicki, W., T. Kupiec, M.S. and R. Pawlowski (2003) Validation of *Cytochrome b* sequence analysis as a method of species identification Journal of Forensic Sci. 48(1) : 1-5.

Broeck, W. V. D., P. Mortier and P. Simoens (2001) Scanning electron microscopic study of different hair types in various breeds of rabbits Folia Morphol. 60(1) : 33-40.

Brunner, H. and B. Coman (1974) The identification of mammalian hair. Inkata Press, Melbourn Australia : 1-23.

Chakraborty, R. and J. K. De (2002) Structure of mid dorsal guard hairs of hunting leopard, *Acinonyx jubatus venaticus* (Griffith) and Lesser panda, *Ailurus fulgens* F. Cuvier (Mammalia : Carnivora). Rec. Zool. Surv. India : 100(Part 1-2) : 131-136.

Chakraborty, R. and J. K. De (2005) Identification of dorsal guard hairs of nine Indian species of the family viverridae (Carnivora : Mammalia). Rec. zool. Surv. India. 104 (Part 3-4) : 13-21.

Chakraborty R. and J. K. De (1995) Structure and pattern of cuticular scales on mid dorsal guard hairs of marbled cat, *Felis marmorata charltoni* gray (mammalian : carnivore : felidae). Rec. Zool. Surv. India. 95(1-2) : 66-67.

Chattha, S. A., K. M. Anjum, M. Altaf and M. Z. Yousaf (2011) Hair mounting technique : helpful in conservation of carnivores. Fuuast J. Biol. 1(2) : 53-59.

Dahiya, M. S. and S. K. Yadav, (2013) Scanning Electron Microscopic Characterization and Elemental Analysis of Hair : A Tool in Identification of Felidae Animals J Forensic Res. 4(1) http://dx.doi.org/10.4172/2157-7145.1000178.

Davis, A. K. (2010) A technique for rapidly quantifying mammal hair morphology for zoological research. Folia Zool. 59(2) : 87-92.

De, J. K. and R. Chakraborty (2012) Identification of dorsal guard hairs of nine species of the family bovidae (Artiodactyla: Mammalia). Rec. zool. Surv. India. 112(2) : 39-52.

Deedrick, D. W. and S. L. Koch (2004) Microscopy of Hair Part II: A Practical Guide and Manual for Animal Hairs. Forensic science communications 6(3) : 1-18 https://archives.fbi.gov/archives/about- us/lab/forensic-science-communications/fsc/july2004/research/2004_03_research02.htm.

Dharaiya N. and V. C. Soni (2012) Identification of hairs of some mammalian prey of large cats in Gir Protected Area. India Journal of Threatened Taxa. 4(9) : 2928–2932| www.threatenedtaxa.org.

Farag M. F., M. H. Ghoniem, A. H. Abou-Hadeed and K. Dhama (2015) Forensic Identification of some Wild Animal Hair using Light and Scanning Electron Microscopy. Adv. Anim. Vet. Sci. 3(10) : 559-568.

Felix, G. A., U. Piovezan, J. Quadros, R. S. Juliano, F. V. Alves, and M. C. S. Fioravanti (2014) Thricology for identifying mammal species and breeds : its use in research and agriculture. Arch. Zootec. 63(R): 107-116.

Gharu, J. and S. Trivedi (2015a) Cuticle Scale Patterns, Medulla and Pigment in Hairs of Some Carnivores. Journal of Chemical, Biological and Physical Sciences. 5(1) : 538-544.

Gharu, J. and S. Trivedi (2015b) Comparison of cuticle scale patterns, medulla and pigment in hairs of domestic goat, sheep, cow and buffalo from Rajasthan (India). Journal of Chemical, Biological and Physical Sciences 5(1) : 570-577.

Gharu J. and S. Trivedi (2013) Hair cuticle scale patterns in Hanuman Langur (*Semnopithecus entellus*) and Grey Slender Loris (*Loris lydekkerianus*). Biological Forum – An International Journal 5 (2) : 11-15.

Gharu J. and S. Trivedi (2015c) Hair cuticle scale patterns, medulla and pigment in Equidae. Journal of Chemical, Biological and Physical Sciences. 5 (2) : 1441-1446.

Guan, Z., Y. Zhou, J. Liu, X. Jiang, S. Li, S. Yang and A. Chen (2013) A simple method to extract DNA from hair shaft using enzymatic laundry powder PLOS ONE. 8(7): 1-7.

Hanfee, F. (1998) Wildlife Trade: A Handbook for Enforcement Staff. WWF Tiger Conservation Program. 4: 56.

Hausman,L.A.(1920b)Structural Characteristics of the Hair of Mammals. The American Naturalist.54(635):496-523. URL: http://www.jstor.org/stable/2456345 Accessed: 15-10-2016.

Hausman L. A. (1920a) The microscopic identification of commercial fur hairs. The Scientific Monthly. 10(1) : 70-78. URL:http://www.jstor.org/stable/6890.

Homan, J. A. and H. H. Genoways (1978) An analysis of hair structure and its phylogenetic implications among heteromyid rodents. J. Bfornrn., 59(4):74-760.

Hsieh, H. M., L. H. Huang, L.C. Tsai, Y.C. Kuo, H.H. Meng, A. L. James and C.I. Lee (2003) Species identification of rhinoceros horns using the *Cytochrome b* gene. Forensic Science International. 136: 1–11.

Joshi, H. R., S. A. Gaikwad, M. P. S. Tomar and K. Shrivastava (2012) Comparative Trichology of Common Wild Herbivores of India. Advances in Applied Science Research. 3 (6) : 3455-3458.

Kamalakannan, M. (2017a) Characterisations of hair of Hoolock Gibbon *Hoolock hoolock* (Harlan, 1834) (Hylobatidae: Primates: Mammalia). Journal of Entomology and Zoology Studies. 5(2) : 986-988.

Kamalakannan, M. (2017b) Microscopic hair characteristics of Namdhapa Flying Squirrel *Biswamoyopterus biswasi* Saha, 1981 (Sciuridae: Rodentia: Mammalia). Journal of Entomology and Zoology Studies. 5(3) : 363-364.

Kamalakannan, M. (2017c) Comparative study of dorsal guard hair of large Indian Civet (*Viverra zibetha*) and Small Indian Civet (*Viverricula indica*) (Carnivora: Viverridae: Mammalia). International Journal of Current Microbiology and Applied Sciences. 6(5) : 1391-1394.

Kamalakannan, M. (2017d) Identification of dorsal guard hairs of Nilgai *Boselaphus tragocamelus* (Pallas, 1766) (Bovidae: Artiodactyla: Mammalia). Bull. Env. Pharmacol. Life Sci. Vol 6(5) : 95-98.

Kamalakannan, M. (2017e) Identification of hairs of *Ratufa bicolour* (Sparrman, 1778), *Ratufa indica* (Erxleben, 1777) and *Ratufa macroura* Pennant, 1769. Journal of Entomology and Zoology Studies. 5(2) : 983-985.

Kamalakannan, M., J. K. De and C. K. Manna (2013) Identification of dorsal guard hairs surface structure of Indian Chevrotain *Moschiolaindica* Gray. 1852 (Tragulidae : Artiodactyla : Mammalia). Biolife 1(4) : 155-158.

Kamalakannan, M. and J. K. De (2017a) Comparative hair morphology of the Indian Otter species *Aonyx cinera*, *Lutra lutra* and *Lutrogale perspicillata* (Mustelidae : Carvivora : Mammalian). Journal of Entomology and Zoology Studies. 5(3) : 826-828.

Kamalakannan, M. and J. K. De (2017b) Hair morphology of Striped Hyena *Hyaena hyaena* (Linnaeus, 1758). International Journal of Current Microbiology and Applied Sciences. 6(5) : 1438-1441.

Khan, A., J. Maryam, T. Yaqub and A. Nadeem (2014) Human Hair Analysis among Four Different Castes Having Potential Application in Forensic Investigation. Forensic Res. 5 : 215. doi:10.4172/2157-7145.1000215.

Kirk, P. L., S. Magagnose, and D. Salisbury (1949) Casting of hairs-its technique and application to species and personal identification. Journal of Criminal Law and Criminology. 40(2) : 236-241.

Kitpipit, T. and P. Thanakiatkrai (2013) Tiger hair morphology and its variations for wildlife forensic investigation. Maejo Int. J. Sci. Technol. 7(03) : 433-443.

Koch, S. L. (2004) Microscopy of Hair Part I: A Practical Guide and Manual for Animal Hairs. Forensic science communications. 6(1) : 1-28 https://archives.fbi.gov/archives/about-us/lab/forensic-science- communications/fsc/july2004/research/2004_01_research01.htm.

Koppikar, B. R. and J. H. Sabnis (1976) Identification of hairs of some Indian Mammals. Journal of the Bombay Natural History Society. 73 (1) : 5-20.

Kshirsagar, S. V., B. Singh and S. P. Fulari (2009) Comparative Study of Human and Animal Hair in Relation with Diameter and Medullary Index. Indian Journal of Forensic Medicine and Pathology. 2 (3) :105-108.

Kuhn R. A. (2009) Comparative analysis of structural and functional hair coat characteristics, including heat loss regulation, in the Lutrinae (Carnivora: Mustelidae). Ph. D. Thesis submitted to University of Hamburg, Hamburg, Germany

Kumar S., Stecher G., Li M., Knyaz C., and Tamura K. (2018). MEGA X: Molecular Evolutionary Genetics Analysis across computing platforms. *Molecular Biology and Evolution* 35:1547-1549.

Lee E., T. Y. Choi, D. Woo, M. S. Min, S. Sugita and H. Lee (2014) Species identification key of Korean mammal hair. J. vet Med Sci. 76(5) : 667-675. DOI : 10.1292/jvms.13-0569.

Lungu, A., C. Recordati, V. Ferrazzi and D. Gallazzi (2007) Image analysis of animal hair :

Morphological features useful in forensic veterinary medicine. LUCRĂRI STIINłIFICE MEDICINĂ VETERINARĂ VOL. XL : 439-446.

Lyne, A. G. and T. S. McMahon (1950) Observations on the surface structure of the hairs of Tasmanian monotremes and marsupials. Pap. And Proc. Roy. Soc. Tasmania: 71-84.

Marinis, A. M. D. and P. Agnelli (1993) Guide to the microscope analysis of Italian mammals hairs: Insectivora, Rodentia and Lagomorpha. Bolletino di zoologia. 60 : 225-232 DOI: 10.1080/11250009309355815.

Marinis, A. M. D. and A. Asprea (2005) How Did Domestication Change the Hair Morphology in Sheep and Goats? Human Evolution. DOI 10.1007/s11598-006-9010-0.

Marinis, A. M. D. and A. Asprea (2006) Hair identification key of wild and domestic ungulates from southern Europe. Wildlife Biology. 12(3) : 306-320.

Mukherjee, P., C. Pasi, S. Patel and S. Dhurwey (2016) Comparative trichological analysis of common domestic mammals of Jabalpur district. Int. J. Adv. Res. Biol. Sci. 3(2) : 93-98.

Nameer, P.O. (2015) A checklist of mammals of Kerala. India Journal of Threatened Taxa. 7(13): 7971–7982.

Negi, P., A. Baberia, K. Yadav, M. Singh Sankhla and R. Singh (2017) Comparison of different animal species hairs with respect to their medullary index for the individual identification and comparison from the animals of local village of Palam Vihar, Gurugram, Haryana. International Journal of Recent Research and Applied Studies, 4, 12(6) : 34-36.

Pepper, C. M. and D. H. Sharon (1977) A new technique for cross sectioning free hairs. The Journal of Investigative Dermatology, 68(2) : 111-112.

Prates, L., F. Ballejo and A. Blai (2016) Analysis of hair remains from a hunter-gatherer grave from Patagonia taxonomic identification and archeological implications. Journal of Archeological Science. 8 : 142-146.

Rajaram, A. and R. K. Menon (1986) Scanning electron microscope study of the hair keratins of some animals of the Indian subcontinent – A preliminary report. J. Bombay Nat. Hist. Soc. 83 : 427-429.

Rana, J., S. B. Banubakode, N. C. Nandeshwar, R. Charjan, U. P. Mainde and S. K. Patel (2017) Comparative morphological study of tactile hair of tiger and leopard. Global Journal of Bio-science and Biotechnology. 6(3) : 500-503.

Ridgway, R. (1886) A nomenclature of colors for naturalists, and compendium of useful knowledge for ornithologists. University Press : John Wilson and Son, Cambridge. Boston Little, brown and company.

Riggot, J. M. and D. E. H. Wyatt (1980) Scanning electron microscopy of hair from different regions of the body of the rat J. Anat. 130(1) : 121- 126.

Sahajpal, V., S. P. Goyal, R. Jayapal, K. Yoganand, and M. K. Thakar (2008) Hair characteristics of four Indian bear species. Science and Justice. 48 : 8–15.

Sahajpal, V., S. P. Goyal, R. Raza and R. Jayapal (2009a) Identification of mongoose (Genus: Herpestes) species from hair through band pattern studies using discriminate functional analysis (DFA) and microscopic examination. Science and Justice. 49 : 205–209.

Sahajpal, V., S. P. Goyal, M. K. Thakar and R. Jayapal (2009b) Microscopic hair characteristics of a few bovid species listed under Schedule-I of Wildlife (Protection) Act 1972 of India. Forensic Science International. 189 : 34–45.

Sahajpal, V. and S. P. Goyal (2010) Identification of a forensic case using microscopy and forensically informative nucleotide sequencing (FINS): A case study of small Indian civet (*Viverricula indica*). Science and Justice. 50 :94–97.

Sahajpal, V., S. P. Goyal, K. Singh and V. Thakur (2016) Dealing wildlife offences in India ; Role of the hair as physical evidence. International Journal of Trichology. 1(1) : 18-26.

Sarkar, P. S., J. K. De and C. K. Manna (2011) Identification of dorsal guard hairs of five species of the family Cercopithecidae (Primates: Mammalia). Current Science 100(10) : 1725-1728.

Sarkar, P. S. and J. K. De (2013) Tricho-taxonomic study of Dorsal Guard Hairs of Indian Species of Rodents Belonging to Subfamily- Sciurinae (Sciuridae: Rodentia: Mammalia). Biological Forum – An International Journal. 5(1) : 1-10.

Sato, H., H. Matsuda, S. Kubota and K. Kawano (2006) Statistical comparison of dog and cat guard hairs using numerical morphology. Forensic Science International. 158 : 94-103.

Shelley, W. B. and S. Ohman (1969) Technique for cross sectioning hair specimens. The Journal of Investigative Dermatology 52(6) : 533-536.

Soni, V. C., Y. Ashalatadevi and D. Nishit (2004) Hair structure is an ideal criterion to identify various species of genus Panthera. Journal of Tissue Research. 4(1) : 161-163.

Snedecor, G. W. and W. G. Cochran (1996) Statistical Methods, 6th Edn., Oxford and IBH Publishing House, Calcutta.

Takayanagi, K., H. Asamura, K. Tsukada, M. Ota, S. Saito and H. Fukushima (2003) Investigation of DNA extraction from hair shafts. International Congress Series 1239: 759– 764

Tarditi, C. R., R. A. Grahn, J. J. Evans, J.D. Kurushima, and L.A. Lyons (2011) Mitochondrial DNA sequencing of cat hair: An informative forensic tool Journal of Forensic Sci. 56(Suppl 1): S36–S46. doi:10.1111/j.1556-4029.2010.01592.x.

Taru, P., G. Mukwada and W. Chingombe (2013) Microscopic hair characteristics of south african blue wildebeest (*Connochaetes taurinus*), black wildebeest (*Connochaetes gnou*) and red rock hare (*Pronolagus crassicaudatus*). J. Life Sci. 5(2) : 123-126.

Taylor, R. J. (1985) Identification of the hair of Tasmanian mammals. Papers and proceedings of the Royal Society of Tasmania. 119 : 69-82.

Teerink, B. J. (1991) Hair of West Europian Mammals : Atlas and Identification Key. Cambridge university Press, Cambridge PP. 1-223.

Toth, M. A. (2002) Identification of Hungarian mustelidae and other small carnivores using guard hair analysis. Acta Zoologica Academiae Scientiarum Hungaricae. 48(3) : 237-250.

Tridico, S. R., M. M. Houck, K. P. Kirkbride, M. E. Smith and B. C. Yates (2014) Morphological identification of animal hairs: Myths and misconceptions, possibilities and pitfalls. Forensic Science International. 238 : 101–107.

Verma, A. M., S. Khan, P. C. Joshi, A. Bahuguna and K. Dev (2016) Wildlife forensic techniques: DNA extraction from hard biological matter of Chital (*Axis axis*), molecular analysis, use of SEM and EDXin species identification. Journal of Biotechnology and Biochemistry. 2(7) : PP 104-110.

Vernall, D. G. (1961) A Study of the Size and Shape of Cross Sections of Hair from Four Races of Men. American Journal of Physical Anthropology 19(4) : 345-350. https://doi.org/10.1002/ajpa.1330190405.

Vinayak, V., Chitralekha, S. Kaur, A. Kadyan and A. Rai (2012) Forensic trichology and its importance in crime cases. Nature and Science. 10(9) : 116-120.

Wan Q.H. and S.G. Fang (2003) Application of species-specific polymerase chain reaction in the forensic identification of tiger species Forensic Science International. 131: 75-78.

Zafarina, Z. And S. Panneerchelvam (2009) Analysis of hair samples using microscopical and molecular techniques to ascertain claims of rare animal species. Malaysian Journal of Medical Sciences. 16(3) : 35-40.